Topics in Atmospheric and Oceanographic Sciences

Hans-Jörg Isemer · Lutz Hasse

The Bunker Climate Atlas of the North Atlantic Ocean

Volume 1: Observations

With 181 Charts and 31 Figures

Springer-Verlag
Berlin Heidelberg New York Tokyo

Dipl.-Met. Hans-Jörg Isemer
Professor Lutz Hasse

Institut für Meereskunde an der Universität Kiel
Abt. Maritime Meteorologie
Düsternbrooker Weg 20
2300 Kiel 1, FRG

ISBN 3-540-15568-6 Springer-Verlag Berlin Heidelberg New York Tokyo
ISBN 0-387-15568-6 Springer-Verlag New York Heidelberg Berlin Tokyo

Typesetting and printing: Beltz Offsetdruck, Hemsbach/Bergstr.
Bookbinding: J. Schäffer OHG, Grünstadt.
2132/3130-543210

Preface

Marine Meteorology has a long tradition, and studies of surface meteorological conditions have been published repeatedly since the end of the last century. Recently, the demand has grown for more detailed descriptions. This stems both from the public's interest in climatic change and from our growing ability to analyse atmospheric and oceanic processes with the aid of numerical models. These models require input data on a regular, finely spaced grid; the increased amount of oceanic data available permits us to provide detailed charts both of surface meteorological conditions and of the air–sea interaction.

The present climate atlas of the North Atlantic Ocean is based on data originally evaluated by Andrew F. Bunker of Woods Hole Oceanographic Institution. He analysed observations from the ships of the Voluntary Observing Fleet in many parts of the world oceans to calculate the various components of the heat budget at the air–sea interface. When Bunker died in 1979, he left the major part of his data and results in an unpublished state. Since he had expended considerable effort to validate the data and calculate air–sea fluxes by the so-called "individual method", it was considered worthwhile to make this unique set of climate data available to the scientific community. Bunker's analysed fields for the North Atlantic Ocean are presented in this atlas. It deals with the surface climate of the North Atlantic Ocean from the equator to 65°N, in the period 1941 to 1972. By the use of monthly and annual mean charts and other diagrams, the annual cycles of various oceanic and atmospheric surface parameters are presented. Volume 1 contains the observed meteorological quantities. For the most part they represent the basic information required to understand Volume 2, which presents derived parameters such as energy budget terms and ocean transports.

This North Atlantic Ocean atlas is unprecedented in the size of the data base and in spatial resolution; it is based on objective analysis and the most recent understanding of the parameterisation of derived quantities such as fluxes of heat and momentum. Meteorologists and oceanographers, as well as other marine scientists, may use the charts for climatological investigations and as a background for studies of actual events. The data are also available on magnetic tape, for use, for example, as boundary conditions for numerical modelling.

This atlas has been produced as a part of the research of the Sonderforschungsbereich *Warmwassersphäre des Atlantiks* at the Institut für Meereskunde, Kiel, and is sponsored by the Deutsche Forschungsgemeinschaft.

Professor Wolfgang Krauß, as the chairman of the Sonderforschungsbereich, initiated the project. We are grateful for his permanent interest in this atlas.

The primary work was done by the late Andrew F. Bunker. His name is included in the title of this atlas in appreciation of his basic contribution to this publication. Professor Henry Stommel and Roger Goldsmith were very helpful during the stay of one of the authors at Woods Hole Oceanographic Institution.

We thank Dr. Heinz Fechner for many substantial discussions on interpolation problems and Dr. Fred Dobson of Bedford Institute of Oceanography, who made helpful comments on the project and kindly reviewed the text.

Members of the staff and several students of the Institut für Meereskunde helped in various stages of the project. Especially the work of Doris Michaelis, who retyped most of the data from papercopies, and Ute Hargens and Elisabeth Rudolph, who assisted in preparing the computer plots, is gratefully acknowledged.

Kiel, Summer 1985

H.-J. Isemer
L. Hasse

Contents

Introduction

Great efforts have been made in recent years to analyse and understand the behaviour of the atmospheric and oceanic climate on a global scale. Goals, procedures and requirements were summarized in 1979 by the International Council of Scientific Unions (ICSU), at the time when the World Climate Research Program (WCRP) was established; they have recently been supplemented (WCRP 1984). The annual cycle of various oceanic and atmospheric parameters at the air–sea interface, highly resolved in space, is basic information for the study of interannual variability and long-term climatic change (stream two and three of the WCRP). The actual incentive for this atlas has stemmed from a research programme at the Institut für Meereskunde, Kiel, which is designed to investigate the *warm-water sphere* of the Atlantic Ocean. Mean fields of atmospheric and air–sea interaction parameters, defined on a 1-degree grid, are needed for the study of both dynamic and thermodynamic processes of the North Atlantic Ocean with various numerical models. Not only measured parameters but also derived quantities, such as energy budget components and wind stress data, are required.

Several atlasses of the climate of the North Atlantic Ocean have already been published (e.g. Brit. Met. Off. 1948, DHI 1967, Meserve 1974). But at least for the North Atlantic Ocean no comprehensive publication presenting a homogeneous set of meteorological observations and derived quantities is available today. In most cases contouring techniques have been subjective, spatial resolution coarse, and results have not been tailored to numerical calculations. Hastenrath and Lamb (1977, 1978) provide a comprehensive high-resolution atlas of surface conditions and energy budget terms, but only for the tropical oceans.

The first comprehensive worldwide atlas of energy budget terms at the air–sea interface, including the North Atlantic Ocean, was published by Budyko (1963); its merits and drawbacks have been discussed by Bunker and Worthington (1976). Budyko (1978) published an updated version of his atlas; mean annual maps of energy budget terms may also be found in Budyko (1974). Esbensen and Kushnir (1981) have recently published an energy budget atlas for the world ocean; they use a 4×5-degree grid net. The present work differs from all of these in techniques applied and spatial resolution.

In an extensive heat budget calculation, the late Andrew F. Bunker derived climatological mean monthly values for the period 1941 to 1972, defined on irregularly shaped areas over the North Atlantic Ocean (Bunker 1975, 1976, Bunker and Worthington 1976). Only parts of Bunker's results have been published; the bulk of his calculated fields has remained in printout files at Woods Hole Oceanographic Institution and has not been readily available for further scientific research. It is worthwhile to continue the work on his special data set for the following reasons: (1) Bunker did a careful quality check of the entire basic data set, which consists of observations from the ships of the Voluntary Observing Fleet, V.O.F. (WMO 1977); (2) he calculated means of derived quantities by the "individual" method, using only those observations which passed error checking procedures; (3) the Bunker data set is the most comprehensive one for the North Atlantic Ocean and is frequently quoted in the literature in connection with heat budget calculations or meridional heat transport problems (e.g. Hastenrath 1982, Dobson et al. 1982, Bryan 1983).

Bunker's data set was read from printouts and stored in computer files. The data were interpolated from Bunker's original irregularly shaped areas onto a 1-degree grid, and are now available on tape in the original and interpolated versions. The basic philosophy of the work has been to leave Bunker's result unchanged as far as possible. Consequently measured and observed parameters, which are presented in the first volume, have not been corrected at all, although the resulting fields may be questionable in some critical areas where the data base is small and the information is not significant in the climatological sense. Also, corrections for biases are not applied. Data analysis, interpolation methods and construction of the charts have been carried out using objective techniques. The fields of the various parameters and their annual variation are displayed as monthly and annual maps. Charts of the range of the annual cycle and graphs of the ocean-wide zonal averages are presented as well as diagrams showing the annual cycle for characteristic areas in the North Atlantic Ocean.

The Bunker Data Set

Weather logs from merchant ships and other marine reports are available from the National Climatic Center, Asheville, U.S.A. in unified formats in the so-called TDF-11 data deck. In a study of the surface energy fluxes of the North Atlantic Ocean Bunker (1976) obtained 12 million observations from 66 Marsden Squares in the North Atlantic Ocean and, after an additional quality check, processed nearly 8 million of them, covering the period 1941 to 1972. Since the TDF-11 data deck is a collection of marine weather reports from several national weather services and organisations (Table 1), the data coverage for the period mentioned is the best available.

Bunker calculated means and, for selected parameters, standard deviations of observed and derived quantities for each calendar month within subdivisions, called gerrymanders, of the North Atlantic Ocean from the equator up to 80°N. The boundaries of these sometimes irregularly shaped gerrymanders (see Fig. 1) were chosen to (1) produce hydrographically homogeneous areas, and (2) provide an optimum number of ship observations within each gerrymander in each month (Bunker 1976). The original Bunker climatological data set contains 52 parameters and is available on computer printouts at the Woods Hole Oceanographic Institution (WHOI), Woods Hole, Massachussetts, U.S.A. A detailed description is given in Goldsmith and Bunker (1979).

Table 1. Sources of marine surface observations in the TDF-11 tape deck of the National Climatic Center with periods of records, which Bunker used for the North Atlantic Ocean from 1941 to 1972. (After Goldsmith and Bunker 1979)

Source	Period
U.S. Navy marine observations	1945–1951
U.S. merchant marine	1949–1963
Japanese ship observations	1941–1961
International marine observations	1963–1972
U.S. Navy MAR marine observations	1941–1945
Great Britain marine observations	1941–1956
U.S.S.R. marine synoptic observations	1957–1958
Japanese whaling fleet observations	1946–1956
Netherlands marine observations	1941–1955
U.S. Navy ship logs	1942–1945
Deutsche Seewarte marine observations	1949–1954
Danish marine observations	1941–1956

Data coverage extends up to 80°N. However, in most areas north of 65°N a complete annual cycle cannot be presented, due to a lack of ship observations mainly in winter and spring, while even during summer and fall a significant climatological field presentation is hindered by the small data base. In general, the number of ship observations within single gerrymanders varies from zero to more than 10,000 for different calendar month (Fig. 2). Weare and Strubb (1981) found for undisturbed tropical environments that at least 11 randomly distributed marine observations per single month are required "to be confident at the 95% level that one has observations which are uniformly sampled in space and time." For higher latitudes in winter this number increases to 50–75 observations per single month (Weare and Strubb 1981). After Bunker (1976), long-term monthly gerrymander means are reliable – compared with ocean weather ship data – if they are calculated from at least 500 observations. But only 261 of the total of 502 gerrymanders (52%) fulfill this condition in all calendar months, while for 28% of all gerrymanders the total number of observations is less than 500 in each calendar month. This means that a construction of monthly maps for most parts of the North Atlantic Ocean would be impossible, if Bunker's threshold were used to eliminate less-than-significant information.

As an alternative approach to the problem we calculated confidence intervals for the monthly gerrymander mean values using a "Student's t" test from standard deviations of selected parameter in the data set. The confidence level was chosen to be 95%. For this investigation we assumed mean values to be climatologically significant, if confidence intervals were smaller than (1) 1°C for sea surface temperature and air temperature, (2) 3 hPa for sea level air pressure and (3) 3 m s^{-1} for the wind components. The results show that north of 65°N nearly all gerrymanders contain insufficient information. In the region between 65°N and 60°N the number of gerrymanders with climatologically significant information is larger than those without, so it was decided to present charts from the equator up to 65°N. In the latter area the total number of gerrymanders is 474. Nevertheless there are five regions south of 65°N with confidence intervals exceeding the chosen thresholds, so that contours especially in the area between 60°N and 65°N should be interpreted

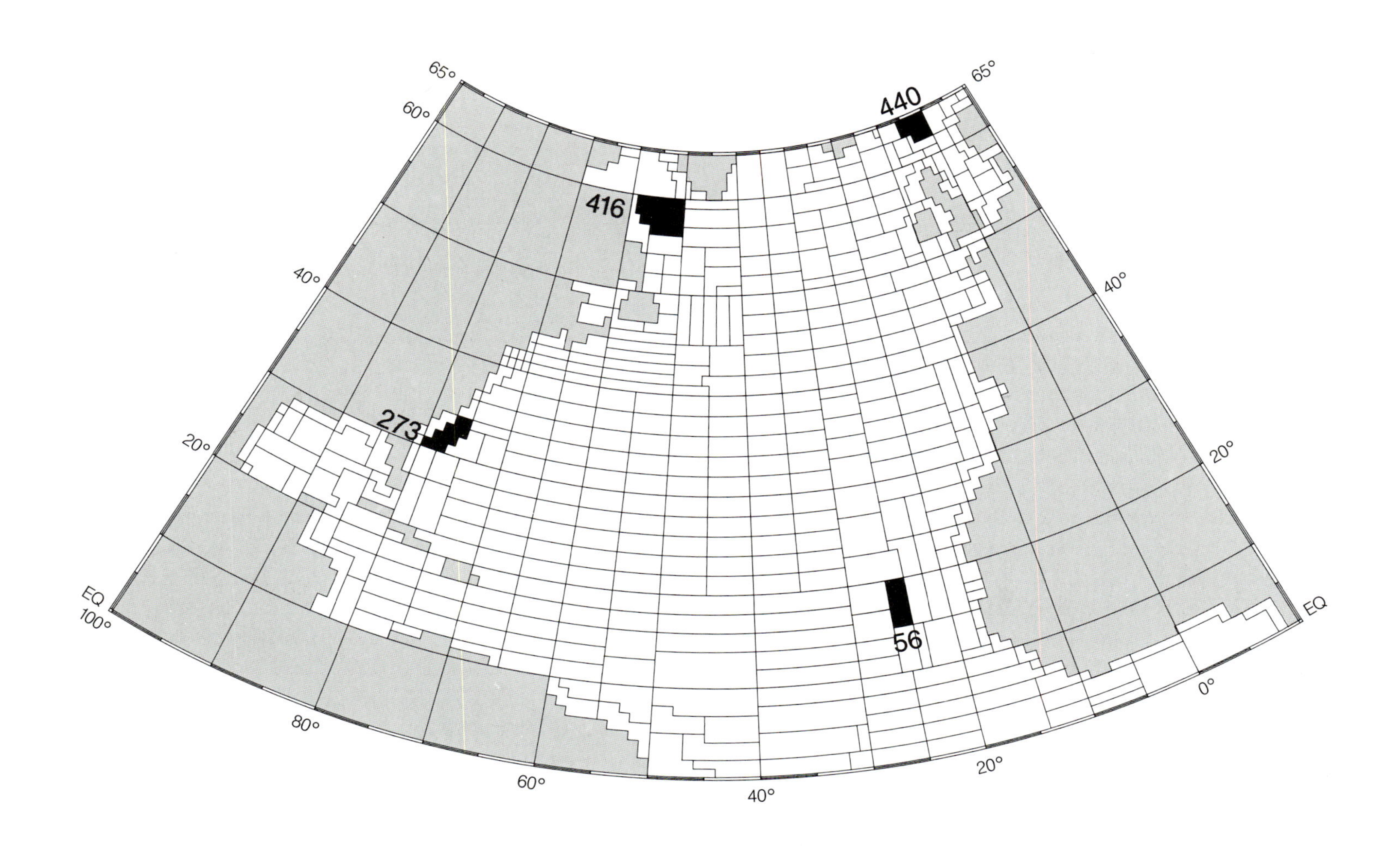

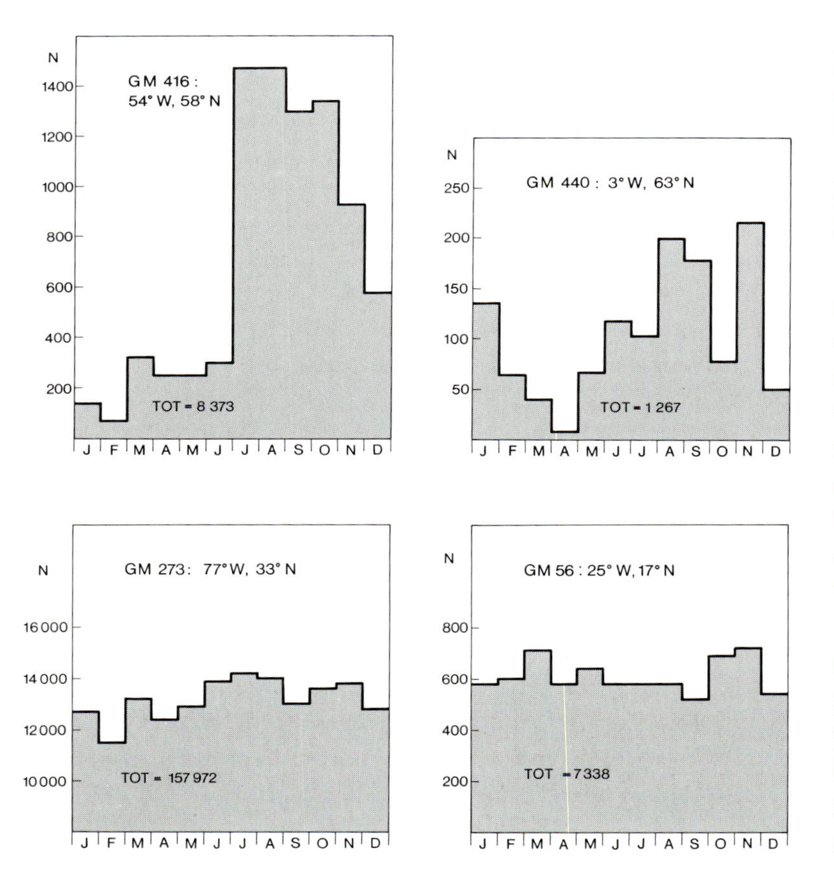

Fig. 1. The gerrymander configuration for the North Atlantic Ocean. (After Bunker 1976). The annual variation of number of ship observations in the four gerrymanders, which are marked black, is given in Fig. 2

Fig. 2. The annual variation of number of ship observations in four selected gerrymanders (see Fig. 1). TOT = total annual sum; GM = gerrymander

and discussed with care. The reader is advised to use the appended overlay which marks these regions. The area north of 60°N and west of 60°W was excluded from presentation. In some month results from the Labrador current region off the Canadian coast between 53°N and 60°N are also doubtful: especially in May all parameters show quite unrealistic gradients there, because in the narrow gerrymander at the Canadian coast (see Fig. 1) only eight observations were made in the period 1941 to 1972.

The number of ship observations per gerrymander was converted to number per unit area and is shown for each gerrymander in Charts 1–5. The area of one or two grid-points at 35°N or 65°N respectively is 10,000 km^2. The seasonal distribution of observations if fairly uniform for most of the area, except in northern latitudes, where shipping is hindered by ice during winter and spring.

Interpolation Method and Construction of Maps and Diagrams

The gerrymander configuration used by Bunker prohibits straightforward use of the fields for objective analysis and numerical modelling. Hence, a special interpolation method was used to produce fields on a regular 1-degree grid. First a preliminary field is constructed by ascribing the gerrymander means to the respective 1-degree grid points within each gerrymander, thus building up a "step"-field containing zero gradients inside each gerrymander and unrealistic steps at the boundaries. Next, local quadratic, two-dimensional polynomials are fitted to each gerrymander and all of its neighbours. The procedure takes the information for the field configuration within each gerrymander (which is described by a constant in the "step"-field version) from its surroundings, permitting quadratic behaviour in both horizontal directions. The following points describe the basic procedures, conditions and attributes of the interpolation scheme used:

1. The grid points of the "step"-field are the data points for the polynomial fit (representing the area-averaged character of the initial data field). By doing so we assume uniform spatial distribution of the basic data within each gerrymander, as no further information is available in the data set.

2. The mean of the central gerrymander is conserved by constraining the mean of the interpolated gridpoint values to be equal to the original gerrymander mean. This determines the first coefficient of the polynomial.

3. One fit is made for each gerrymander, using all of its neighbouring gerrymanders. At least five additional, independent pieces of information are needed to solve a quadratic system. In general, the gerrymander configuration gives eight neighbouring values, so the coefficients of the polynomial are determined by a least squares fit. If there are less than five surrounding gerrymanders (this occurs most often in coastal regions), the order of the polynomial is reduced.

4. Calculations with the procedure occasionally showed unsatisfactory results, especially when the magnitude of the difference (the "step") between the data of the central gerrymander and two or more of the neighbouring gerrymanders is very different. A slight remedial weighting was introduced to give the gridpoint values of gerrymanders with larger steps more influence.

5. The resulting field over the North Atlantic Ocean is composed of a collection of 474 individual polynomial pieces. A low pass filter which consists of two runs of the two-dimensional Hann filter (e.g. Holloway 1958, Shapiro 1970) was adopted to remove numerical noise and to guarantee reasonable first derivatives: "reasonable" in this context means that the isolines should not reproduce the contours of the gerrymanders themselves.

6. The interpolation procedure is the same for all parameters. The filter changes the conservation of the mean within single gerrymanders only very slightly. It leaves the boundary values at the equator, 65°N, and the coastlines unchanged.

7. As is obvious from the above, the interpolated 1-degree scale data are not independent.

From the interpolated data on the 1-degree grid various maps and diagrams were constructed:

- maps of monthly and annual means,
- maps of monthly and annual standard deviations,
- maps of the range of the annual cycle,
- time latitude diagrams of zonally averaged values, and
- diagrams of the annual cycle in characteristic regions of the North Atlantic Ocean.

The annual field was derived from the means of the monthly values at every gridpoint, weighted by the number of days per calendar month. When monthly changes are small or may be deduced from other parameters, only four monthly maps, i.e. the maps for the respective central month of each meteorological season (January, April, July, October) are presented.

Not all of the 52 parameters in the original Bunker data set are presented here. For the monthly and annual maps, as well as for the maps of the annual range, we chose the following parameters: sea surface temperature, air temperature, air- minus sea surface temperature difference, mixing ratio, total and low cloudiness, precipitation frequency, sea level air pressure and scalar wind speed. The maps of the resultant wind field are constructed from the interpolated mean wind components and are presented as combined pictures of arrows for direction and isolines for the absolute value of the resultant wind. For the sake of clarity an arrow is plotted only at every second grid point with the end of the

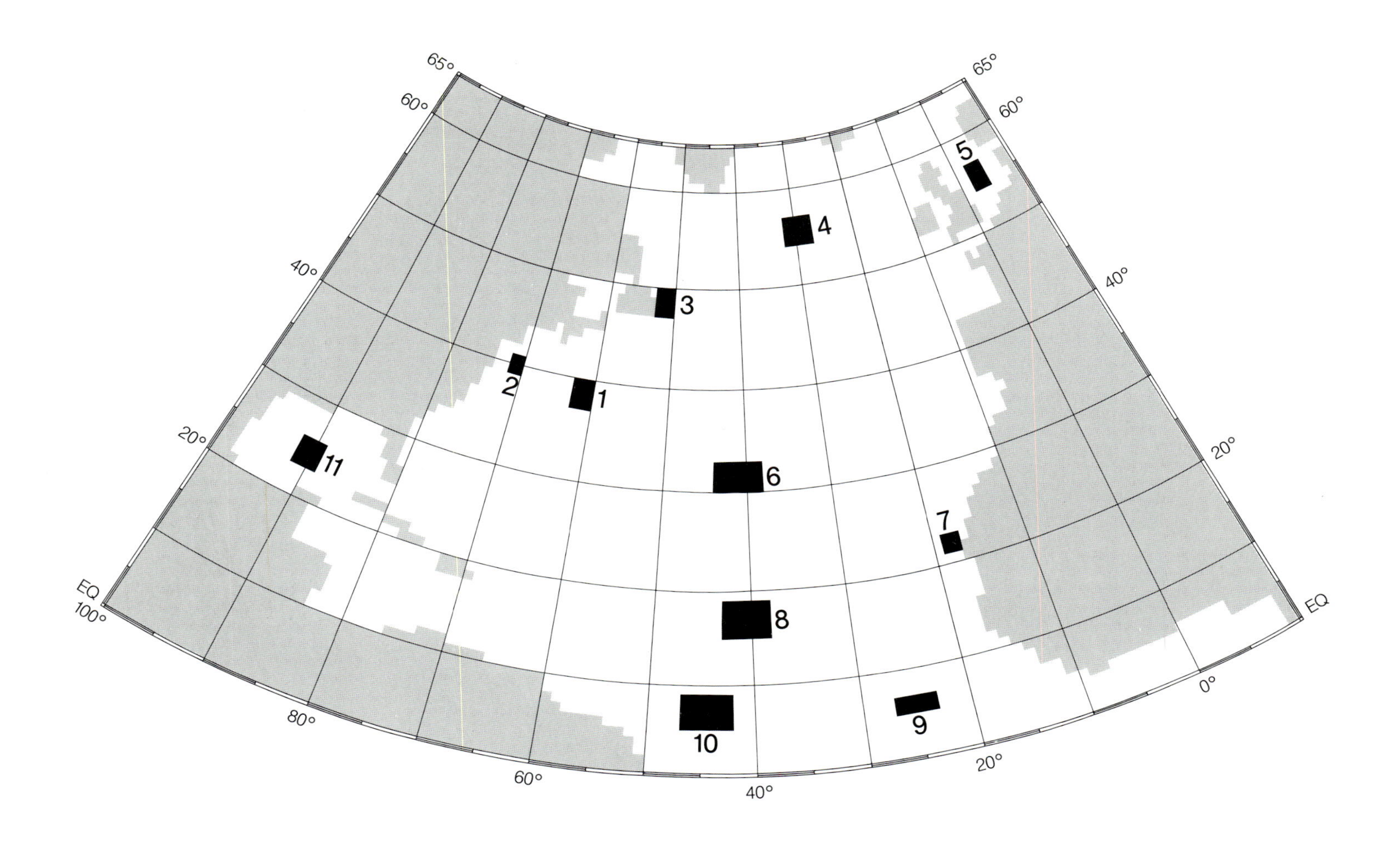

Fig. 3. Selected characteristic areas of the North Atlantic Ocean. The annual cycles of various parameters in these areas are given in Figs. 20 to 31

shaft at the grid point. The following parameters were not in the original data set but are calculated from interpolated fields: relative humidity, mixing ratio minus mixing ratio at sea surface temperature (this parameter is presented, because together with wind speed it controls the latent heat flux) and directional steadiness and divergence of the wind. The latter was calculated from the interpolated mean wind components using a centred difference scheme which takes into account the poleward convergence of the meridians. Steadiness of the wind is calculated from the absolute value of the resultant wind, divided by the mean scalar wind speed. The annual field of directional steadiness is obtained from the annual fields of the resultant wind and scalar wind speed.

Bunker computed standard deviations from single ship observations for selected parameters. They represent the day-to-day variations about the long-term monthly mean within each gerrymander. These standard deviations contain a contribution from spatial variations of the field within the gerrymander. The procedure which we used to obtain standard deviations on a regular 1-degree grid is the following: Bunker's standard deviations were converted to variances. The squared differences between the original "step"-field values and the interpolated values at every grid point were calculated and averaged for each gerrymander. These averages are the spatial variances within each gerrymander and were substracted from the variances given by Bunker. Next, the interpolation procedure was applied to these residual variance fields. The gridpoint values were finally reconverted to standard deviations. Standard deviation fields are given for the central month in each season and for the year. The latter is the average of the monthly values, and does not contain the contribution from the annual cycle. Standard deviations for the following parameters are presented: sea surface temperature, air- minus sea surface temperature difference, mixing ratio, sea level air pressure and the wind components. Standard deviation of mixing ratio, computed from single ship observations, is not given in the Bunker data, but he evaluated mean dew point temperature and its standard deviation. From the interpolated fields of the latter and sea level air pressure we computed the standard deviation of mixing ratio as half the difference between the mixing ratios for mean dew point temperature plus and minus one standard deviation.

The range of the annual cycle for each gridpoint was calculated as the difference between the extremes of the monthly means. No further smoothing was applied to these fields.

The contour maps of the fields of monthly and annual means and standard deviations as well as of the annual range are objectively produced by linear interpolation between the gridpoints. The conic projection used, with intersections at 20°N and 55°N, is a true areal representation with maximum angular deviation less than 6 degrees between 10°N and 60°N. Only a rough design for the coast lines, based on a 1-degree grid net, is used to enclose the area where data are available.

Time-latitude diagrams are presented for zonally averaged variables in order to give a condensed presentation of the annual variation as a function of latitude. These diagrams are given for the above-mentioned parameters and also for the wind components. Such ocean-wide averages were formed for 1-degree latitude strips of every month, wherever data are available over the North Atlantic Ocean, including areas containing possibly non-significant information. As a consequence of zonal averaging, the information about longitudinal variations is missing from these diagrams. The fields are smoothed with a single pass of the one-dimensional Hann filter used in the latitudinal direction.

As an alternate approach, annual variations of selected, characteristic areas of the North Atlantic Ocean are presented for a set of 14 parameters. The position of these areas is shown in Fig. 3. The short title of the areas indicates the type of climate regime for which the area was taken as reprensentative. Areas 1, 2, 3, 7 and 10 are nearly coincident with the respective gerrymanders Bunker (1976) used for a similar description. Total averages over the North Atlantic Ocean from the equator to 65°N (area 12) are shown in the same way in Fig. 31. The presentation of such large-scale averages is less meaningful for some parameters (e.g. wind direction), but is given here for sake of completeness. The monthly mean values were computed from the 1-degree gridpoint data by area weighted averaging. As a visual aid they were connected by straight or dashed lines (the presentation of mean wind direction in this manner is of course doubtful). The corresponding figures for the energy budget terms will be given in Volume 2 of this atlas.

Random and Systematic Errors

Random errors, biases or trends of climatological means calculated from V.O.F. ship data may be induced through instrumental or observational errors, or changes in instrumental equipment and observing practices. Additionally, long period area means can be biased by uneven distribution of the basic observations (1) in the time of the day as well as of the month and (2) in space. This has been shown in a comprehensive work by Weare and Strubb (1981). In the following we quote some random and systematic errors of the basic data, which are known to effect long-term monthly averages.

In the period 1941 to 1972 most measurements of sea surface temperatures aboard merchant ships were made at the water intake in the ships' engine room, the actual depth of the inlet varying between 2 and 10 m below the surface. There have been some detailed investigations about the difference between the intake temperature and the bucket temperature, which is normally measured between the surface and 1-m depth. Walden (1966) found by a comparison of simultaneous measurements from a collection of V.O.F. ship data in the North Atlantic Ocean, that on average the intake temperature is higher by 0.3 °C than the bucket temperature. This difference undergoes diurnal variations and is shown to be a function of wind speed and latitude. Saur (1963) compared simultaneous measurements in the North Pacific and also found the intake temperature to be systematically higher by 0.7°C. Tabata (1978) gave a review of the previous literature of the subject: the difference between intake- and bucket sea surface temperature ranges from 0.2° to 0.7 °C. The intake temperature can also differ from the skin temperature (e.g. Hasse 1970) of the ocean surface as measured by radiation instruments from satellites (Tabata and Gower 1980). A recent review about satellite measurements in comparison with in situ data is given by Minnet et al. (1984).

Air temperature and wet bulb temperature from shipboard instruments, although tested on research vessels, have been found to be at least 0.2°C too high when compared with buoy data (Augstein et al. 1974). For merchant ships this bias may be even higher. Ramage et al. (1972) also reported an overestimation of these parameters. From a collection of V.O.F. ship data for the period 1980 to 1984 in the North Atlantic Ocean, we found that the majority of the weather reports are made at 12.00 or 18.00 GMT. This implies that the North Atlantic Ocean data set contains by a factor of about two more daytime than nighttime observations, and consequently air- and wet bulb temperature may possibly be even more biased towards high temperatures due to radiation effects. Quayle (1984) compared climatological air temperatures from NOAA data buoys with those from V.O.F. ship data and found the latter to be 0.8 °C higher on average. We therefore assume that mean climatological air temperature may be biased by about 0.5 °C and a similar bias is expected for wet bulb temperature.

Observations of cloudiness are made by eye and climatological values are reliable at best to one okta. First comparisons with satellite images indicate that the ground-observed cloud cover is higher than the results from space measurements (Reed 1977).

Because precipitation cannot be measured reliably over the ocean, precipitation frequency, that is the number of ship observations containing precipitation in any form in the weather log divided by the total number of observations, is commonly calculated from the basic observations (Goldsmith and Bunker 1979). Quayle (1974) found precipitation frequencies from transient merchant ships to be significantly lower than those from ocean weather ships. Systematic percentage errors are as large as 50% and vary seasonally. As this difference cannot be entirely explained by a fair weather bias, errors must be introduced by practices observed on merchant ships. A quantitative correction may be difficult to find because of the subjective nature of the observation of this quantity.

For sea level air pressure measurements from ships, according to our experience a r.m.s. error of 0.6 hPa is typical after elimination of coarse coding and transmission errors; the latter may increase the r.m.s. error by a factor of 2. A frequent source of error seems to be missing or erroneous reduction of pressure from measurement height to sea level. Gradients of sea level air pressure are commonly believed to be more reliable than measurements or estimates of wind from running ships.

In a study on wind speed measurements from V.O.F. ship data in the world ocean Quayle (1980) found that in the North Atlantic Ocean from 1950 to 1971 less than 15% of marine wind reports are based on measurements, while the major part was estimated from the state of the sea surface and then coded in Beaufort equivalent scale. Verploegh (1967) found the observer-induced errors in this latter procedure to be in the same order of magnitude as those of wind speed measurements; random errors are 1.3 m s^{-1}. Generally random errors are assumed to be in the range from 10 to 15% (Dobson 1981). In view of the results of Quayle (1980), the question whether marine wind reports may be influenced by the ship itself seems to be of minor importance, at least for this data set. Dobson (1981) discussed effects of flow distortion and different ship anemometer heights, and recommended no correction of single observations at the present time, since the magnitude and even the sign of possible errors vary from ship to ship (Augstein et al. 1974). Quayle (1980) could not find a significant fair weather bias, supposedly induced by weather routing of merchant ships. While a fair weather bias may be expected due to weather routing, a compensating effect could be induced by ships slowed down in bad weather and also increased transmission of weather observations under bad weather conditions. Bunker converted most of the wind speed data from the old Beaufort equivalent scale, defined by WMO code 1100 (WMO 1971), to meters per second. This code has been in use since 1906. Quayle (1980) analysed non-simultaneous wind speed data on a climatological basis. He compared Beaufort estimates from passing ships and wind speed measurements from ocean weather ships. The results show agreement for wind speeds higher than Bft. 5; in the lower wind speed range estimates from passing ships give smaller wind speeds. However, this difference does not exceed 1 m s^{-1}. Kaufeld (1981) in a similar study using simultaneous data, found a definite discrepancy between the old Beaufort equivalent scale and his own evaluation. This would lead to higher wind speeds below Bft. 8 and lesser speeds above. Systematic differences reach 2 m s^{-1} between Bft. 3 and 5.

The above gives some indication of the difficulties in obtaining reliable and unbiased climatological averages from meteorological observations from V.O.F. ship data. It should be noted by the user of this atlas that we present the data as processed by Bunker without adding any corrections for biases. A short discussion of the influence of biases and uncertainties of parametrisations on calculation of energy exchange components will be given in Volume 2.

The Climate of the Marine Surface Layer over the North Atlantic Ocean

In the following section we present a short description of the main features of the climate of the marine surface layer over the North Atlantic Ocean, as visible in the maps and diagrams contained in this volume. A more general description, including the upper air circulation and the climatology of synoptic systems, has recently been given by Tucker and Barry (1984). Instead of describing details of the monthly fields chart by chart, we will give a more schematic and introductory overwiew. In this way we can avoid quoting all the numbers on the charts and diagrams and giving detailed explanations for all observed patterns.

For an overview, it is conducive to start from known features of climate. There is the meridional gradient of temperature and its annual variation originating from the corresponding meridional and annual variation of the insolation. Stronger gradients of temperature are found in winter than in summer. It is well known that equator pole temperature gradients lead to baroclinic instability. Hence, in the dynamic fields (and in the fields of parameters closely related to the atmospheric circulations, e. g. cloudiness) we find the typical structure of three regimes: the tropical Hadley cell, the prevailing westerlies of the temperate zone and the polar easterlies. The regimes of the atmospheric circulation naturally move with the sun during the course of the year. These general considerations readily explain most of the observed patterns and annual variations of the thermodynamic and dynamic fields in the bulk of the ocean. There are two modifying influences: advection in the atmosphere, and advection in the ocean by currents and upwelling.

It is hardly necessary to say that there is a meridional gradient of both air- and sea surface temperature influencing the fields of all thermodynamic properties. Typical numbers might be taken from the diagrams of zonally integrated values. The spatial distribution of temperatures and humidity shows an approximately sinusoidal annual cycle in the whole North Atlantic Ocean and the thermal equator is seen to migrate meridionally. The shift of the 26°C isotherm of zonally averaged sea surface temperature from 9°N in March to 30°N in August gives an impression of the heating of the large tropical areas. The atmospheric Polar Front is not visible in the surface temperature fields, which is not surprising, since the sequence of midlatitude depressions provides for a smooth meridional temperature variation in the mean. Due to the strong non-linear variation of water vapour pressure with temperature, there is a similar equator pole mixing ratio gradient: from more than 18 g kg^{-1} in the tropics to less than 4 g kg^{-1} in the northwestern part of the North Atlantic Ocean. From June to October maxima in the Caribbean Sea even exceed 20 g kg^{-1}.

Modifications to the general pattern of a smooth equator pole gradient of temperature and humidity are brought about by advection, both in the ocean and in the atmosphere. Such modifications are prominent features in all maps of the temperatures and moisture. In the ocean the general pattern of the meridional gradient is strongly modulated by the presence of the main ocean currents. Strong, narrow currents are found on the west side of the North Atlantic Ocean. Advection by the Florida current and the Gulf Stream, transporting warm surface water from the tropics to northern latitudes and by the cold Labrador current, directed southward and passing around the Grand Banks of Newfoundland, produces a sharp sea surface temperature gradient on the west side of the ocean between 35°N and 45°N. The gradient would probably be even stronger if the Gulf Stream did not show substantial meandering. The footprints of the meandering are found in the monthly maps of standard deviation of sea surface temperature, which show maxima near 5°C just southeast of the Grand Banks. Annual ranges of sea surface temperature are largest off the American coast near 40°N, where a spur of Labrador current water (ca. 4°C) exists in February, while this region is occupied by the Gulf Stream (ca. 22°C) in August. The extension of the Gulf Stream, known as the North Atlantic Current, is assumed to split into several branches towards the north, northeast and southeast after leaving the North American shelf (Dietrich et al. 1975), thus feeding into the subpolar and subtropical gyre. This is also documented in the diverging isotherm configuration of sea surface temperature. Hence, the temperature gradients in the middle and eastern North Atlantic Ocean are small compared with those in some areas in the western part of the ocean. In May, for example, a sea surface temperature gradient of 14°C is found between 40°N and 45°N at 55°W while at 25°W it is only 2°C.

Due to vertical advection of colder water, relatively low sea surface temperatures are found in the coastal upwelling areas off Northwest Africa and to a minor extent off South America in the southern Caribbean. The maps of the resultant wind show that at the coast of Northwest Africa the average wind direction is parallel to the shoreline throughout the year, leading to offshore Ekman transports and upwelling. The upwelling area is associated with the annual north–south migration of the atmospheric circulation system; it moves North and South along the African coast, covering an area between 10°N and 25°N in January and extending northward towards Portugal in July. This is in agreement with more detailed research (Wooster et al. 1975). The southern border of upwelling is marked by a sharp gradient in sea surface temperature. Atmospheric advection works together with the upwelling to produce positive air-sea temperature differences throughout the year (between 1° and 2°C). The strong static stability inhibits formation of convective clouds. Together with a prevalent divergence in the wind field a minimum of total and low cloud cover is conspicuous: from February to May total cloudiness is less than two oktas in the African upwelling area.

Naturally, atmospheric advection can be effective only where there are large property gradients. Hence, areas where a predominant offshore wind carries continental air over the sea should be influenced the most. This is clearly seen, for example, in the charts of annual range of both air temperature and mixing ratio: high ranges are found associated with areas off the continent of North America and close to the coast of Europe. While in the middle of the ocean along 30°W annual ranges of air temperature are small, less than 1°C in the inner tropics to about 7°C at 40°N, they increase to about 18°C near the American coast between 40°N and 50°N, where prevailing west winds carry continental air masses far out over the ocean. Regions of high annual range are also found in the North Sea and between 10°N and 20°N off the African coast. Atmospheric advection may also be effective where oceanic currents induce stronger gradients of atmospheric properties. The proximity of cold and warm ocean currents, together with advection of warm air over the cold Labrador current, are the reasons for the relative frequent occurrence of fog, leading to extreme values of average relative humidity over the Grand Banks of Newfoundland: over 92% in July and over 86% in the yearly mean, while over the bulk of the ocean relative humidity is between 75 and 82% and is one of the least variable of the quantities presented in this atlas.

In some areas atmospheric and oceanic advection supplement each other and produce high air–sea interaction activity, as documented in the charts of air–sea temperature difference. We mention in this context the Florida current and the Gulf Stream area. Their course is well marked by the maxima of negative temperature differences. Extreme negative differences are found in winter even in the monthly mean (nearly −6°C in January) with average winds from the northwest and west. In summer air–sea temperature differences are between −1°C and −2°C. This agrees well with prevailing wind directions from the southwest from July to August. Advection of warm air, on a track parallel to the Gulf Stream axis also leads to positive air–sea temperature differences over the cooler waters of the Labrador current, from April through September. Oceanic and atmospheric advection also lead to maxima of standard deviation of the air–sea temperature difference, extending off the American coast and far out into the ocean.

The mixing ratio difference pattern closely follows the pattern of air–sea temperature differences. The gradient between the Gulf Stream and the Labrador current, where small positive values occur from April to September, is conspicuous near 40°N during the whole year. But, unlike the air–sea temperature difference, there is no strong gradient of mixing ratio difference southeast of the core of the Gulf Stream. Neither are the values in the Gulf of Mexico very different from those in the Gulf Stream region. Also, while the air–sea temperature difference becomes positive in upwelling regions, the mixing ratio difference remains negative with smaller absolute values. Comparison of the zonal averages indicates that, unlike air–sea temperature difference, the mixing ratio difference shows only a weak annual signal between 35°N and 45°N and no positive values at all. The minimum of the mixing ratio difference is situated between 18°N and 26°N from October to December, thus showing a shift in space and time compared to the minimum of the temperature difference in this season.

Ship observations of wind fields have been a major root of marine meteorology. Knowledge of the wind fields of the Atlantic Ocean have to a large extent influenced Hadley's and Ferrel's perception of the three-cell system of the general circulation of the atmosphere. Features like the Intertropical Convergence Zone (ITCZ), the trade winds, the subtropical high and the west wind belt of the temperate zone are easily identifiable in the charts of the dynamic variables.

Although the position and strength of the ITCZ, if taken as a region of deep convection, shows synoptic variation (as is known both from satellite pictures and from the GARP Atlantic Tropical Experiment, GATE 1974), it is well marked in the monthly maps of the resultant wind as a continuous line of change in direction. The latter coincides with a narrow band of minimum of the resultant wind (smaller than 1 m s^{-1} throughout the year) and a somewhat broader area of minimum of the scalar wind speed. In general, the ITCZ is found a few degrees further north in the east of the ocean than in the west. It is situated north of the equator throughout the year, except in the western part of the North Atlantic Ocean from February to May. The features along its axis change considerably in the course of the year. Especially in summer, the centre of the convergence zone between 30°W and 40°W exhibits unorganized patterns of wind direction and a minimum of the value of the resultant wind. To the west of this minimum, the meeting of the northeast trades and the equator-crossing southeast trades is organized in a smooth confluence, directed westward towards the northern coast of South America, while at the east, the trades from both hemispheres recurve into the African continent. The area of wind convergence covers a wider latitudinal range. Extreme values near the axis of the ITCZ reach 12×10^{-6} s^{-1} in January near the equator and 8×10^{-6} s^{-1} in July between 8°N and 10°N. The annual migration of the ITCZ is also seen in the graphs of zonal averages of wind field parameters, especially in the north–south component. It is also well marked in the precipitation frequency maps, although the line of maximum frequency does not fully coincide with the center of the convergence zone. Maxima are found to be near 30% in February and exceed 40% in July and August. The sea level air pressure fields show the ITCZ as a weak "equatorial" trough with minimum values between 1010 hPa in January and 1014 hPa in July. Minima in both the annual range (between 1 and 4 hPa) and in the monthly standard deviations are also characteristic features. The latter do not exceed 3 hPa in the bulk of the area between the equator and 10°N in every calendar month. Standard deviations of sea level air pressure as well as of the wind components grow continuously from the near-equatorial ITCZ region through the trade wind region far into the belt of prevailing westerlies.

The trade winds show high directional steadiness (in some areas more than 95%) even in the zonally averaged values. Unlike scalar wind speed, the absolute value of the resultant wind is largest in the trade wind regime due to its high directional steadiness. In most of the month two maxima are found: the first in the main branch of the trade winds between North Africa and South America (more than 8 m s^{-1}), and the second in the southern Caribbean Seas. The latter exceeds the first in strength except in May and October. Another typical feature of the trade wind region is low precipitation frequency, which is less than 5% in an area extending from the African coast far into the North Atlantic Ocean.

The subtropical high with its centre at the northern edge of the trade winds is a dominant feature of the sea level air pressure field throughout the year. The seasons are marked by a variation in intensity (1018 hPa in winter, more than 1025 hPa in summer) and a north south migration (by about 10° of latitude in the centre of the ocean and at the east side). During spring and summer until July the subtropical high develops from a more zonally oriented belt to a cell with closed isobars centred near 35°N/35°W, with a southwest–northeast directed axis crossing the Azores. From September to January the high shifts southward, broadens and reorganizes into this belt. The centre of the high is well marked in the wind field as an often rather narrow region of minima in both the scalar wind speed and the resultant wind and of direction change or anticyclonic rotation pattern. The relative minimum of cloud cover, marked by the area enclosed by the four-okta isoline, is found southwest of the center of the high pressure area, especially in summer.

The westwind belt of the temperate latitudes reaches 30°N in winter and is confined to areas north of 40°N in July. Maxima of the scalar wind speed reach more than 13 m s^{-1} in winter. From June to August the western part of the ocean between 30°N and 40°N shows winds from the southwest, blowing nearly parallel to the axis of the Florida current and Gulf Stream. The latter area shows high directional steadiness, which reaches 70% in July, while in the other regions of the west wind belt the steadiness varies mainly from less than 10% to as much as 40%. These features reflect the variability which is brought about by the procession of lows and highs: it also appears in the maxima of standard deviations of sea level air pressure and of the wind components. The former reaches maxima of more than 17 hPa in January; in July the pressure variability is smaller (6 hPa to 10 hPa). The wind components show their highest variability between 50°N and 60°N, reaching standard deviations of more than 10 m s^{-1} in January. Except in summer the area of high standard deviations extends around the Grand Banks of Newfoundland and reflects one of the main cyclone tracks in the North Atlantic Ocean. This may also be seen in the charts of scalar wind speed. High annual sea level air pressure ranges dominate in the central North Atlantic Ocean north of 35°N. In the annual mean between 45°N and 55°N the strongest pressure gradient is found. The "Icelandic low" is shown in the annual map as a low pressure centre east of Greenland, while it is not evident in the monthly maps. The precipitation frequency reaches more than 60% (near Ocean Weather Station B in January). A tendency is seen for the maximum to occur in the Greenland–Newfoundland region in the winter month and in the eastern part between Iceland and the British Islands from May to August. Total cloudiness is higher than 6 oktas north of 50°N and reaches maxima of more than 7 oktas southwest (in January) and south (in July) of Greenland.

At the northern edge of the area easterly wind components are occasionally found. Although easterly winds north of the atmospheric polar front are expected as a result of the average polar anticyclone, it should be noted that the persistency of the wind direction north of 60°N is less than 10%. Hence, with relatively few observations in the higher latitudes, these results may be spurious.

The short discussion of the climate of the North Atlantic, as visible in the charts and diagrams of this atlas, shows that Bunker's data set is an excellent base to present a climate atlas of the North Atlantic Ocean. It is possible to give a reasonable description of monthly and annual fields with high spatial resolution. While the high resolution was primarily aimed at for easier use in numerical modelling, it is evident that some features of the meteorological fields appear rather well marked and concentrated in the monthly maps. The variations of the fields of meteorological variables and sea surface temperature given in this volume form the base for a discussion of the fields of surface fluxes of energy and momentum in Volume 2.

Concluding Remarks

The data base for this North Atlantic Ocean atlas consists of 8 million voluntary ship observations from the period 1941 to 1972. They were checked for consistency and averaged by the late Andrew F. Bunker. In order to obtain an optimum spatial resolution, Bunker used partly irregularly shaped averaging areas. These areas are also tailored according to oceanographic near surface structures. Fortunately enough, observational coverage is most dense in areas with strong gradients of sea surface temperature and other properties of the marine surface layer, providing for detailed resolution. From the irregularly shaped areas long-term monthly averages have been transferred to a regular 1-degree grid net, using a special interpolation scheme. The charts presented in this atlas have been obtained by totally objective techniques.

The difficulties in obtaining reliable meteorological observations from shipboard have been treated repeatedly in literature and should not be overlooked. While random errors should not influence mean climatological results, some possible biases, as discussed in the text, may remain. Such biases can be accounted for when calculating derived quantities, like air–sea energy exchange components. In the first volume of this atlas the observations are given as processed by Bunker, systematic correction are not applied.

This atlas presents a description of the annual mean state and the annual cycle of the conditions in the marine surface layer over the North Atlantic Ocean. Maps and diagrams show not only the large-scale features but also surprisingly fine detail. It is the agreement of such details with other knowledge that provides confidence in the results. It is hoped that the atlas will be found useful for climate studies in meteorology and oceanography and will provide background material for a large variety of studies and users in marine science, ocean engineering and other applications.

References

Augstein E, Hoeber H, Krügermeyer L (1974) Fehler bei Temperatur-, Feuchte- und Windmessungen auf Schiffen in tropischen Breiten. METEOR-Forschungsergeb Reihe B 9: 1–10

Br Meteorol Off (1948) Monthly meteorological charts of the Atlantic Ocean. Air Minist, London, MO 483

Bryan K (1983) Poleward heat transport by the ocean. Rev Geophys Space Phys 21: 1131–1137

Budyko MI (1963) Atlas of heat balance of the earth. Acad Sci, Moscow, 69 pp

Budyko MI (1974) Climate and life. Int Geophys Ser, vol 18. Academic Press, London New York, pp 508

Budyko MI (1978) Heat balance of the Earth. Gidrometeoizdat, Leningrad (in Russian)

Bunker AF (1975) Energy exchange at the surface of the western North Atlantic Ocean. Tech Rep, WHOI-75-3, Woods Hole Oceanogr Inst

Bunker AF (1976) Computations of surface energy flux and annual air–sea interaction cycles of the North Atlantic Ocean. Mon Weather Rev 104: 1122–1140

Bunker AF, Worthington LV (1976) Energy exchange charts of the North Atlantic Ocean. Bull Am Meteorol Soc 57: 670–678

Dtsch Hydrogr Inst (1976) Monatskarten für den nordatlantischen Ozean, 4th ed, no 2420. DHI, Hamburg

Dietrich G, Kalle K, Krauss W, Siedler G (1975) Allgemeine Meereskunde. Eine Einführung in die Ozeanographie. Bornträger, Berlin, pp 593

Dobson FW (1981) Review of reference height for and averaging time of surface wind measurements at sea. Marine meteorology and related oceanographic activities. Rep no 3. WMO, Geneva, 64 pp

Dobson FW, Bretherton FP, Burridge DM, Crease J, Kraus EB, Vonder Haar TH (1982) The CAGE experiment, a feasibility study. WCP-22, WMO, Geneva, 95 pp

Esbensen K, Kushnir V (1981) The heat budget of the global oceans: an atlas based on estimates from marine surface observations. Climat Res Inst Rep No 29, 188 figs, Oregon State Univ, 27 pp

Goldsmith RA, Bunker AF (1979) WHOI collection of climatology and air–sea interaction data. Tech Rep, WHOI-79-70, Woods Hole Oceanogr Inst

Hasse L (1970) The sea surface temperature deviation and the heat flow at the sea–air interface. Boundary-Layer Meteorol 1: 368–379

Hastenrath S (1982) On meridional heat transport in the world ocean. J Phys Ocean 12: 922–927

Hastenrath S, Lamb PJ (1977) Climatic atlas of the tropical Atlantic and eastern Pacific Oceans. Univ Wisc Press, Madison, 105 pp

Hastenrath S, Lamb PJ (1978) Heat budget atlas of the tropical Atlantic and eastern Pacific Oceans. Univ Wisc Press, Madison, 104 pp

Holloway JL (1958) Smoothing and filtering of time series and space fields. Adv Geophys 4: 351–389

Kaufeld L (1981) The development of a new Beaufort equivalent scale. Meteorol Rundsch 94: 17–23

Meserve JM (1974) U.S. Navy Marine climatic atlas of the world, vol I. North Atlantic Ocean. NAVIAR 50-1C-528, 386 pp

Minnett PJ, Zavody AM, Llewellyn-Jones DT (1984) Satellite measurements of sea surface temperature for climate research. In: Gautier C, Fieux M (eds) Large–scale oceanographic experiments and satellites. NATO ASI Ser, Reidel Hingham MA, pp 57–85

Quayle RG (1974) A climatic comparison of ocean weather stations and transient ship records. Mar Weather Log 18: 307–311

Quayle RG (1980) Climatic comparisons of estimated and measured wind from ships. J Appl Meteorol 19: 142–156

Quayle RG (1984) Comparisons between ship and buoy climatologies. Mar Weather Log 28: 137–140

Ramage CW, Miller FR, Jeffries C (1972) Meteorological atlas of the international Indian Ocean expedition, vol I. The surface climate of 1963 and 1964. Washington DC

Reed RK (1977) On estimating insolation over the ocean. J Phys Ocean 7: 482–485

Saur JFT (1963) A study of the quality of seawater temperature reported in the logs of ships' weather observations. J App Meterol 2: 417–425

Shapiro R (1970) Smoothing, filtering and boundary effects. Rev Geophys Space Phys 8: 359–387

Tabata S (1978) Comparison of observations of sea surface temperatures at ocean station P and NOAA buoy stations and those made by merchant ships travelling in their

vicinities, in the North Pacific Ocean. J Appl Meteorol 17: 374–385
Tabata S, Gower J F R (1980) A comparison of ship and satellite measurements of sea surface temperatures off the Pacific coast of Canada. J Geophys Res 85: 6638–6648
Tucker G B, Barry R G (1984) Climate of the North Atlantic Ocean. In: Loon H van (ed) Climates of the oceans, world survey of climatology, vol 15. Elsevier Sci Publ New York, pp 193–262
Verploegh G (1976) Observation and analysis of the surface wind over the ocean. K Ned Meteorol Inst Medd Verh 89: 67
Walden H (1966) Zur Messung der Wassertemperatur auf Handelsschiffen. Dtsch Hydrogr Z 19: 21–28
World Clim Res Programme (WCRP) (1984) G/WCRP Plan, ICSU/WMO, 79 pp
Weare B C, Strubb P T (1981) The significance of sampling biases on calculating monthly mean oceanic surface heat fluxes. Tellus 33: 211–224
Wooster W S, Bakun A, McLain D R (1976) The seasonal upwelling cycle along the eastern boundary of the North Atlantic. J Mar Res 34: 131–141
World Meteorol Org (1977) Guide to marine meteorological services. Rep no 471. WMO, Geneva
World Meteorol Org (1979) Reports on marine science affairs. Rep no 3. The Beaufort scale of wind force. WMO, Geneva, 22 pp

Time Latitude Diagrams of Zonal Averages

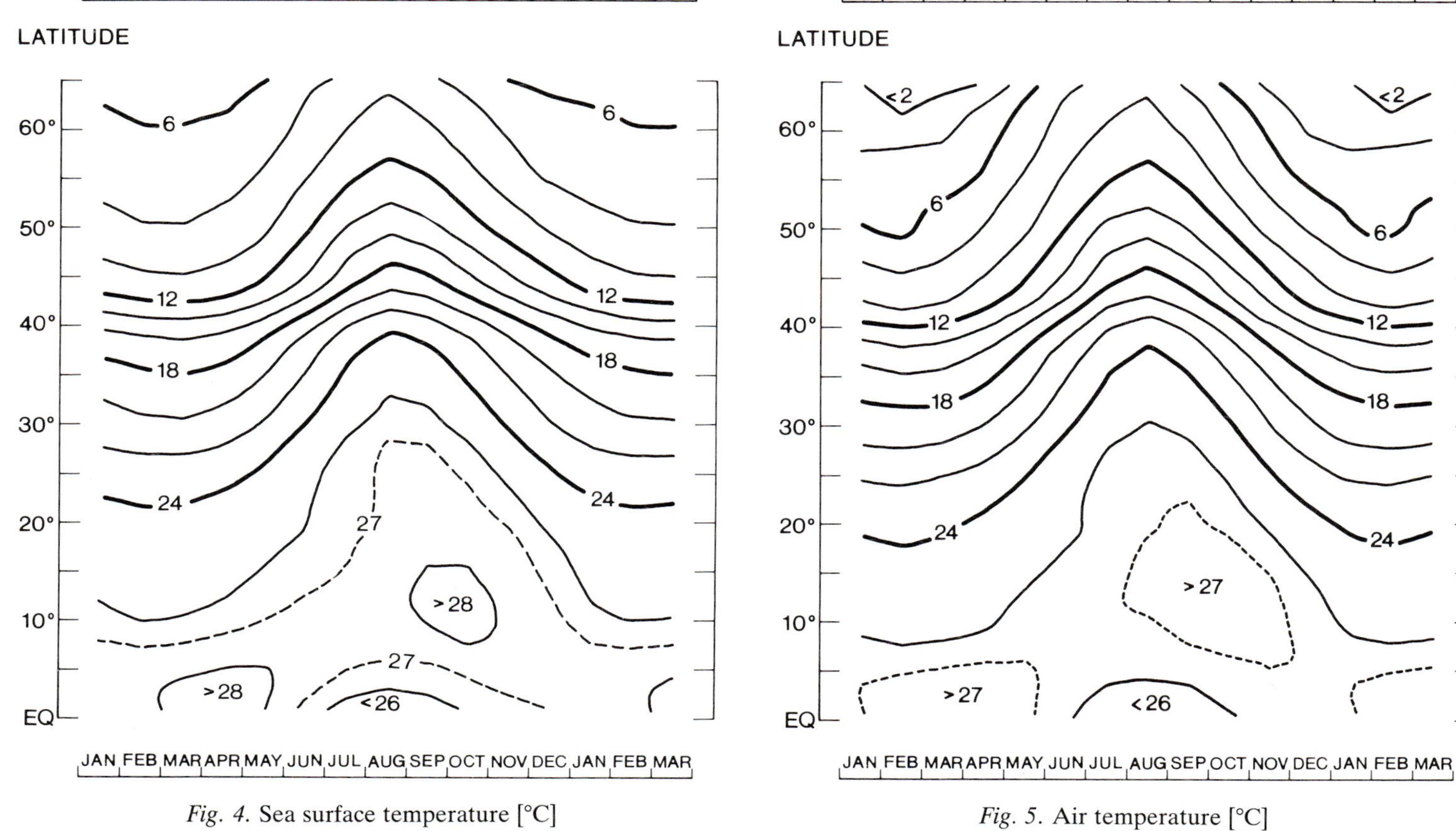

Fig. 4. Sea surface temperature [°C]

Fig. 5. Air temperature [°C]

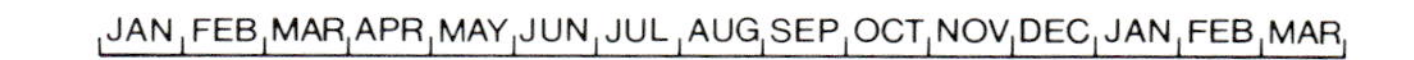

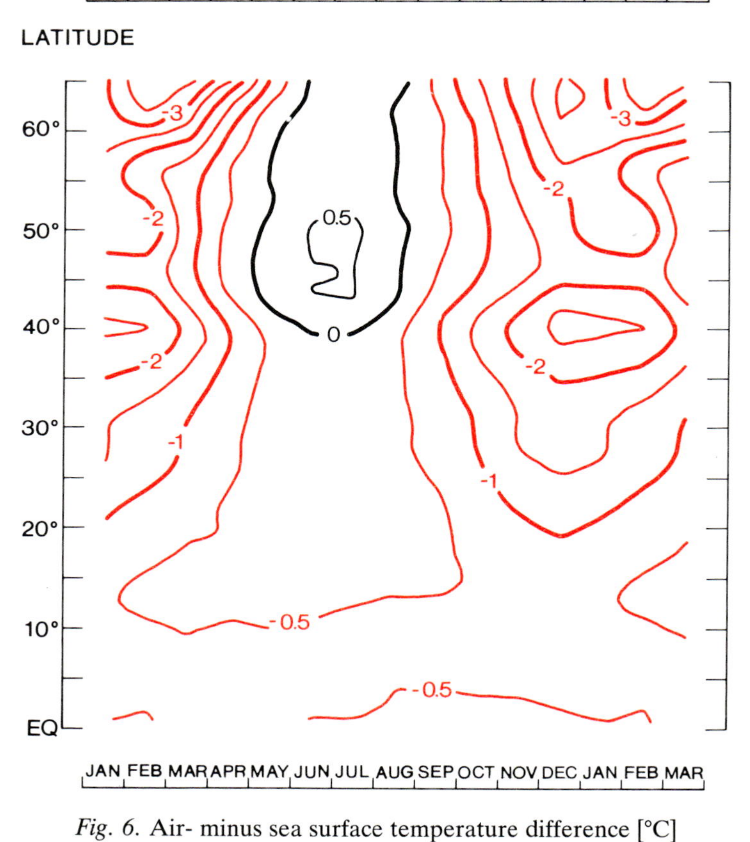

Fig. 6. Air- minus sea surface temperature difference [°C]

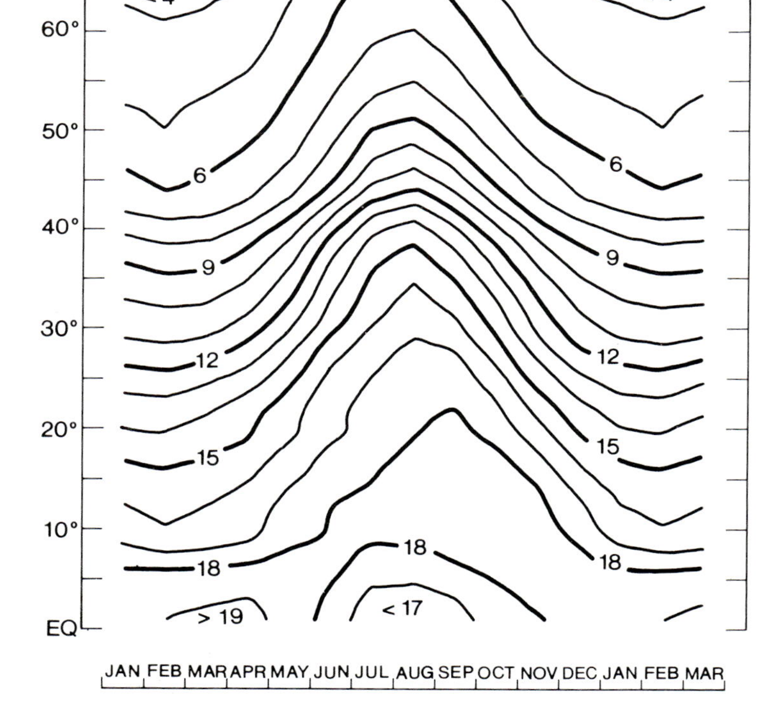

Fig. 7. Mixing ratio [g kg^{-1}]

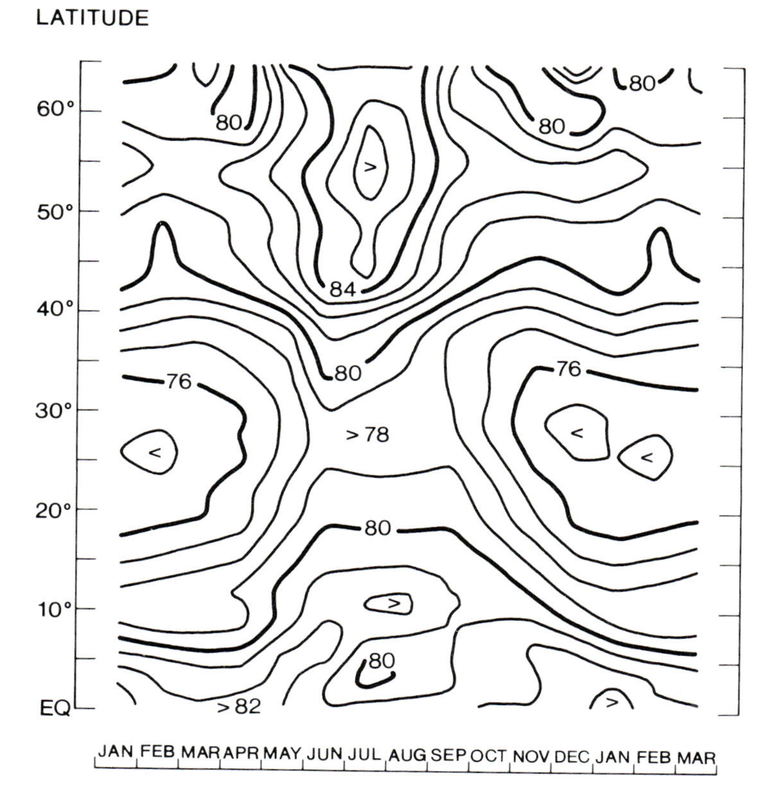

Fig. 8. Relative humidity [%]

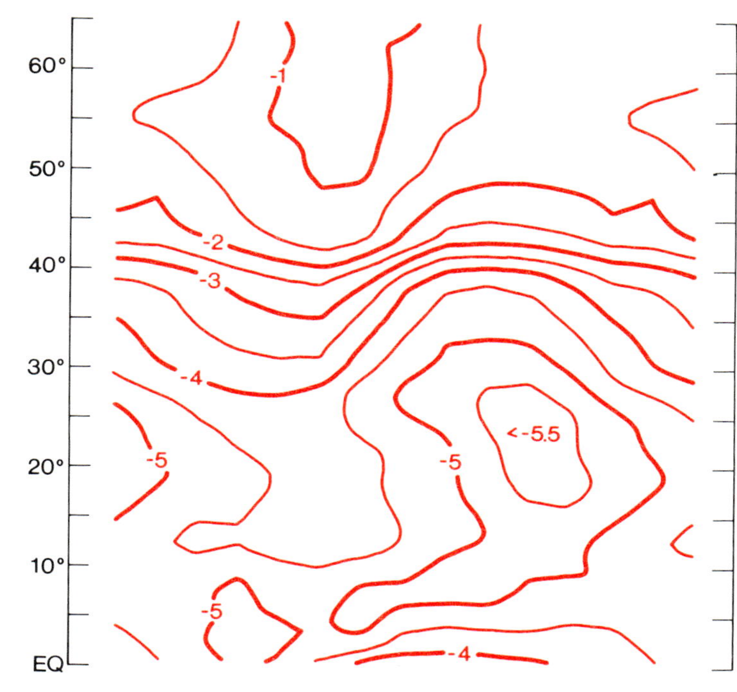

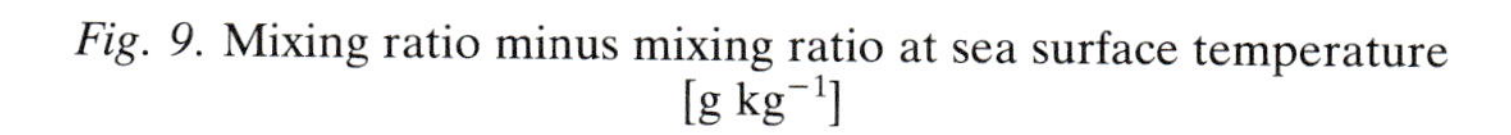

Fig. 9. Mixing ratio minus mixing ratio at sea surface temperature [g kg^{-1}]

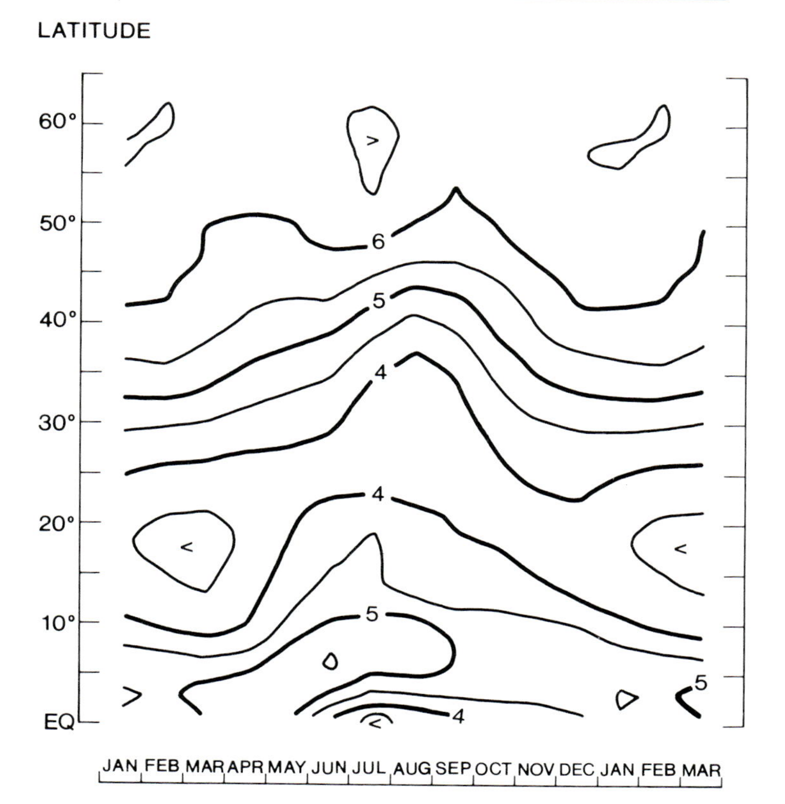

Fig. 10. Total cloud cover [oktas]

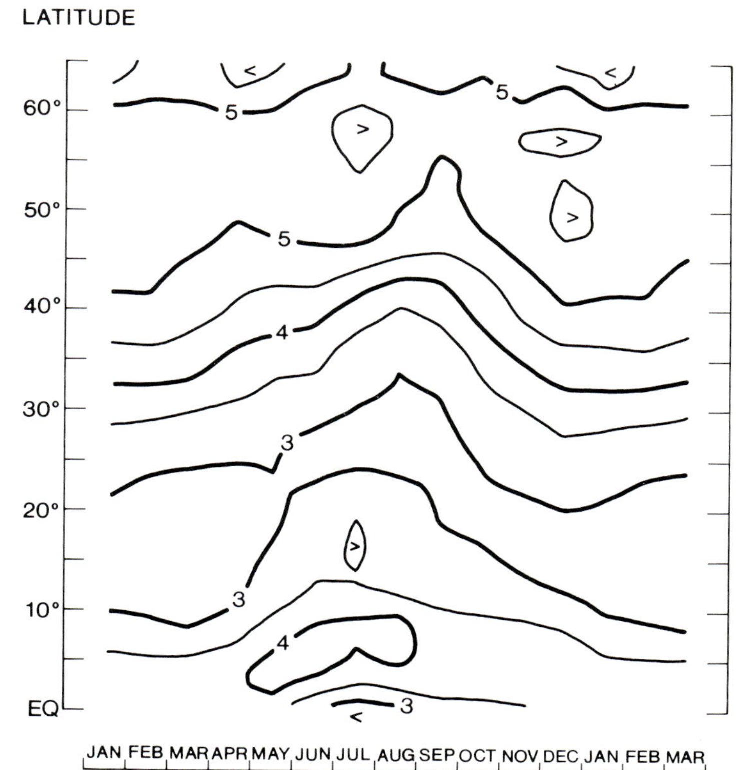

Fig. 11. Low cloud cover [oktas]

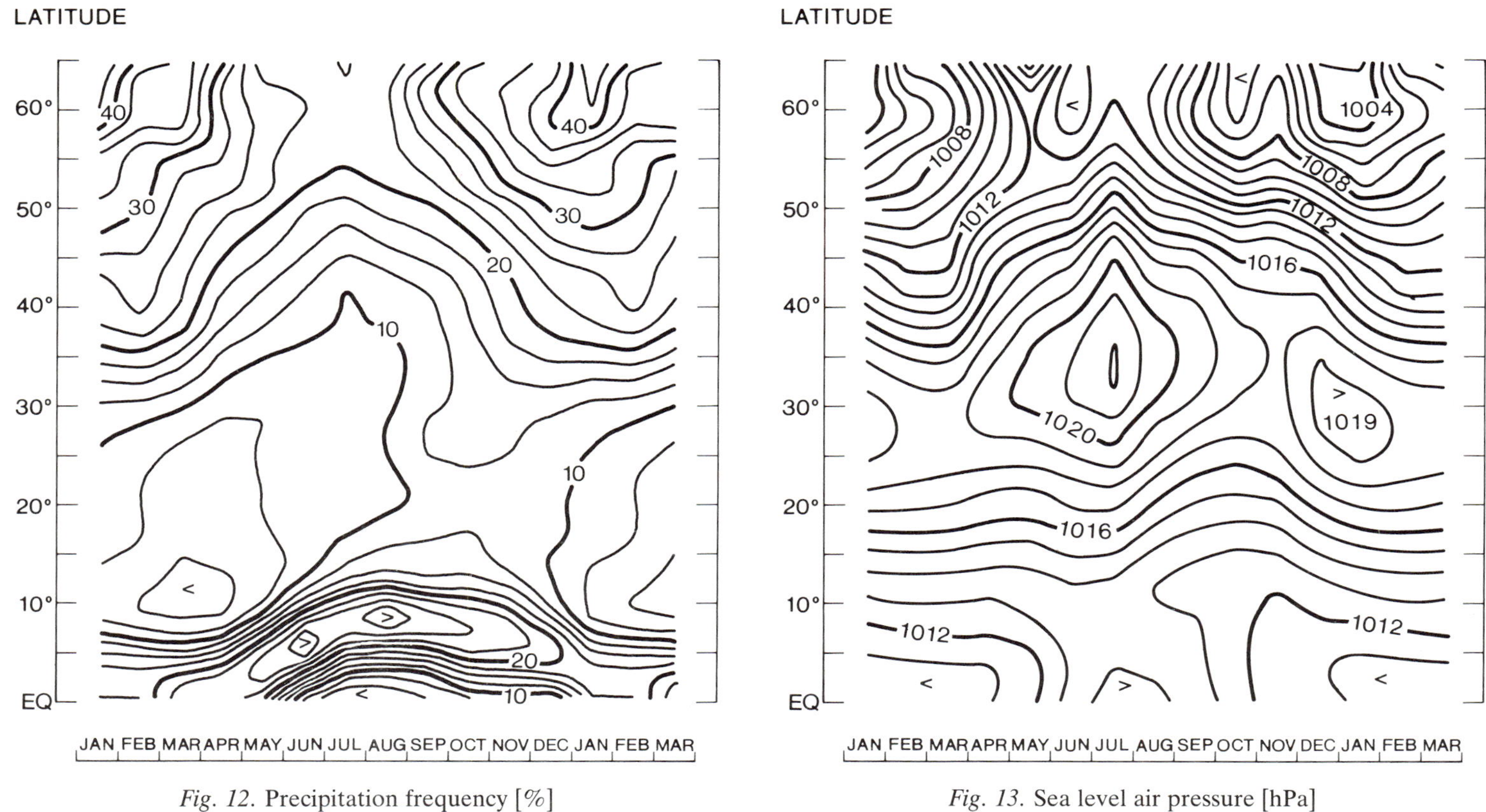

Fig. 12. Precipitation frequency [%]

Fig. 13. Sea level air pressure [hPa]

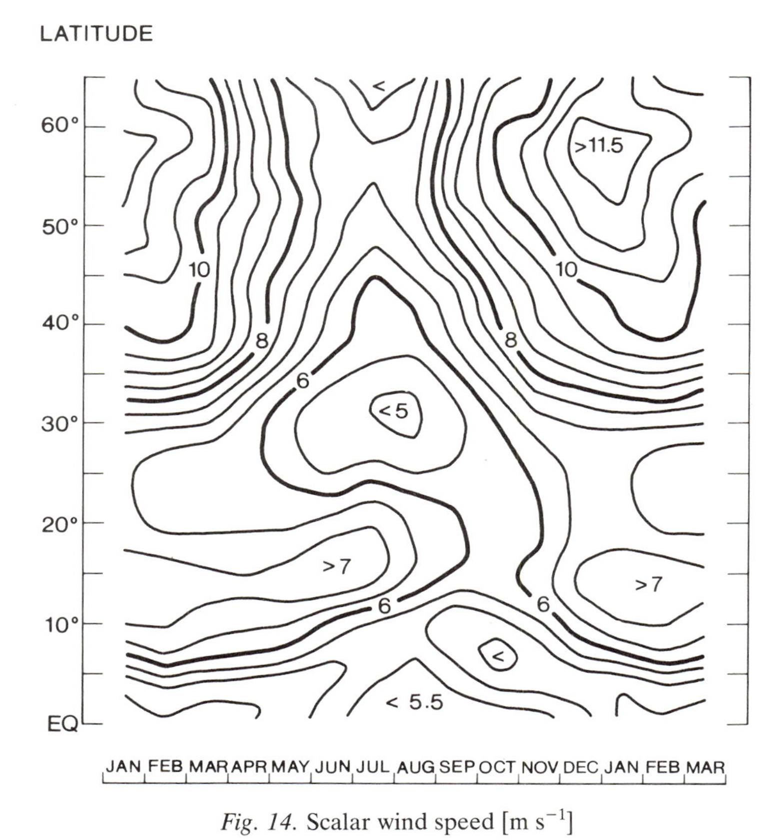

Fig. 14. Scalar wind speed [m s^{-1}]

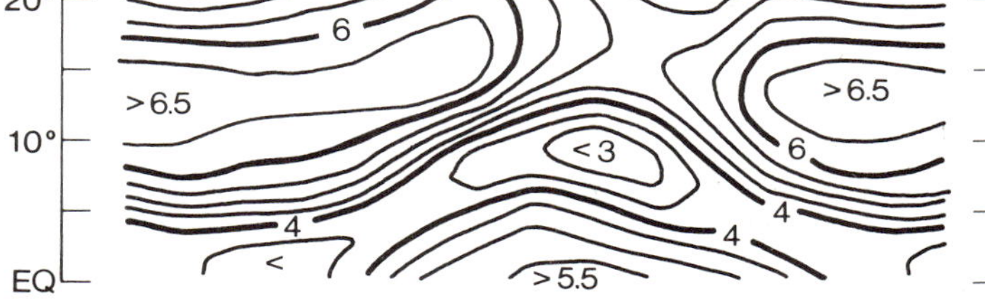

Fig. 15. Absolute value of the resultant wind [m s^{-1}]

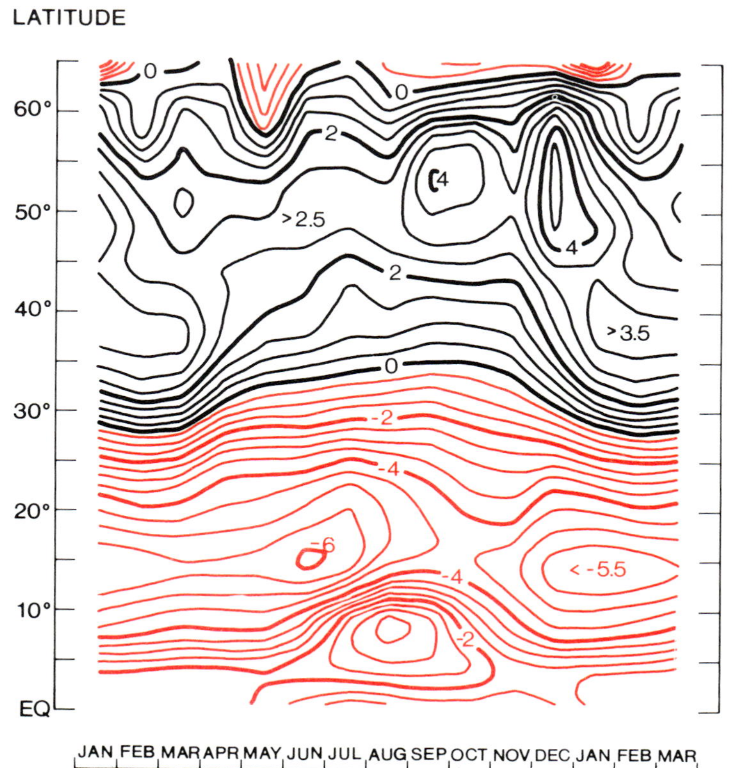

Fig. 16. East-west component of the wind [m s^{-1}]

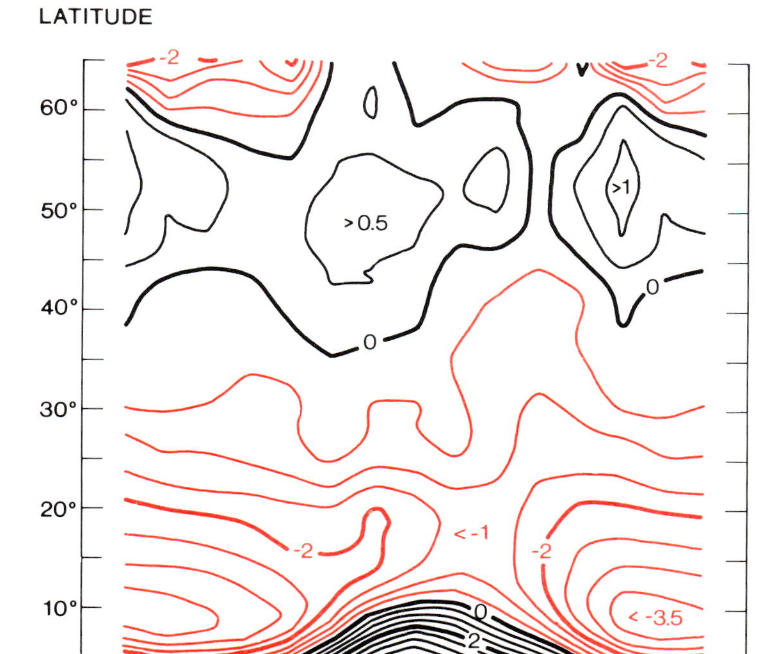

Fig. 17. North-south component of the wind [m s^{-1}]

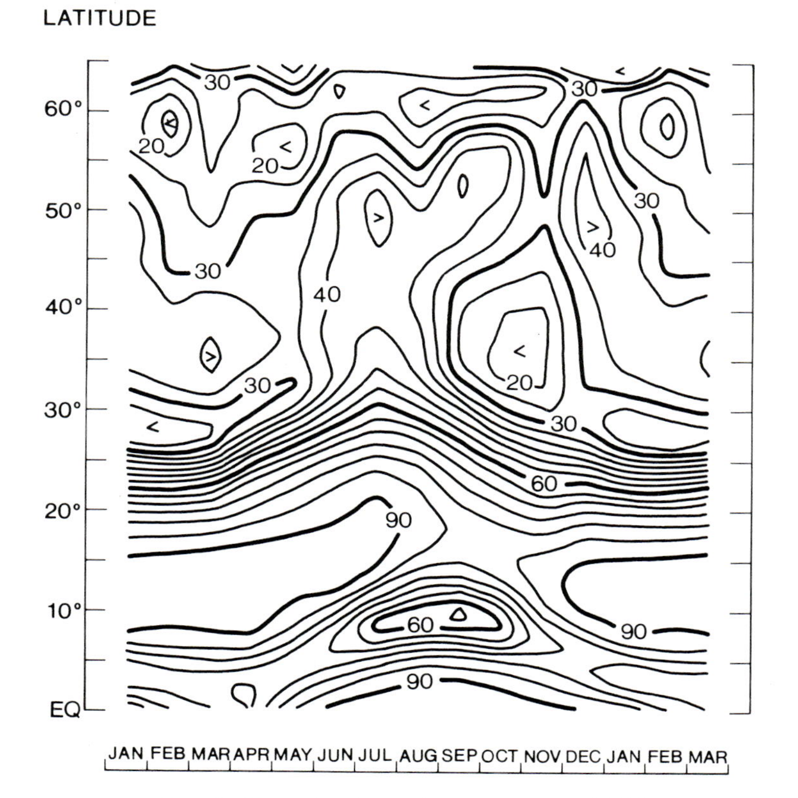

Fig. 18. Directional steadiness of the wind [%]

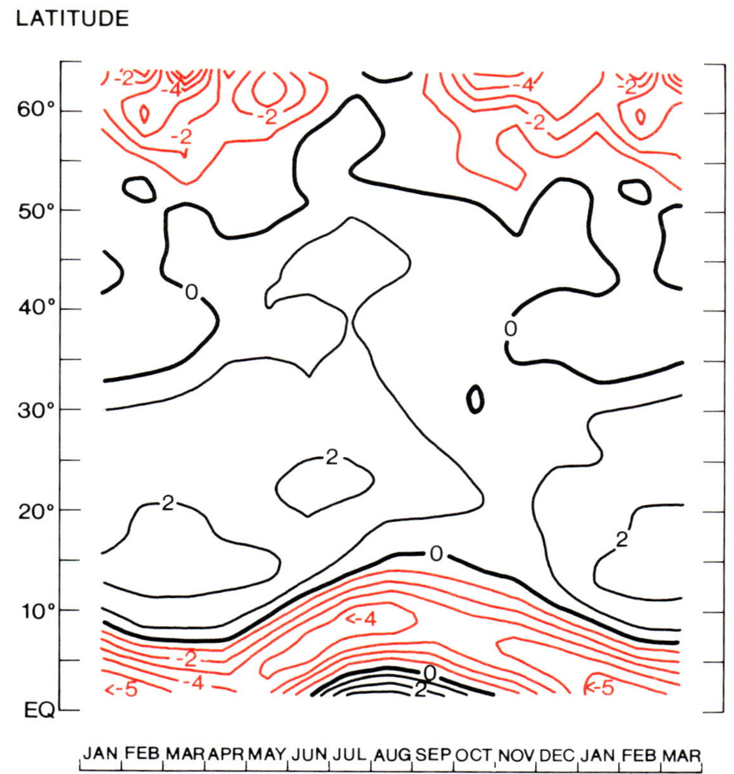

Fig. 19. Divergence of the wind [10^{-6} s^{-1}]

Diagrams of Annual Cycles in Characteristic Areas

No.	Area	Site
1	Gulf Stream	38°N–40°N, 61°W–63°W
2	Middle Atlantic Bight	40°N–41°N, 71°W–72°W
3	Labrador current region at the Grand Banks of Newfoundland	48°N–50°N, 51°W–53°W
4	Midocean westwind drift	55°N–57°N, 28°W–32°W
5	North Sea	55°N–57°N, 5° E– 3°E
6	Subtropical convergence zone southwest of summer subtropical anticyclone	31°N–33°N, 39°W–44°W
7	Upwelling region at the North West African coast	22°N–23°N, 18°W–19°W
8	Trade wind region	16°N–19°N, 39°W–43°W
9	Low latitude central North Atlantic Ocean	6°N– 7°N, 24°W–27°W
10	Low latitude western North Atlantic Ocean	6°N– 9°N, 43°W–47°W
11	Gulf of Mexico	24°N–26°N, 90°W–92°W
12	Total North Atlantic Ocean	1°N–65°N, 10°E–97°W

Parameter	Abbreviation
Sea surface temperature	SST
Air temperature	AIRTEMP
Dewpoint temperature	DEWPOINT
Air- minus sea surface temperature difference	AIRT-SST
Air- minus dewpoint temperature difference	AIRT-DEWP
Mixing ratio	MIXRAT
Mixing ratio minus mixing ratio at sea surface temperature	MIXDIFF
Total cloud cover	CLOUD
Low cloud cover	LCLOUD
Precipitation frequency	PRECIP
Sea level air pressure	PRESSURE
Scalar wind speed	SPEED
Absolute value of the resultant wind	VECWIND
Direction of the resultant wind	DIR

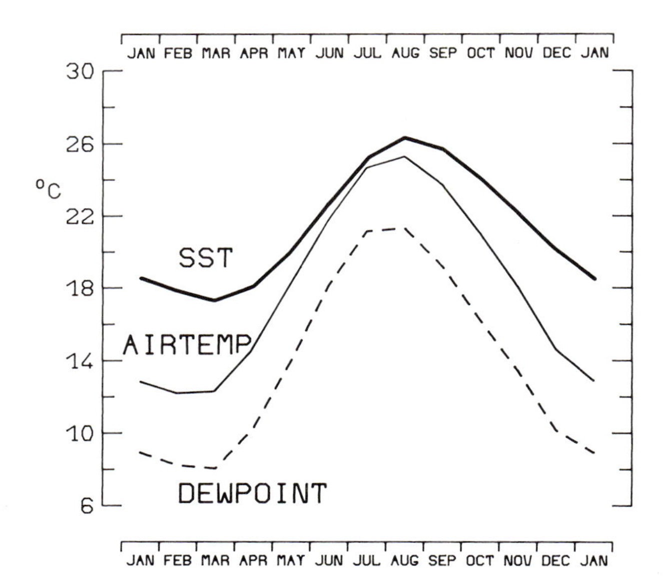

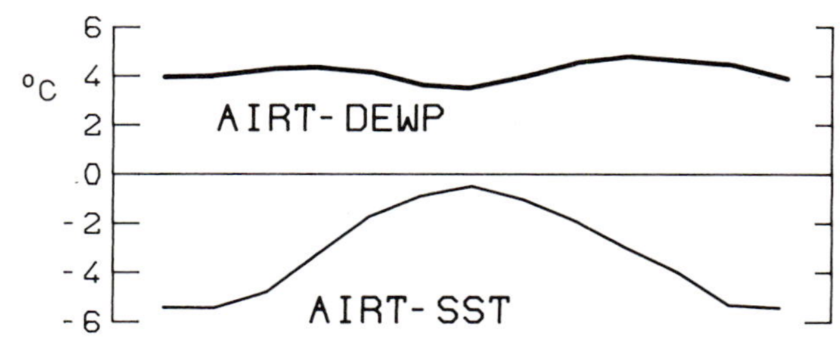

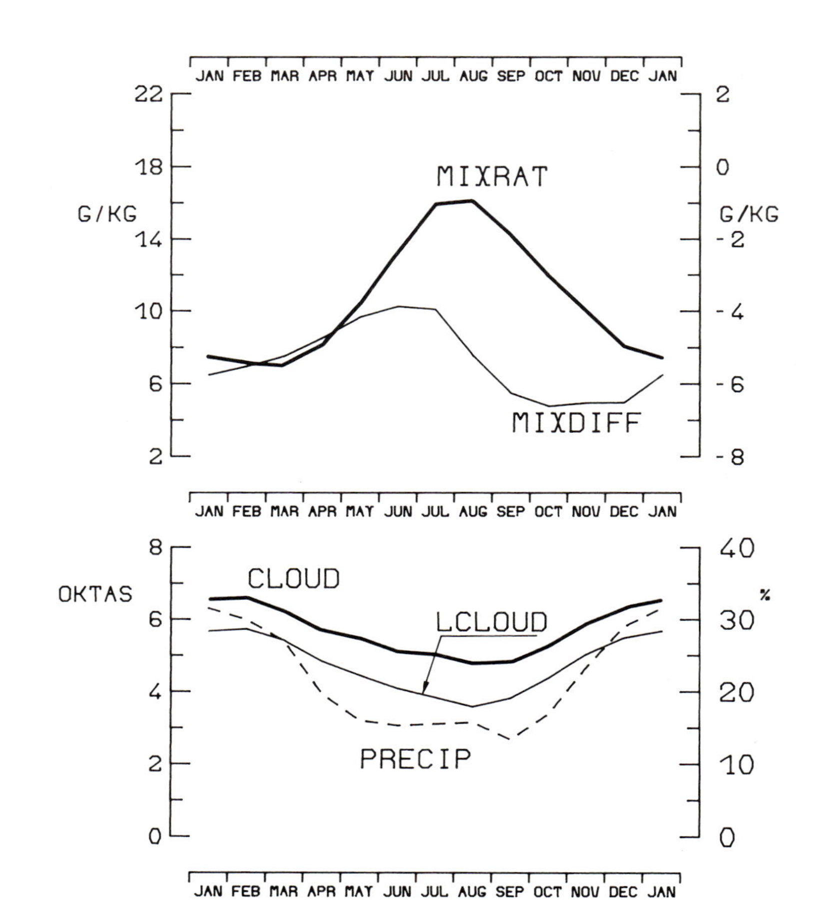

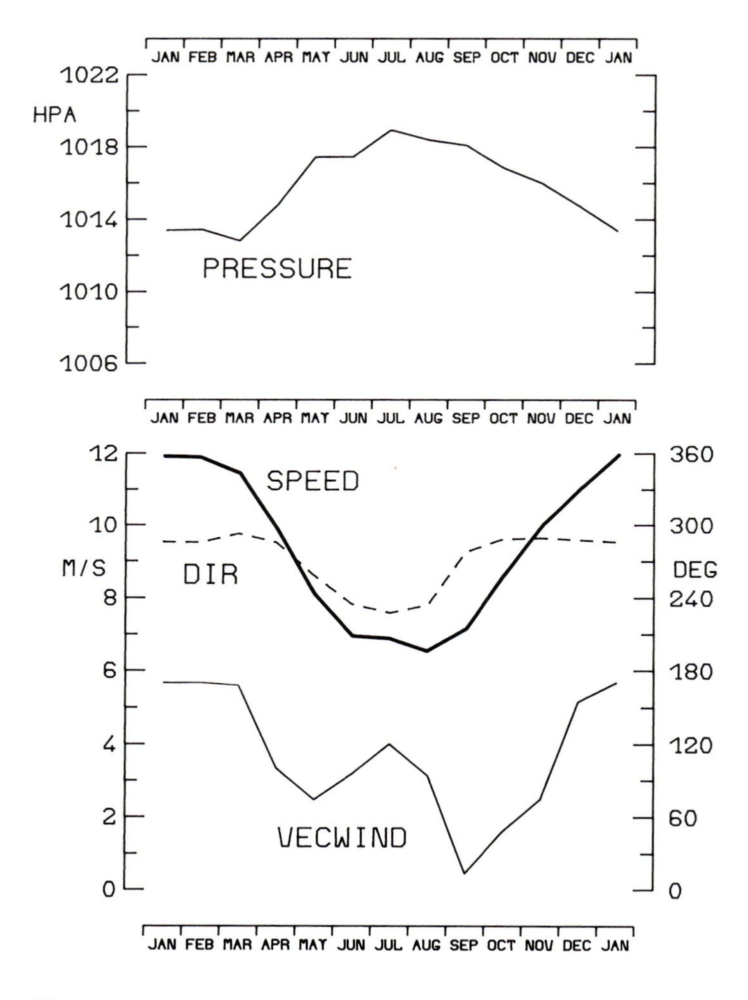

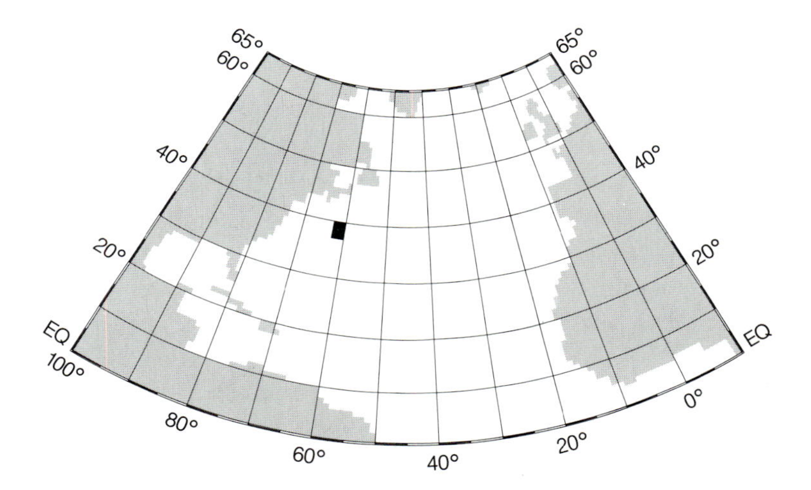

Fig. 20. Gulf Stream

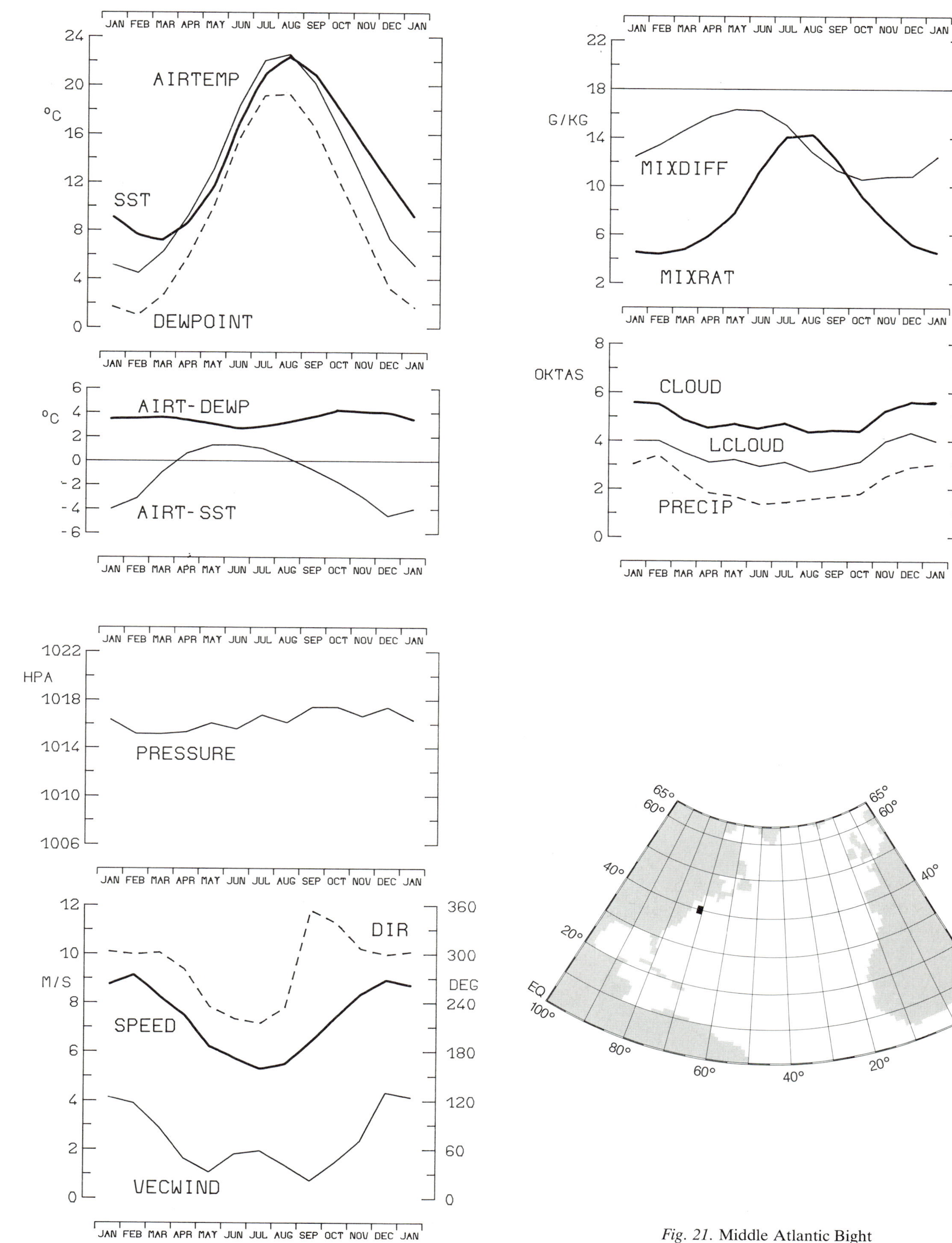

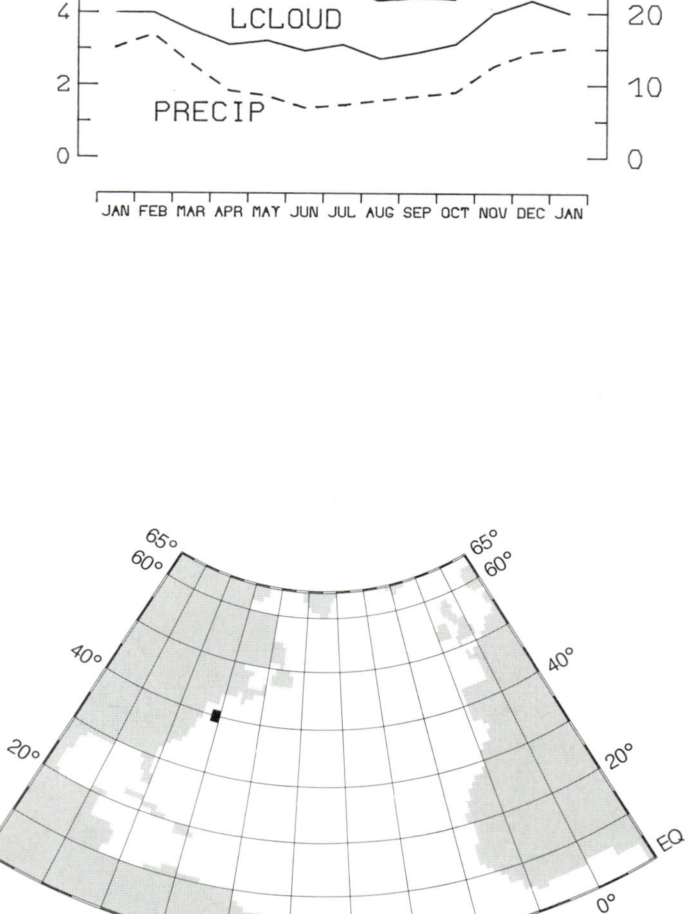

Fig. 21. Middle Atlantic Bight

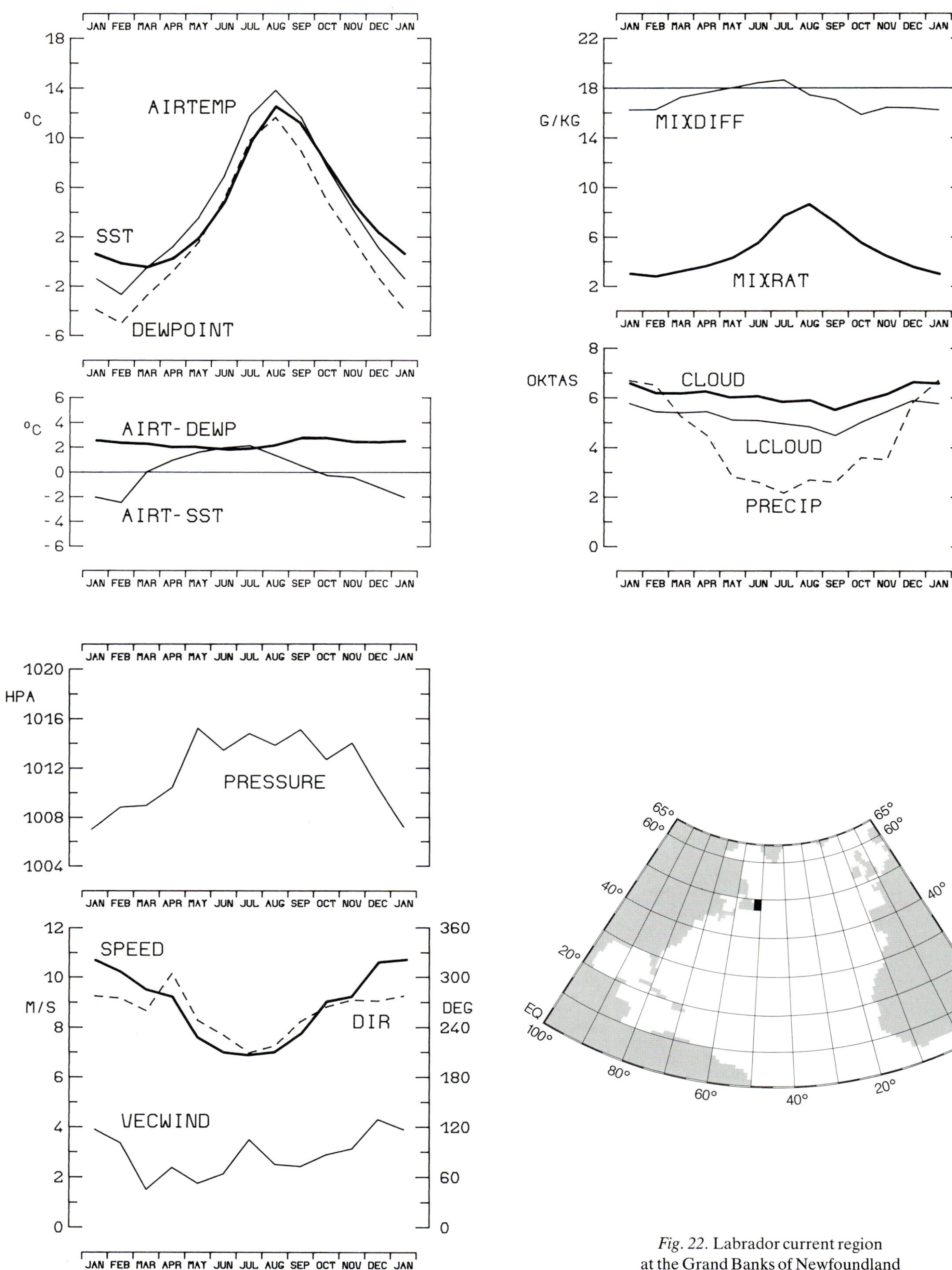

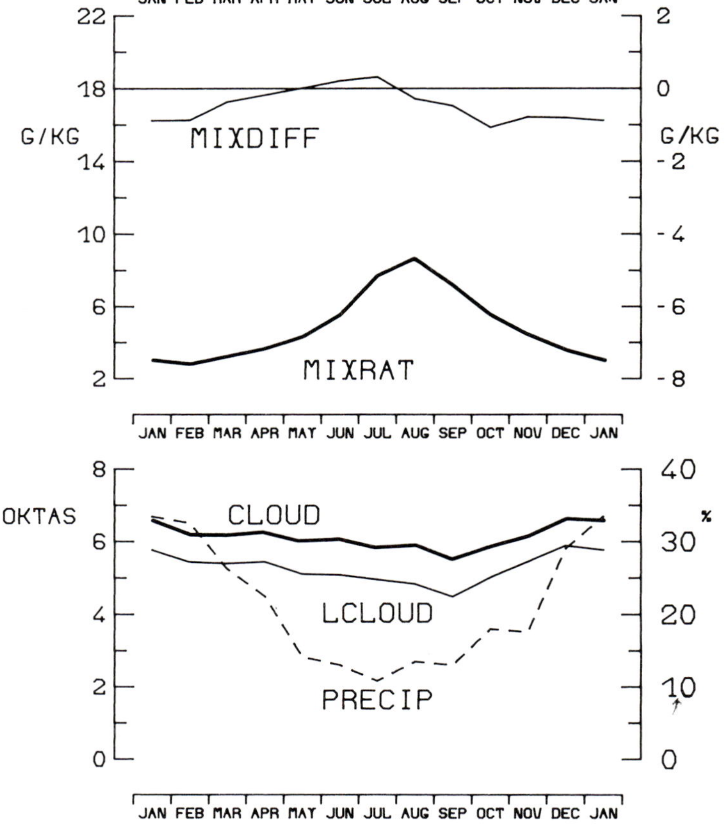

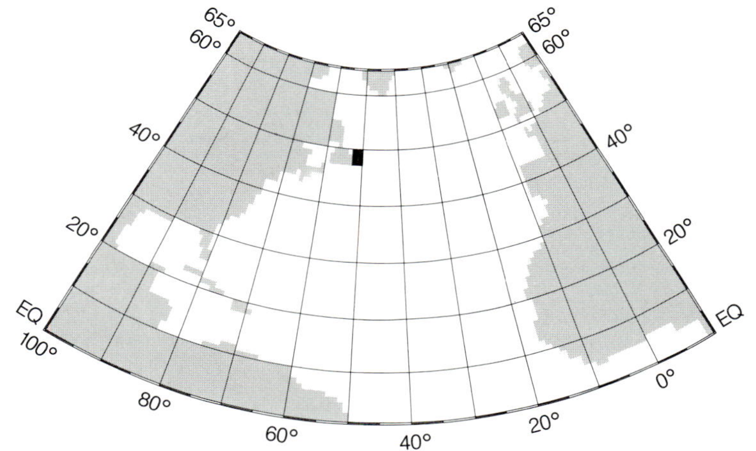

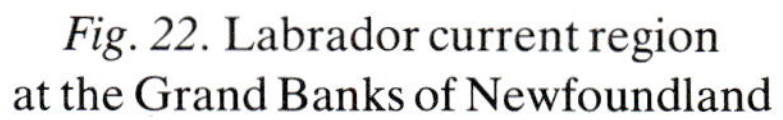
Fig. 22. Labrador current region at the Grand Banks of Newfoundland

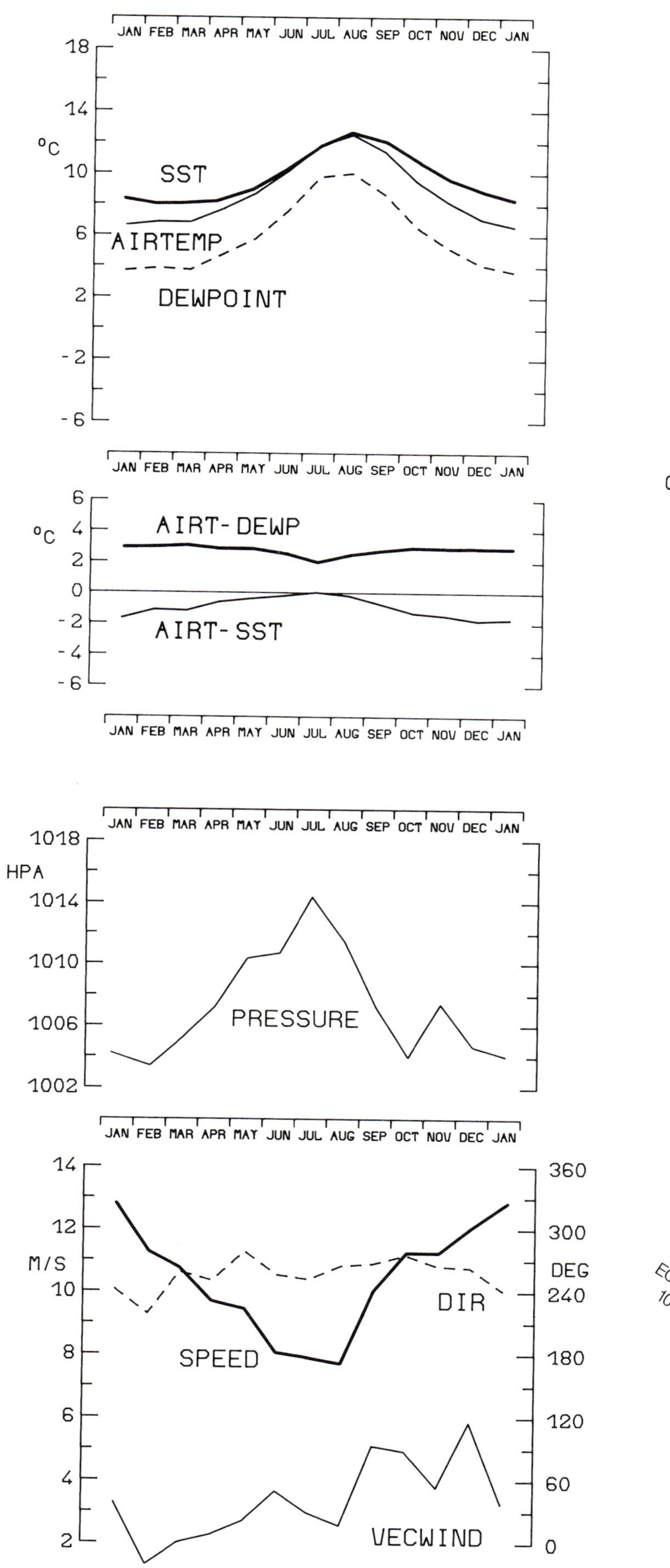

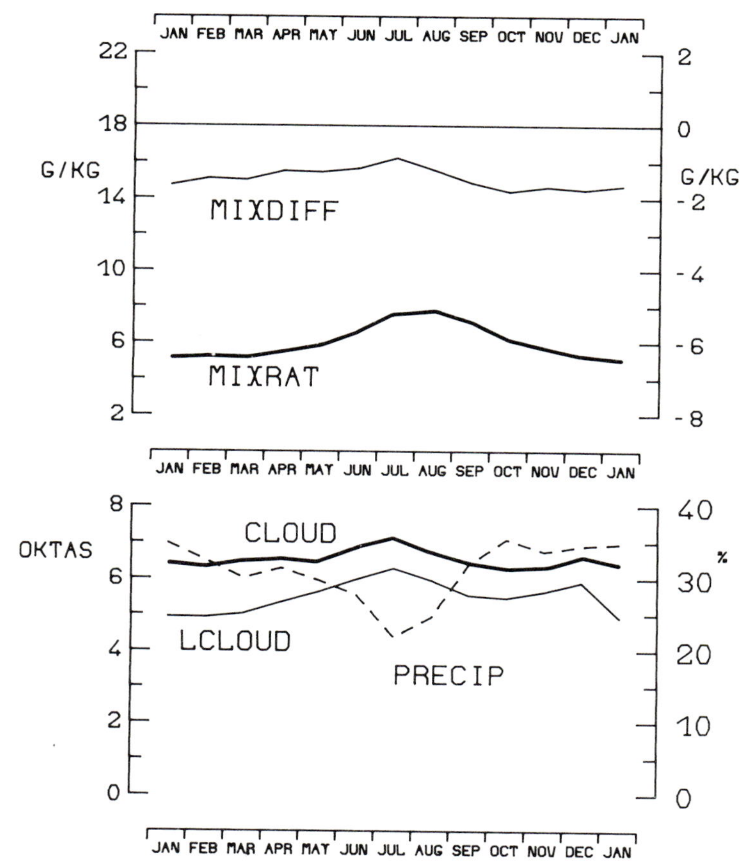

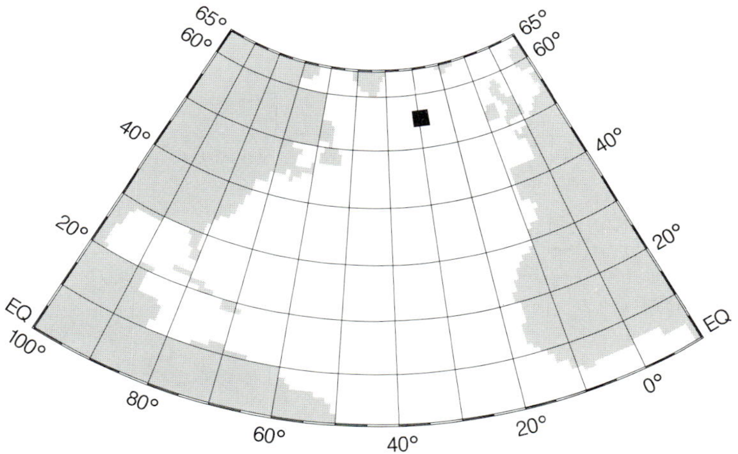

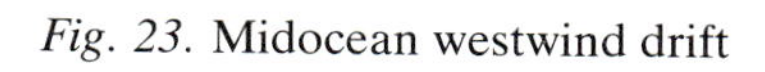
Fig. 23. Midocean westwind drift

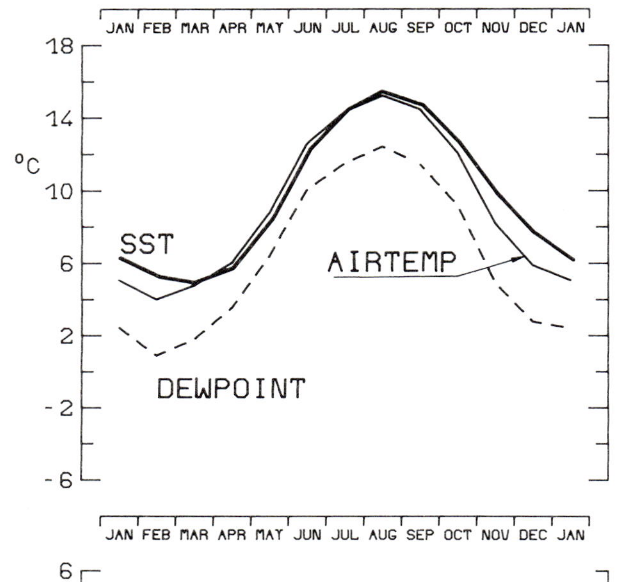

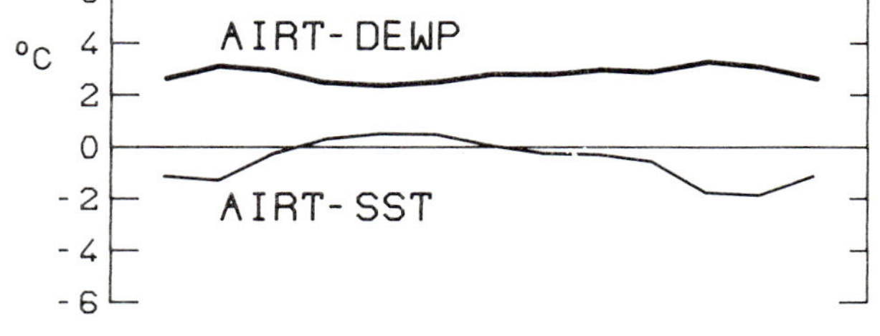

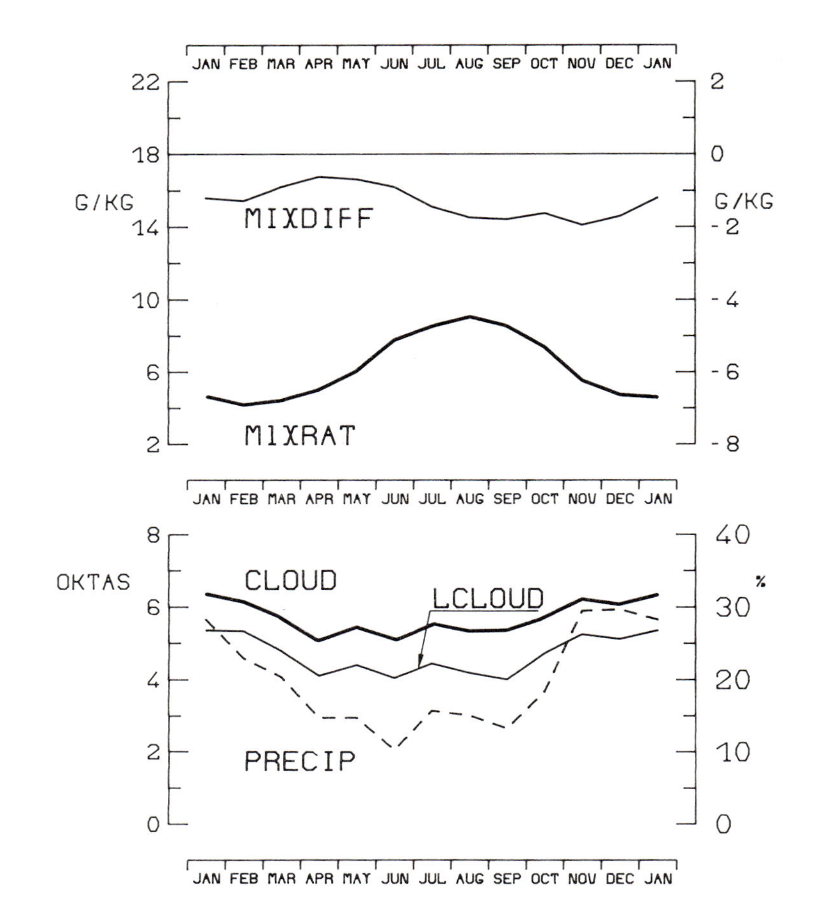

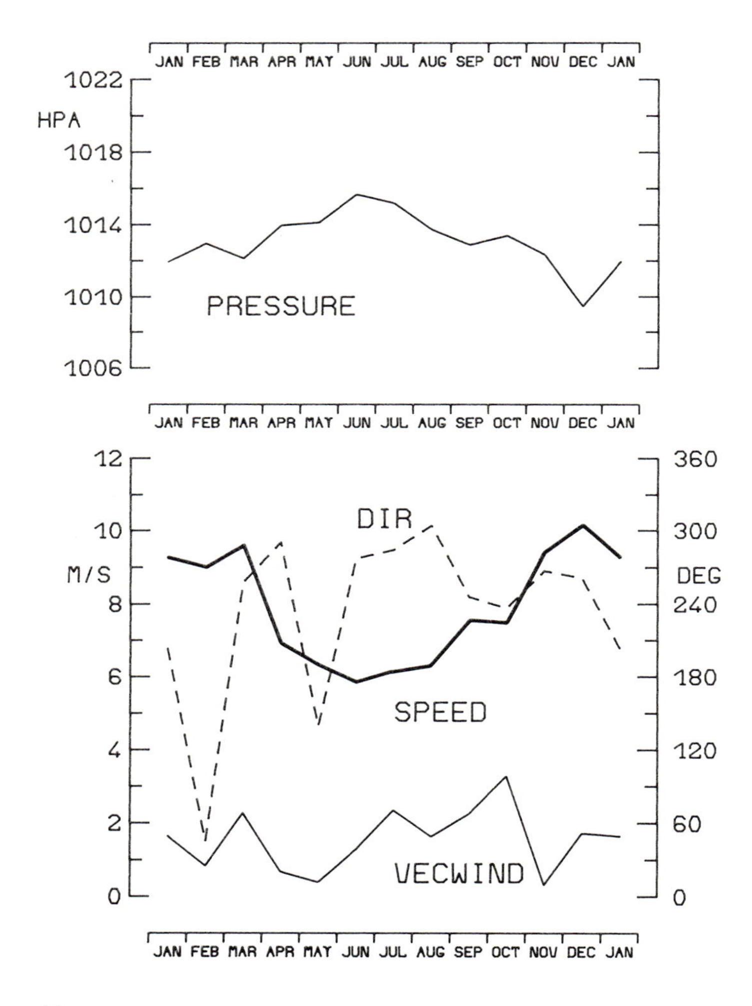

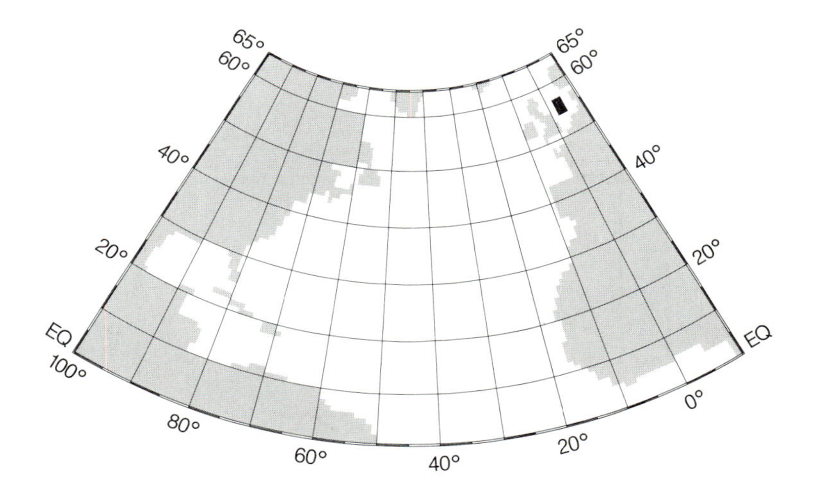

Fig. 24. North Sea

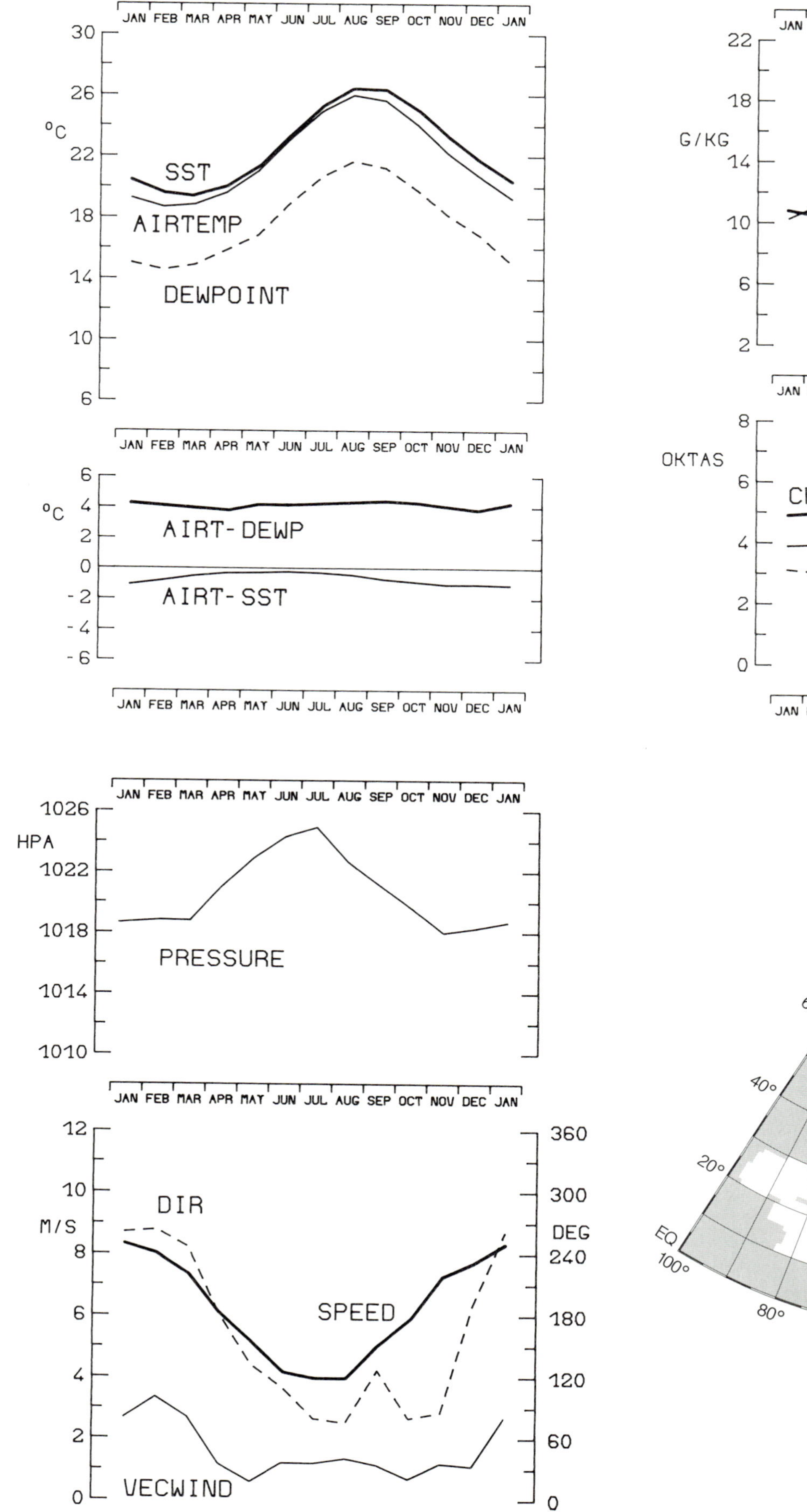

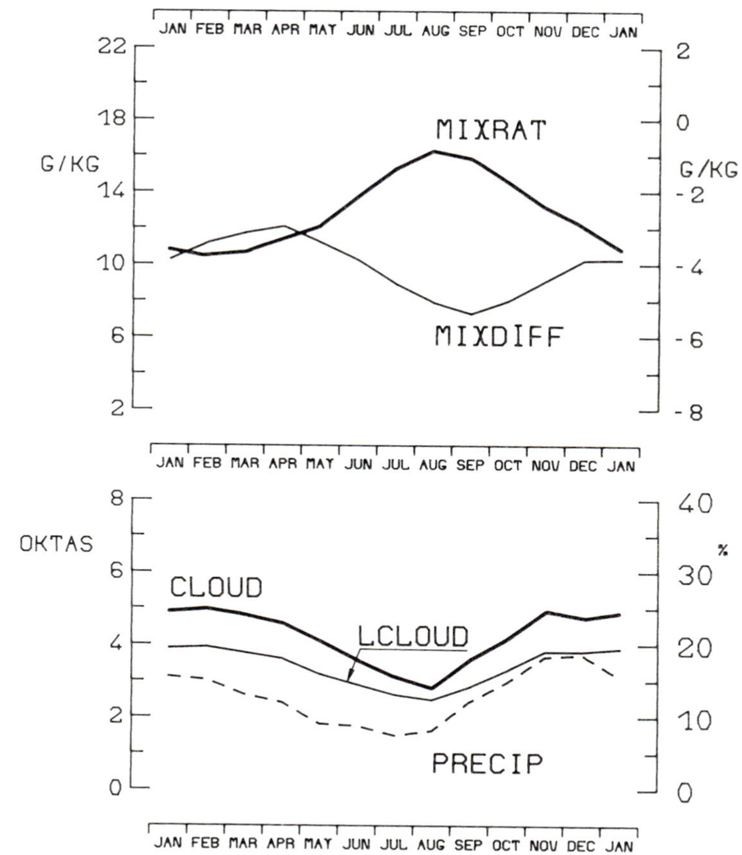

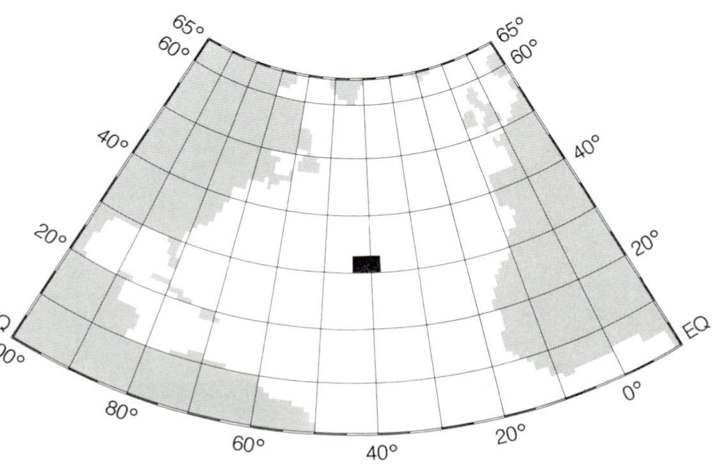

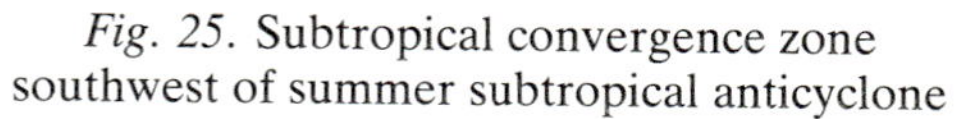

Fig. 25. Subtropical convergence zone southwest of summer subtropical anticyclone

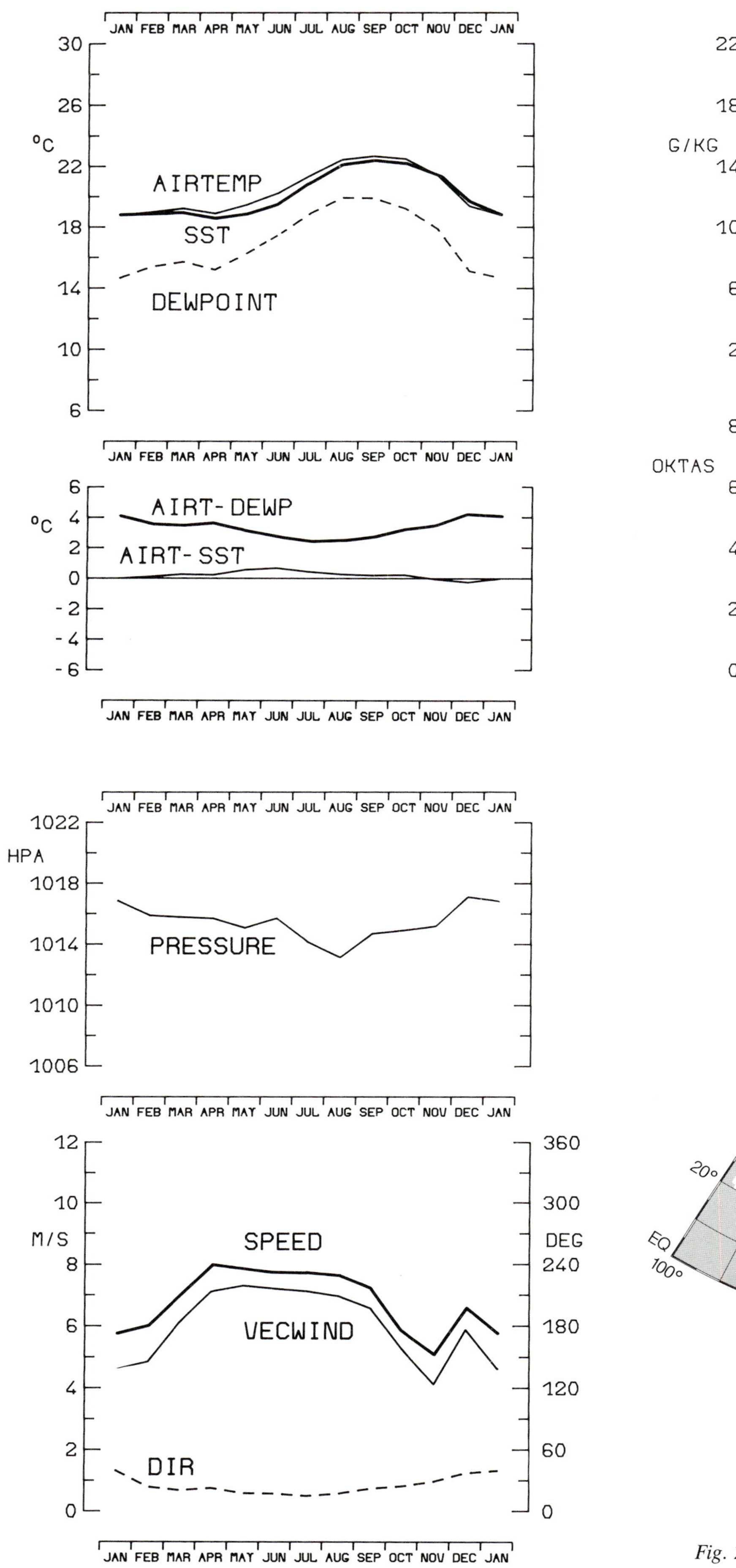

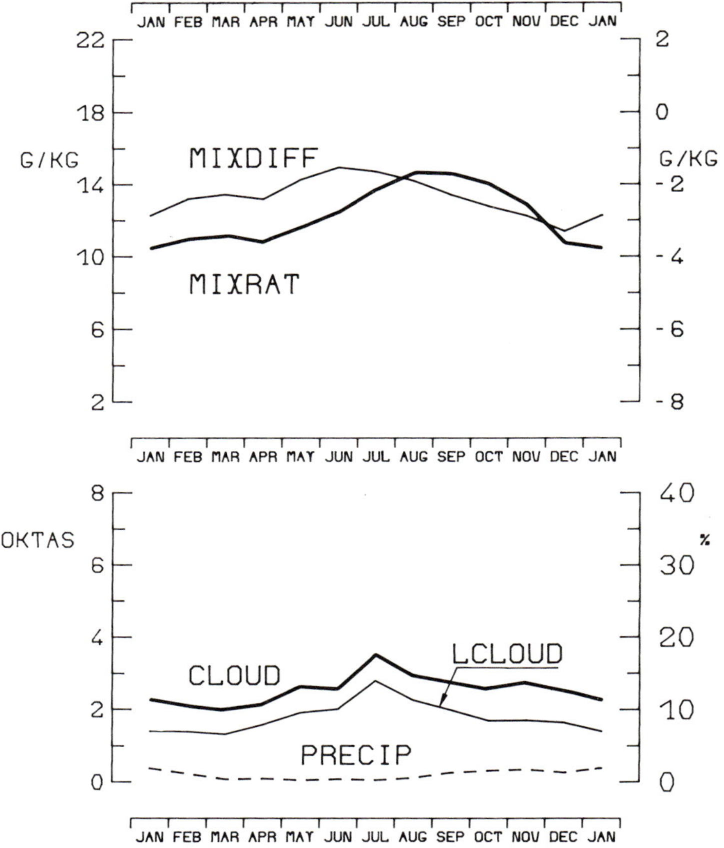

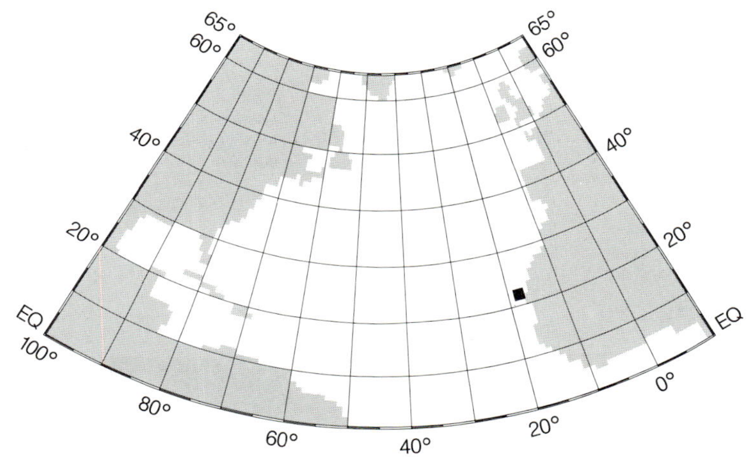

Fig. 26. Upwelling region at the North West African coast

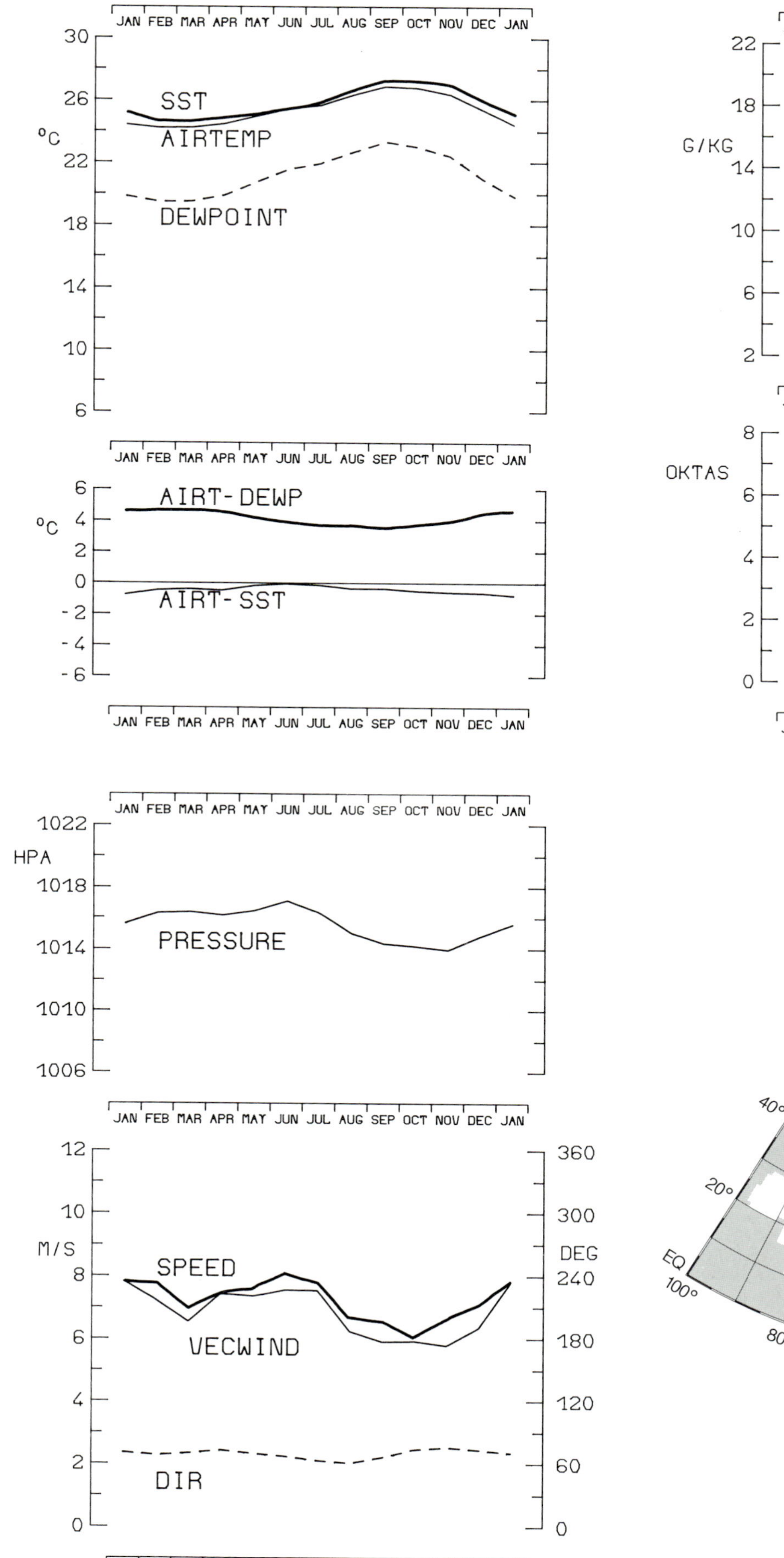

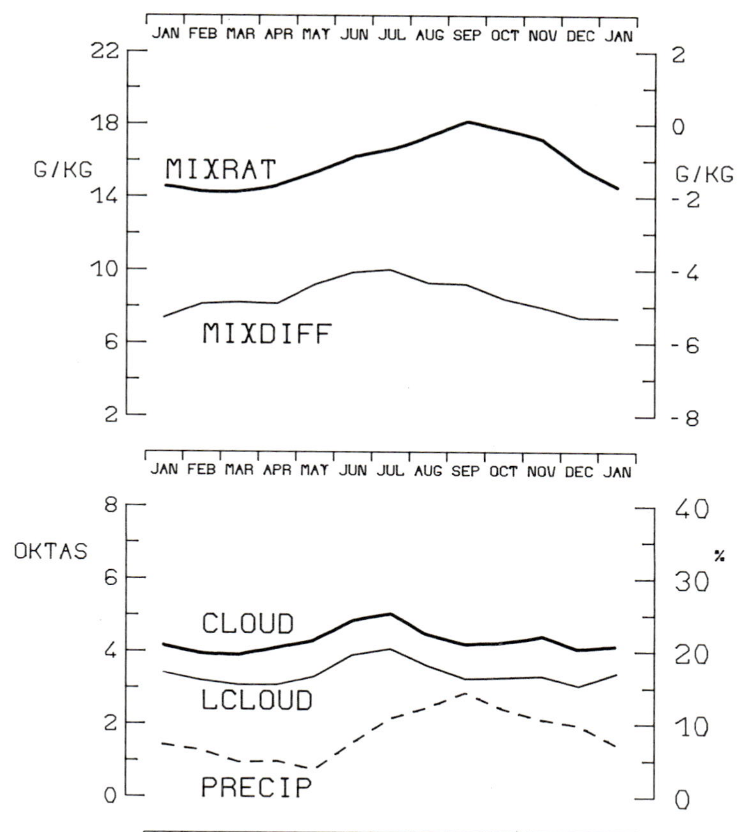

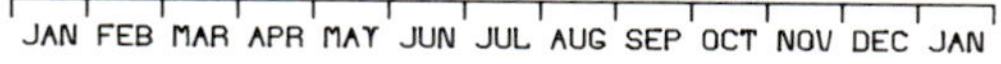

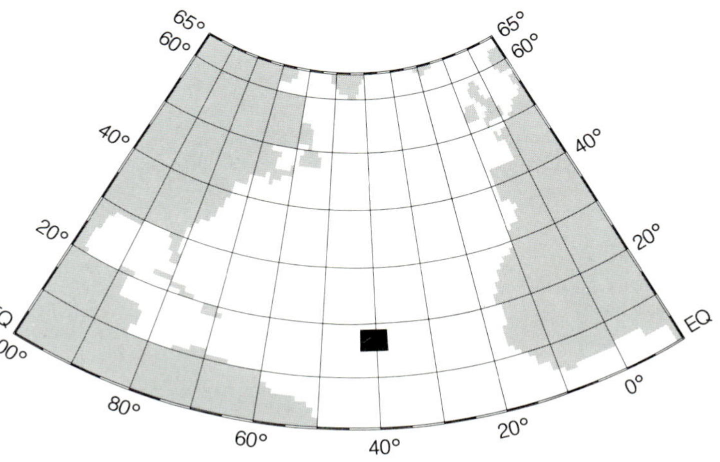

Fig. 27. Trade wind region

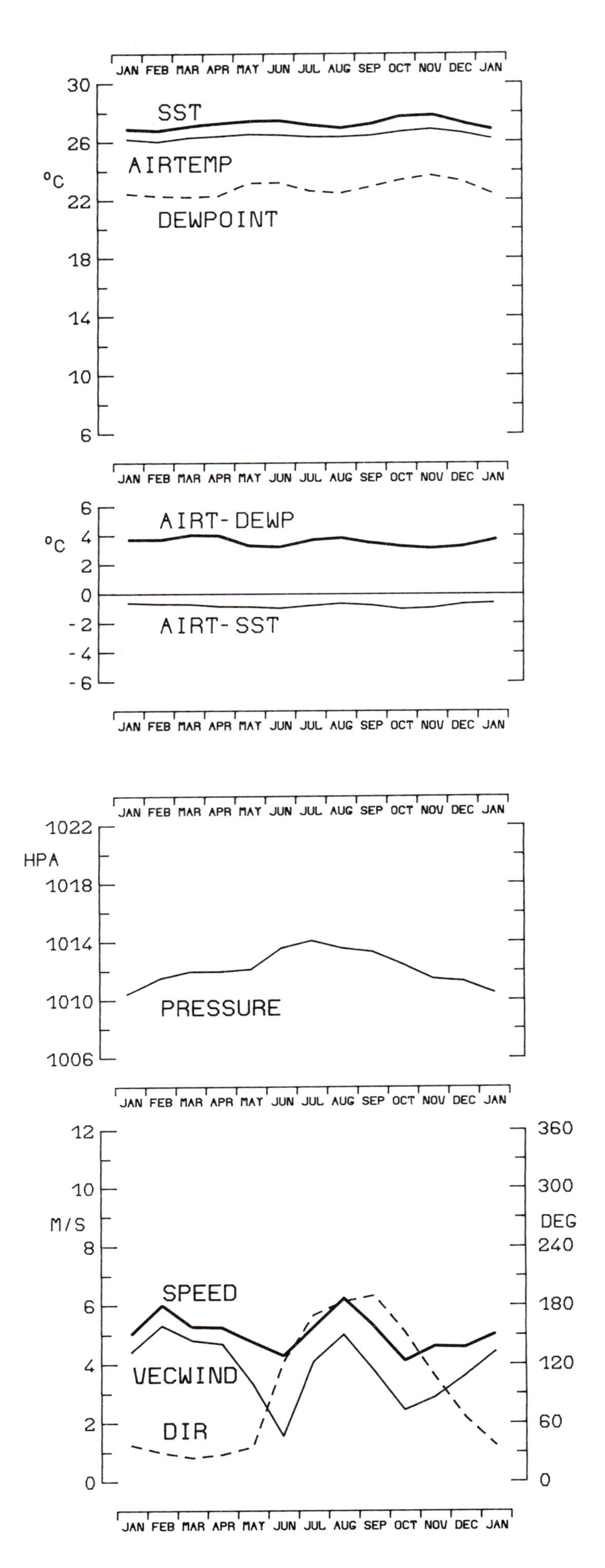

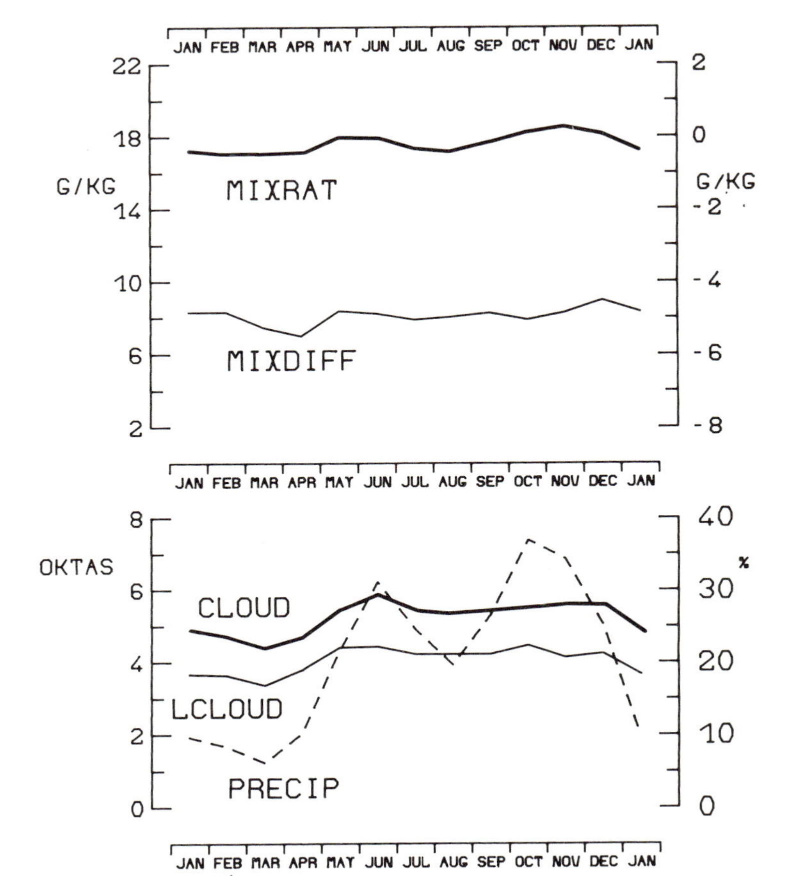

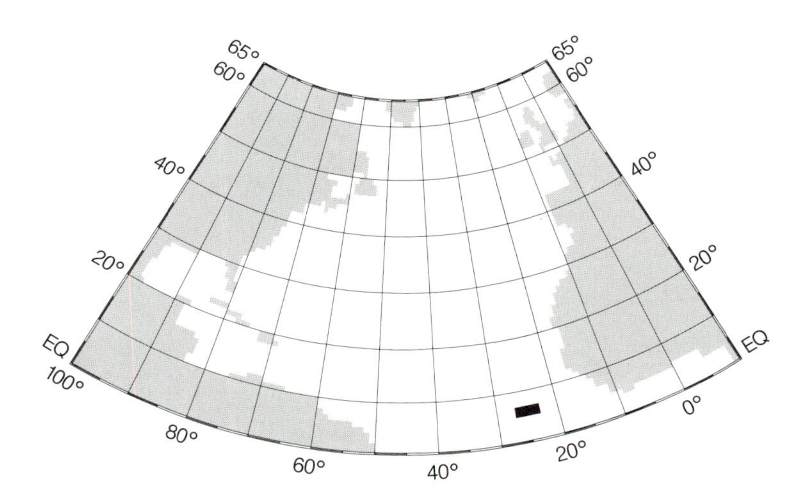

Fig. 28. Low latitude central North Atlantic Ocean

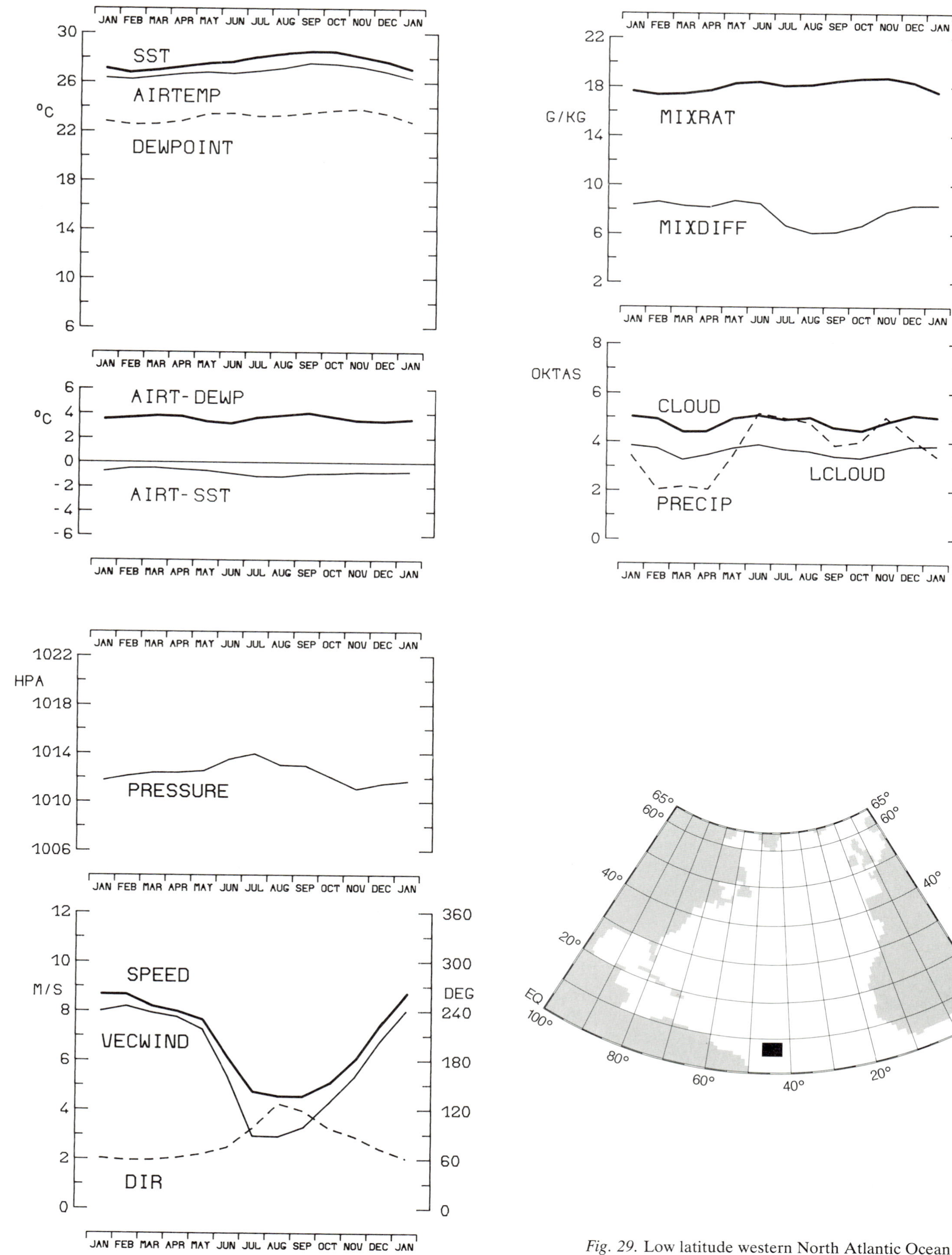

Fig. 29. Low latitude western North Atlantic Ocean

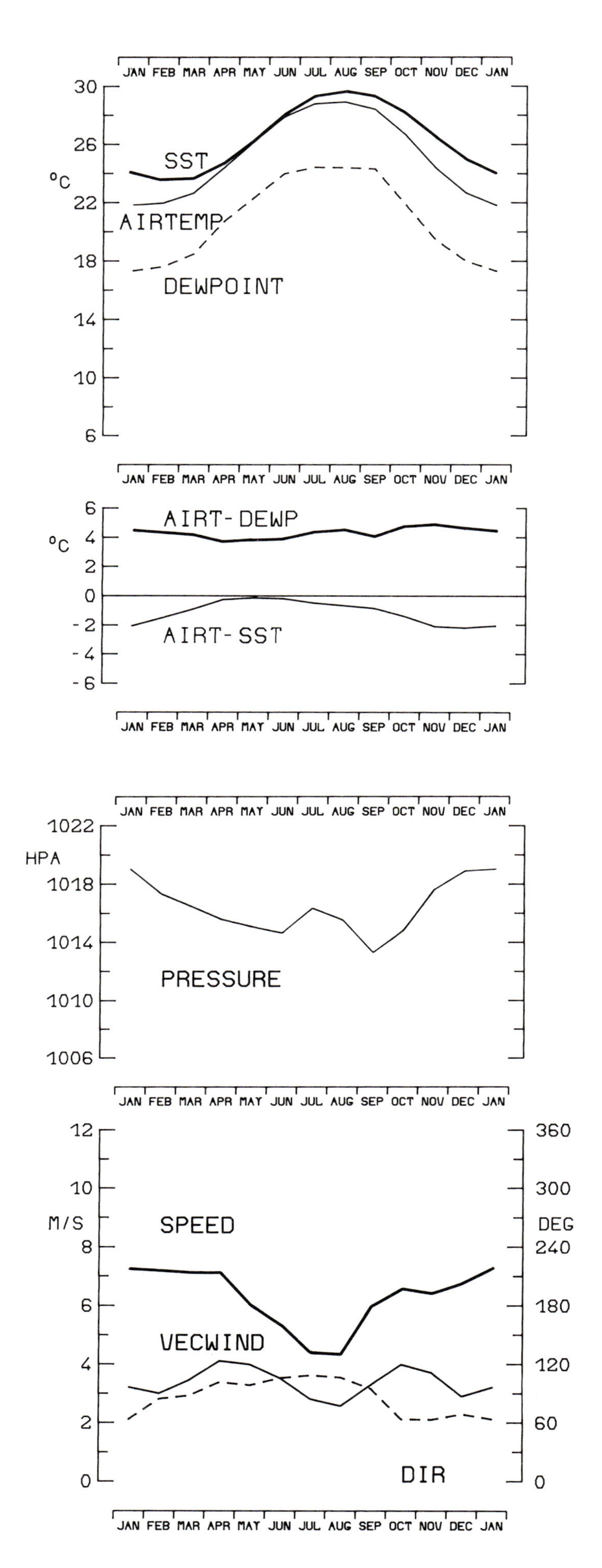

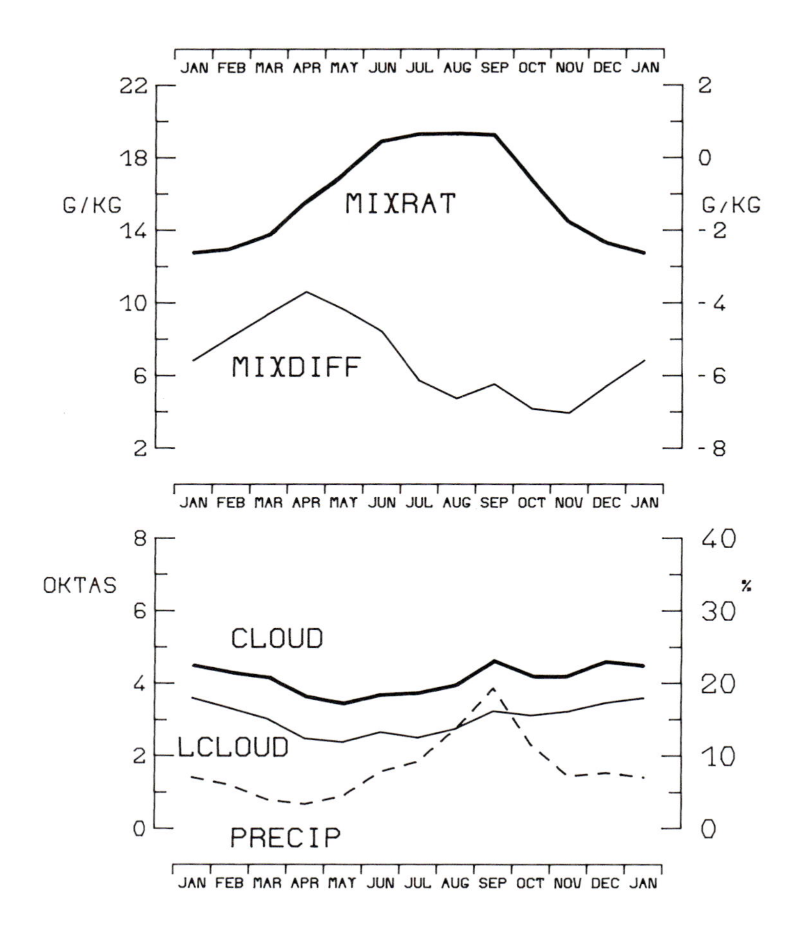

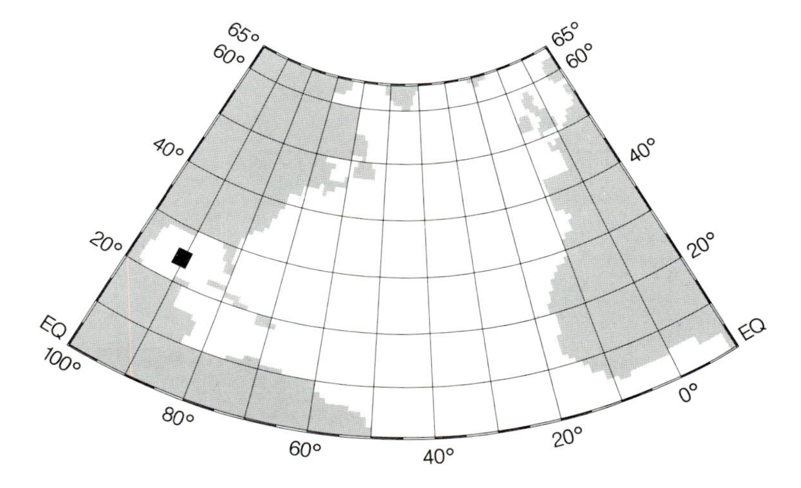

Fig. 30. Gulf of Mexico

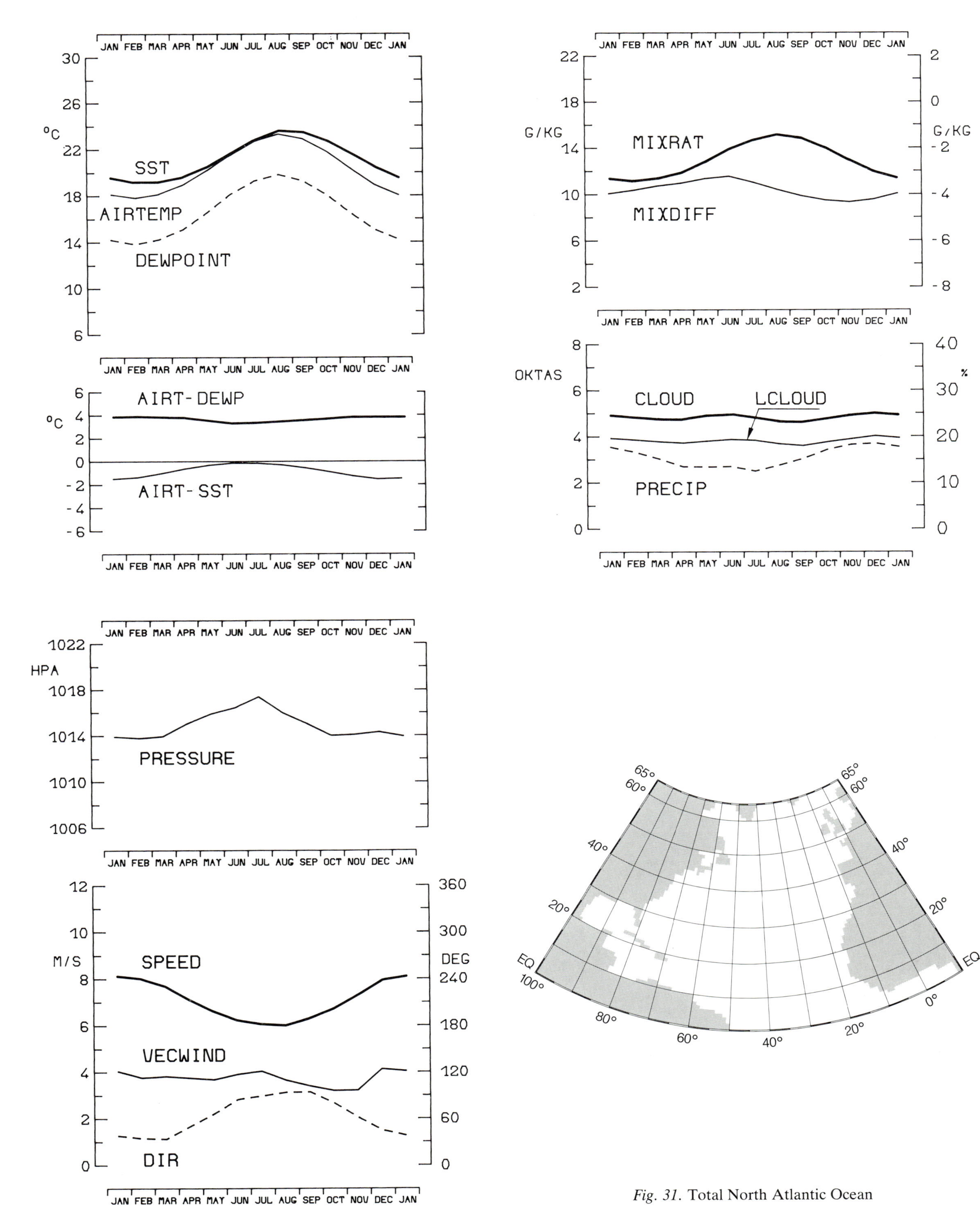

Fig. 31. Total North Atlantic Ocean

Charts of Monthly and Annual Means, Annual Ranges and Standard Deviations

Parameter	*Chart number*
Number of observations	1– 5
Sea surface temperature	6– 19
Standard deviation of sea surface temperature	20– 24
Air temperature	25– 30
Air- minus sea surface temperature difference	31– 44
Standard deviation of the air- minus sea surface temperature difference	45– 49
Mixing ratio	50– 63
Standard deviation of mixing ratio	64– 68
Relative humidity	69– 74
Mixing ratio minus mixing ratio at sea surface temperature	75– 80
Total cloud cover	81– 94
Low cloud cover	95–100
Precipitation frequency	101–114
Sea level air pressure	115–128
Standard deviation of sea level air pressure	129–133
Scalar wind speed	134–147
Resultant wind	148–161
Directional steadiness of the wind	162–166
Standard deviation of the east–west component of the wind	167–171
Standard deviation of the north–south component of the wind	172–176
Divergence of the wind	177–181

Chart 1

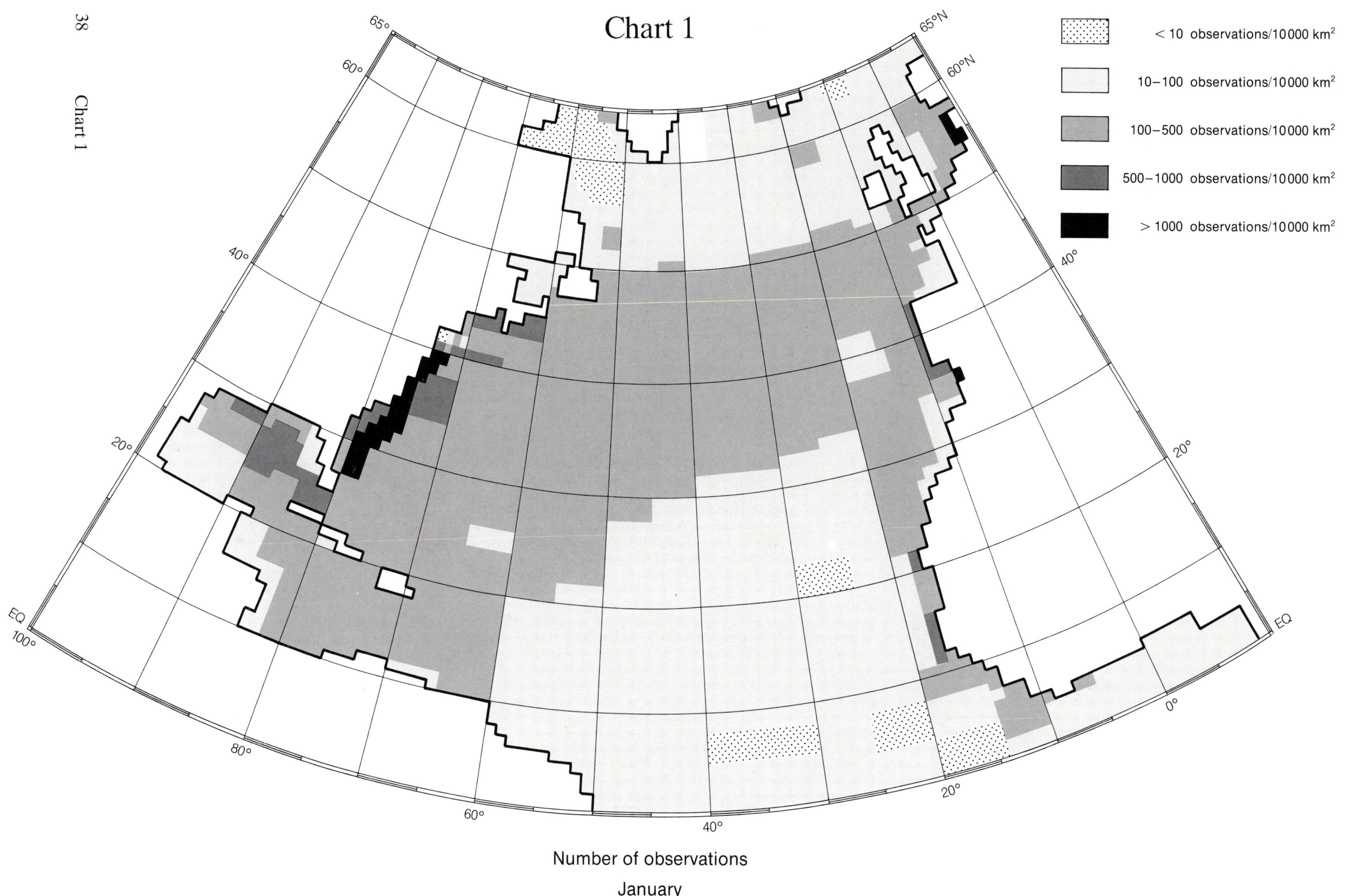

Number of observations
January

Chart 2

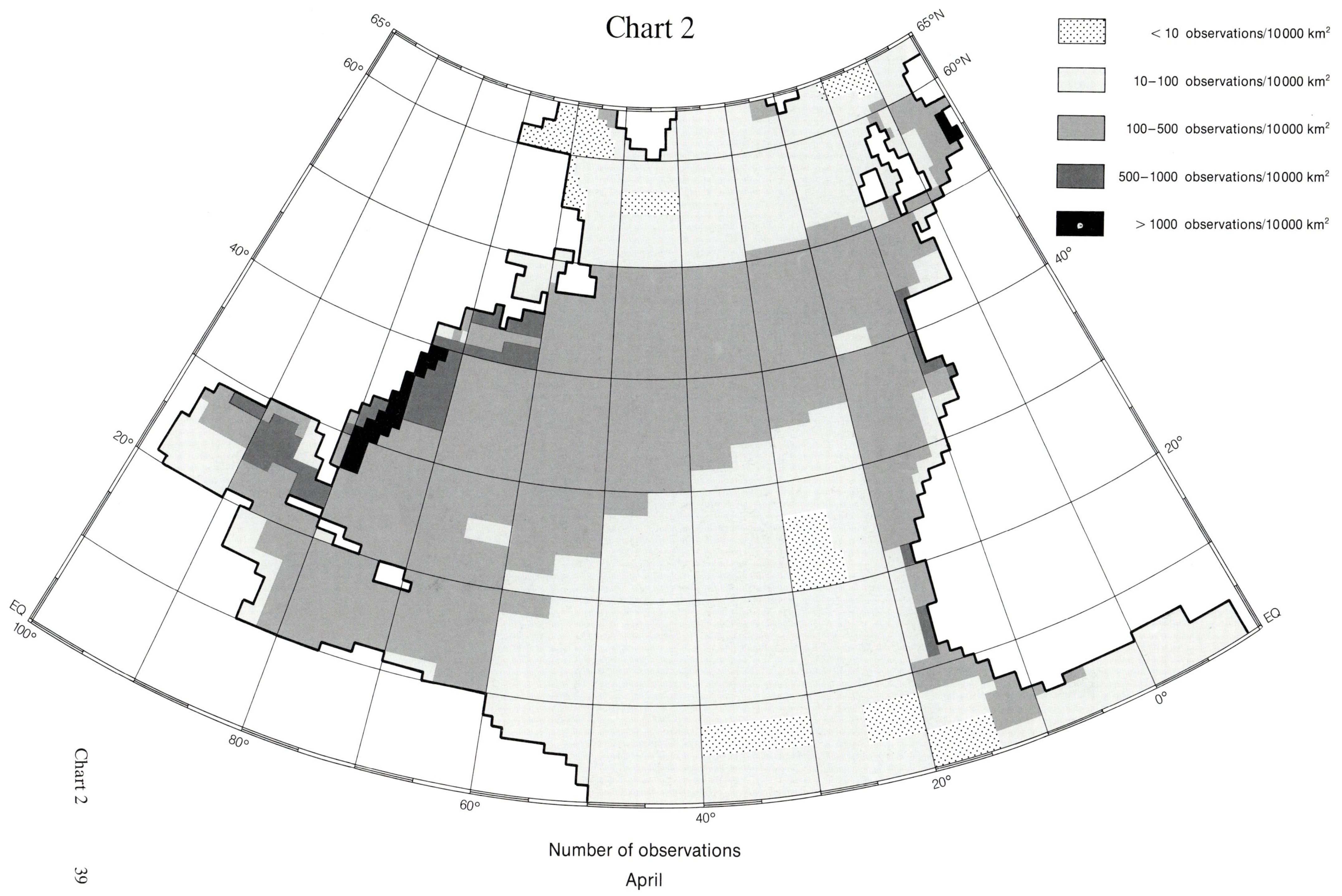

Number of observations
April

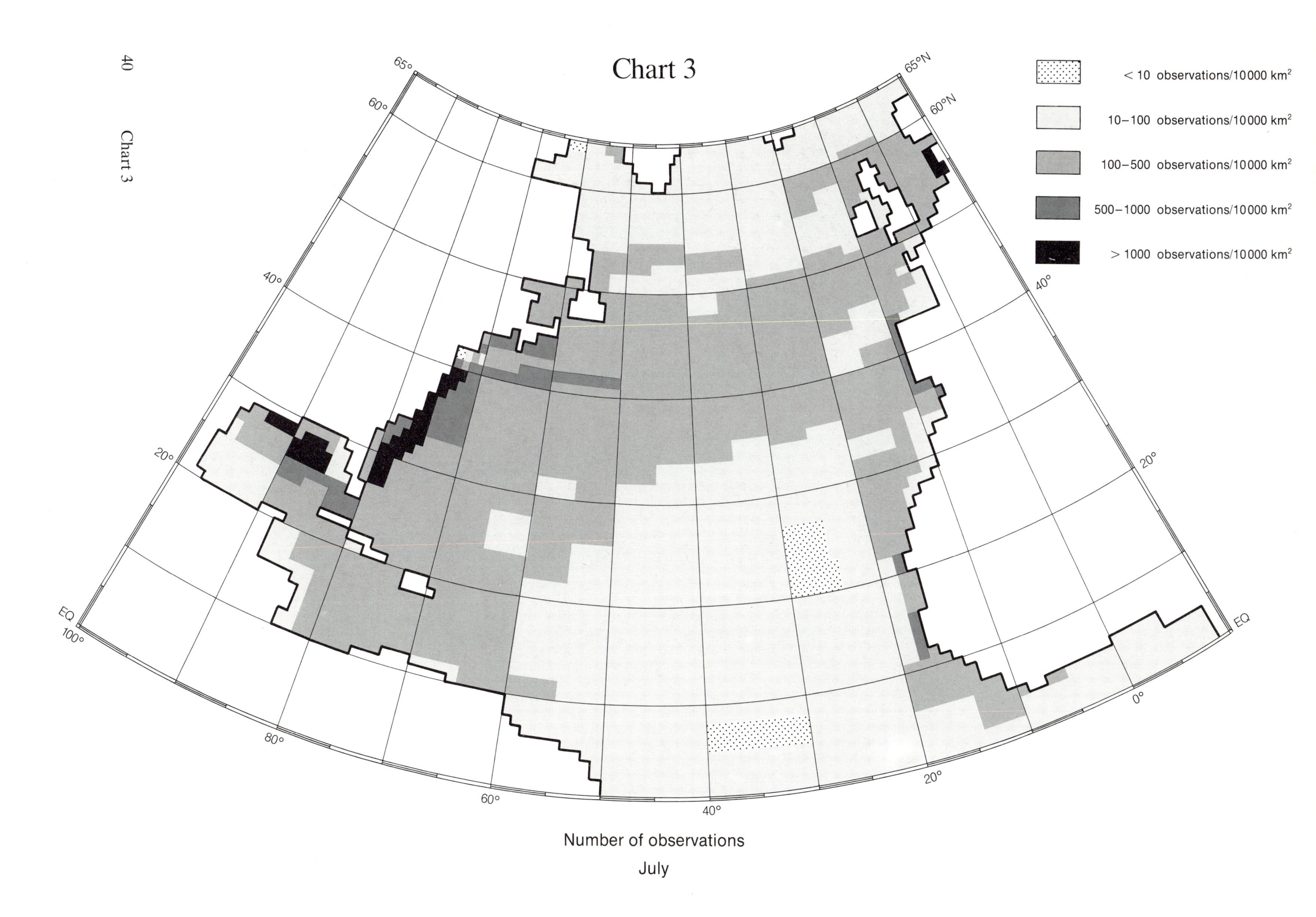

Chart 3
< 10 observations/10000 km²
10–100 observations/10000 km²
100–500 observations/10000 km²
500–1000 observations/10000 km²
> 1000 observations/10000 km²
65°
60°
40°
20°
EQ
100°
80°
60°
40°
20°
0°
EQ
20°
40°
60°N
65°N
Number of observations
July

Chart 4

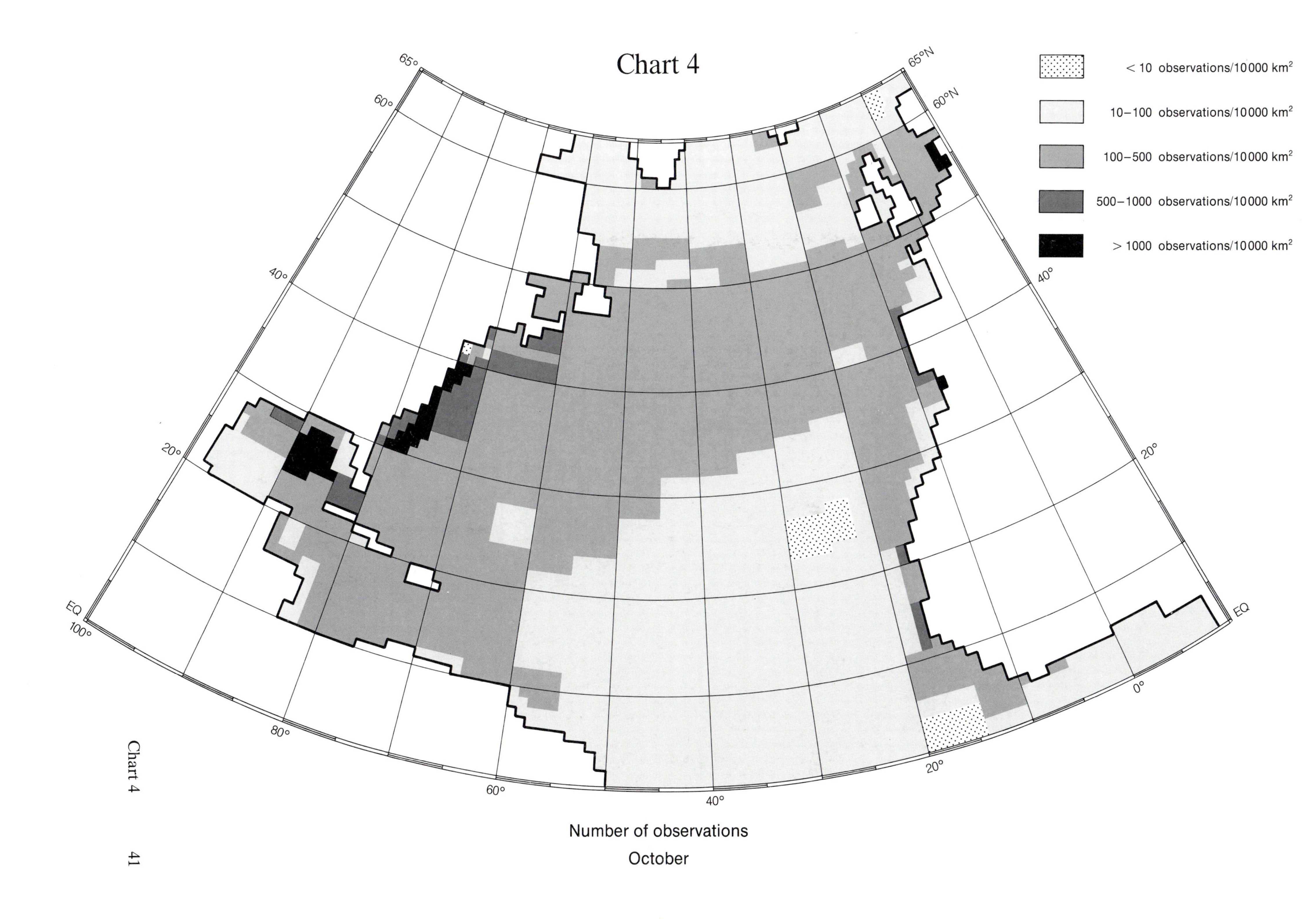

Number of observations
October

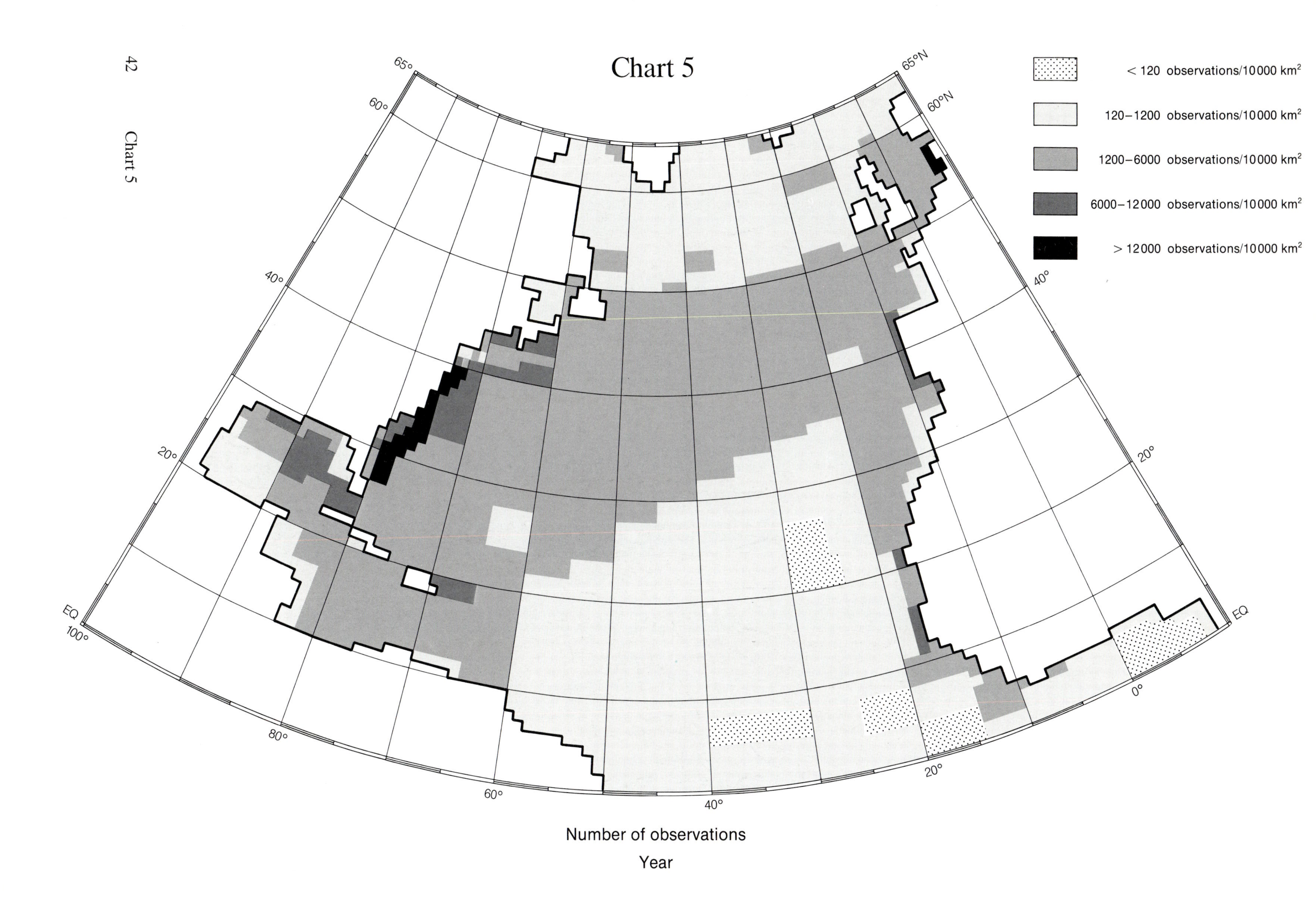

Chart 5
< 120 observations/10000 km²
120–1200 observations/10000 km²
1200–6000 observations/10000 km²
6000–12000 observations/10000 km²
> 12000 observations/10000 km²
65°
60°
40°
20°
EQ
100°
80°
60°
40°
20°
0°
EQ
20°
40°
60°N
65°N
Number of observations
Year

Chart 6

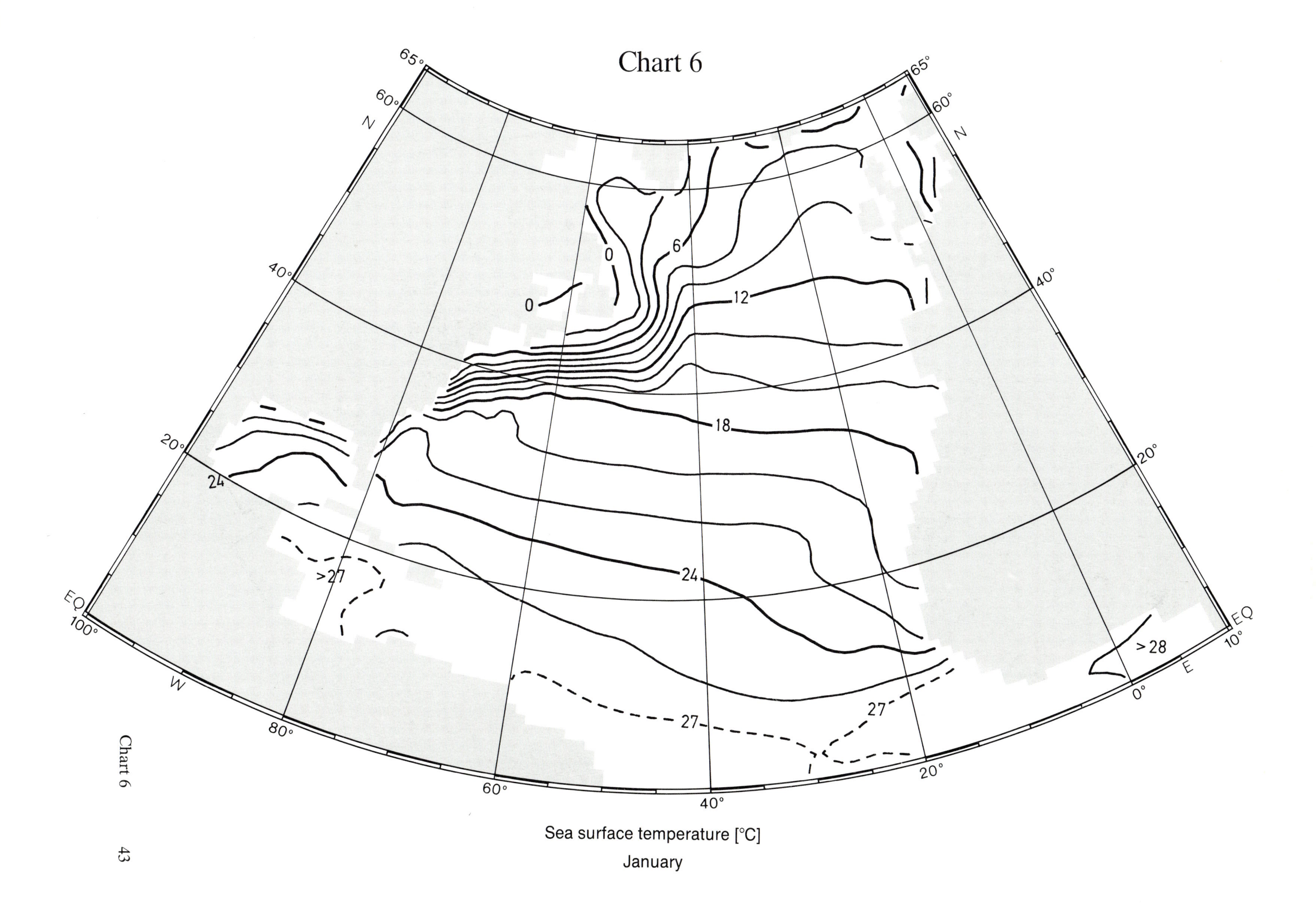

Sea surface temperature [°C]
January

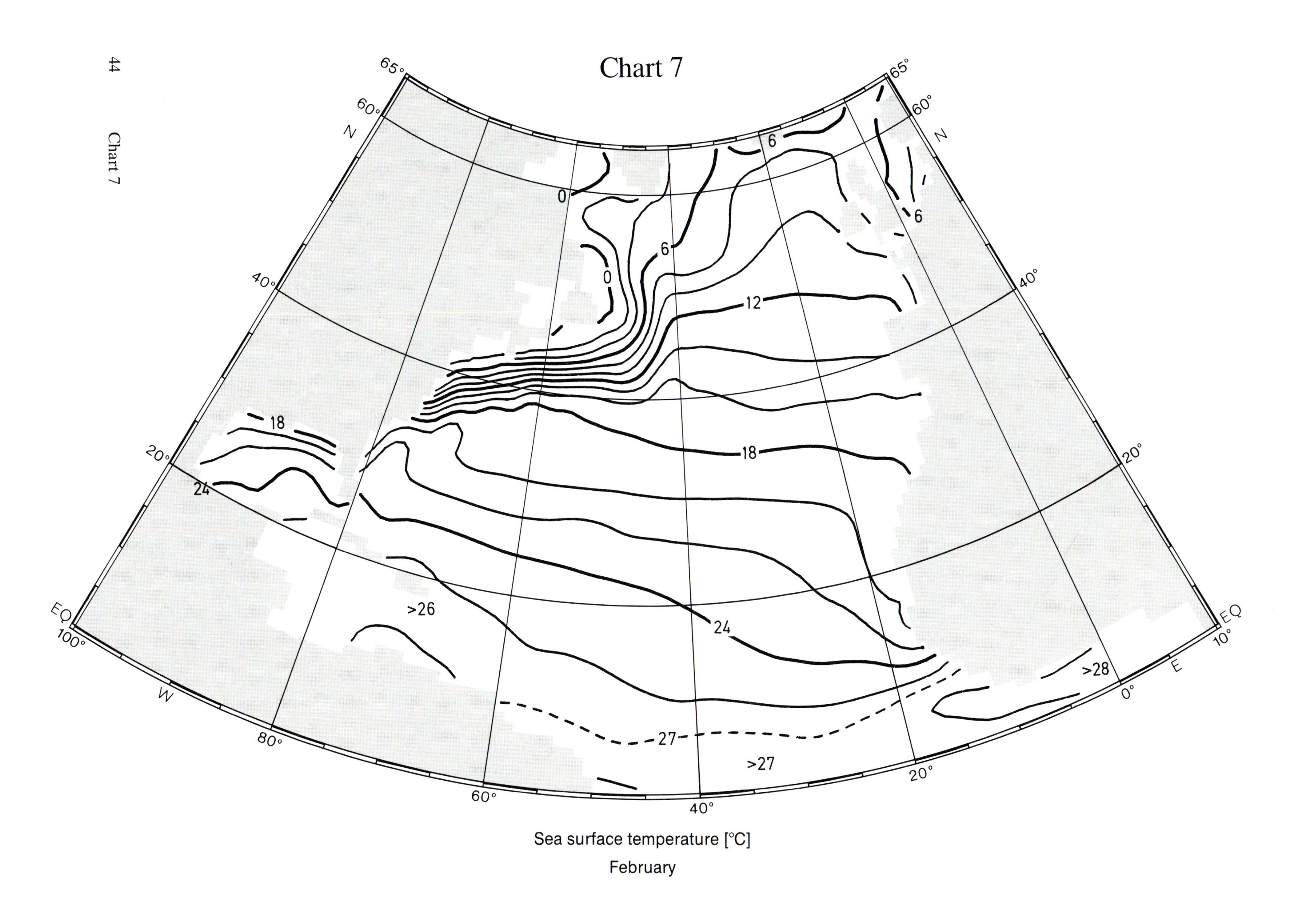
Chart 7
Sea surface temperature [°C]
February
65°
60°
N
40°
20°
EQ
10°
E
0°
20°
40°
60°
80°
100°
W
0
6
12
18
24
>26
27
>27
>28

Chart 8

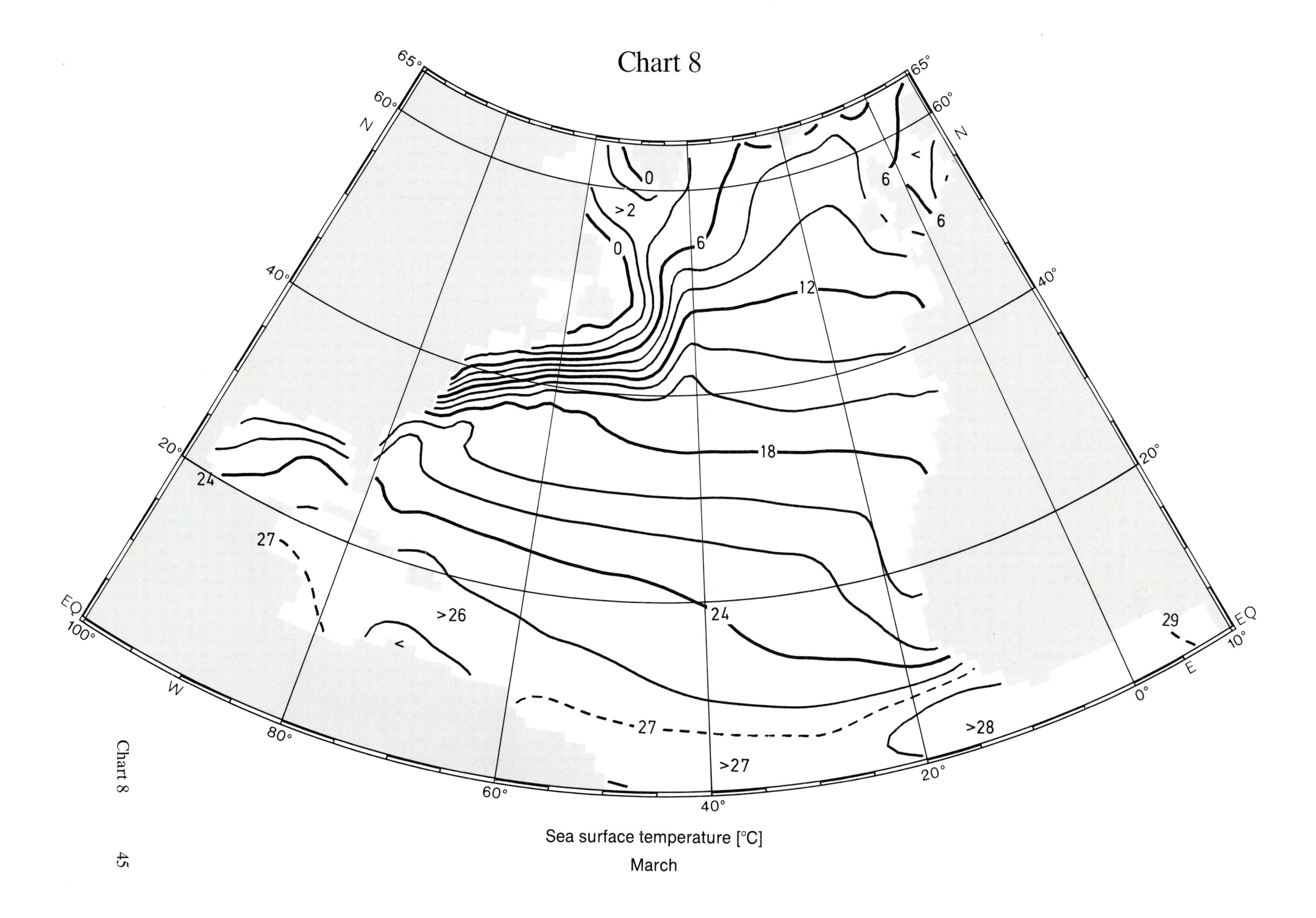

Sea surface temperature [°C]
March

Chart 9

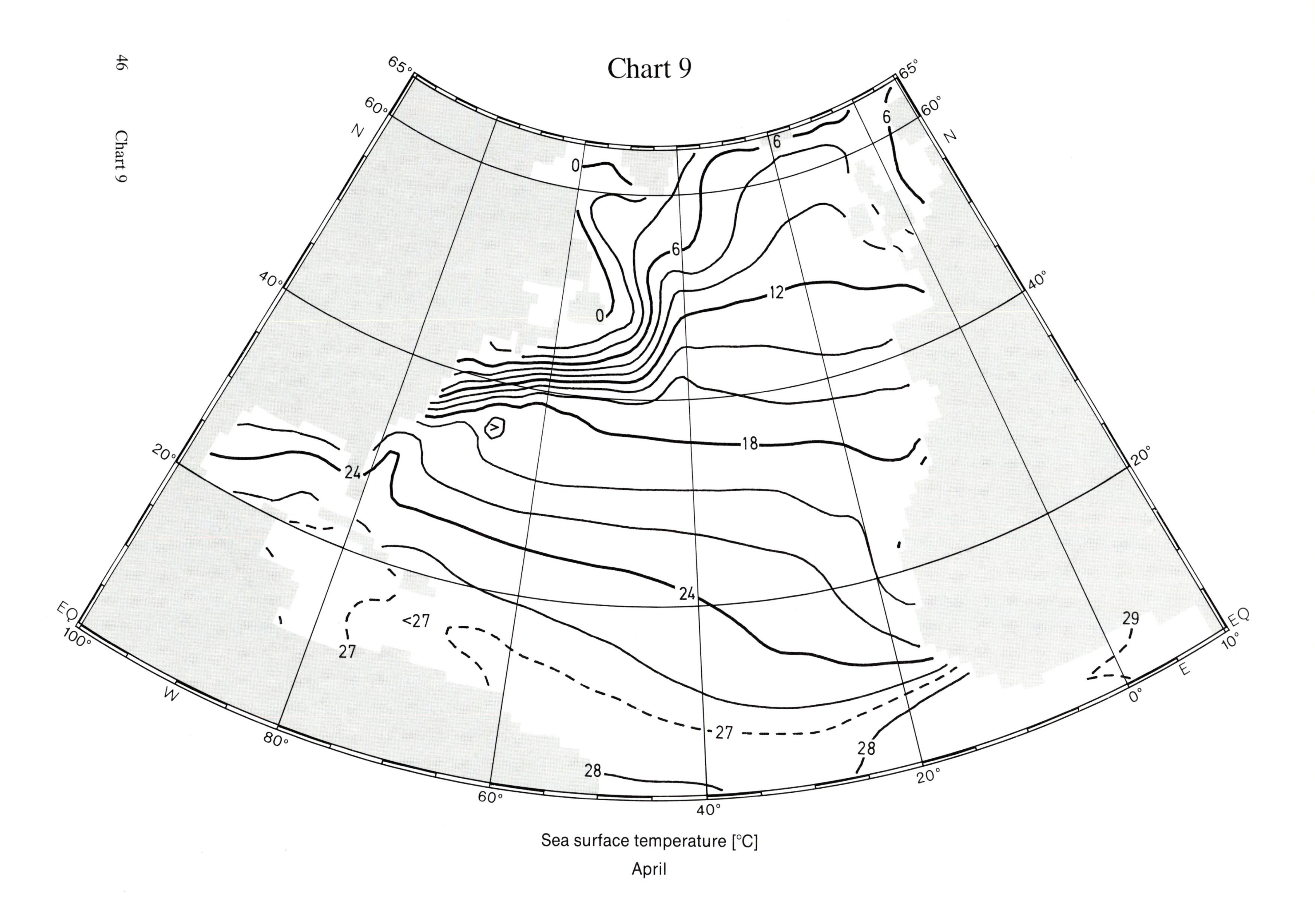

Sea surface temperature [°C]
April

Chart 10

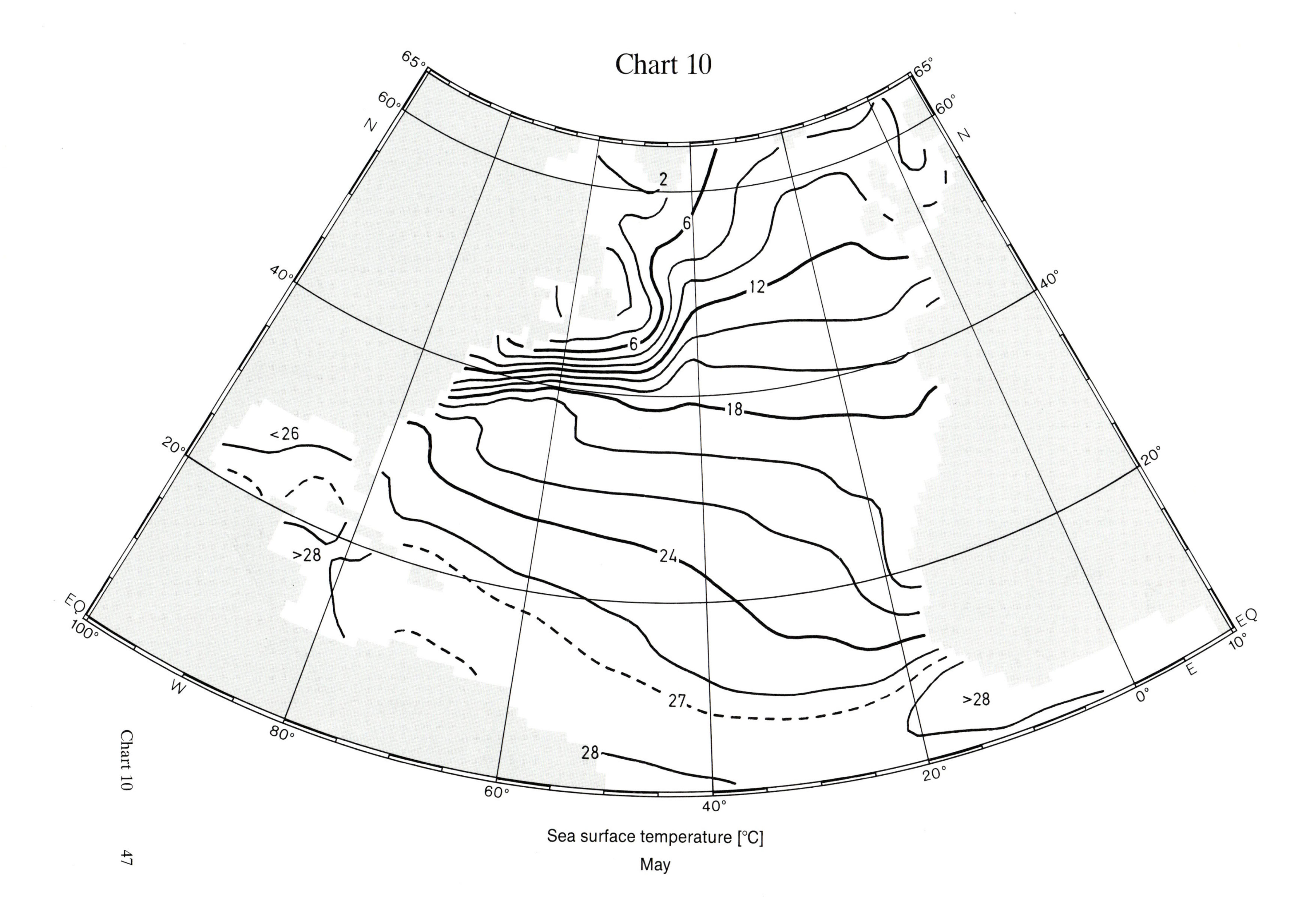

Sea surface temperature [°C]

May

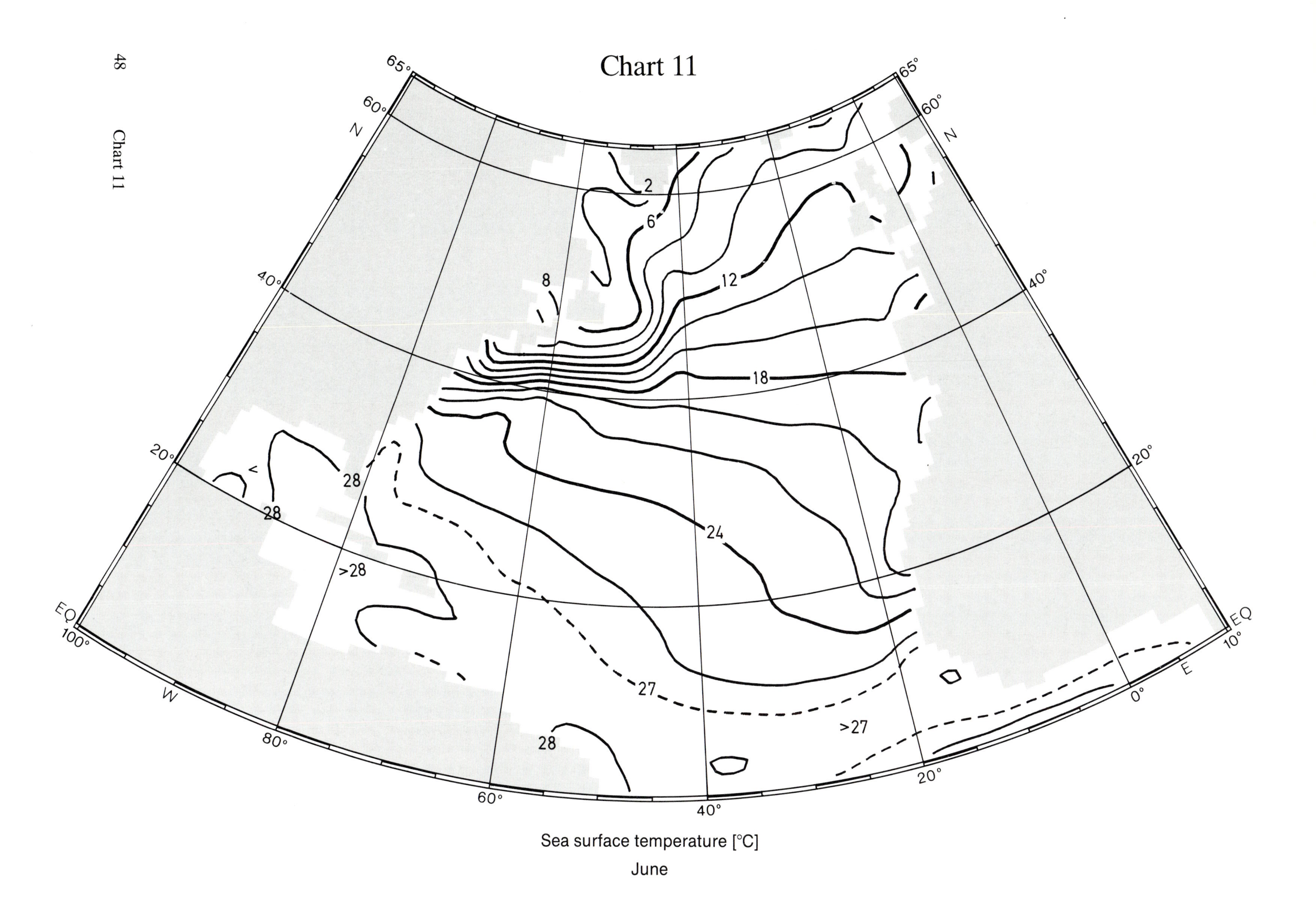
Chart 11
65°
60°
N
40°
20°
EQ
100°
W
80°
60°
40°
20°
0°
E
10°
2
6
8
12
18
24
27
>27
28
>28
Sea surface temperature [°C]
June

Chart 12

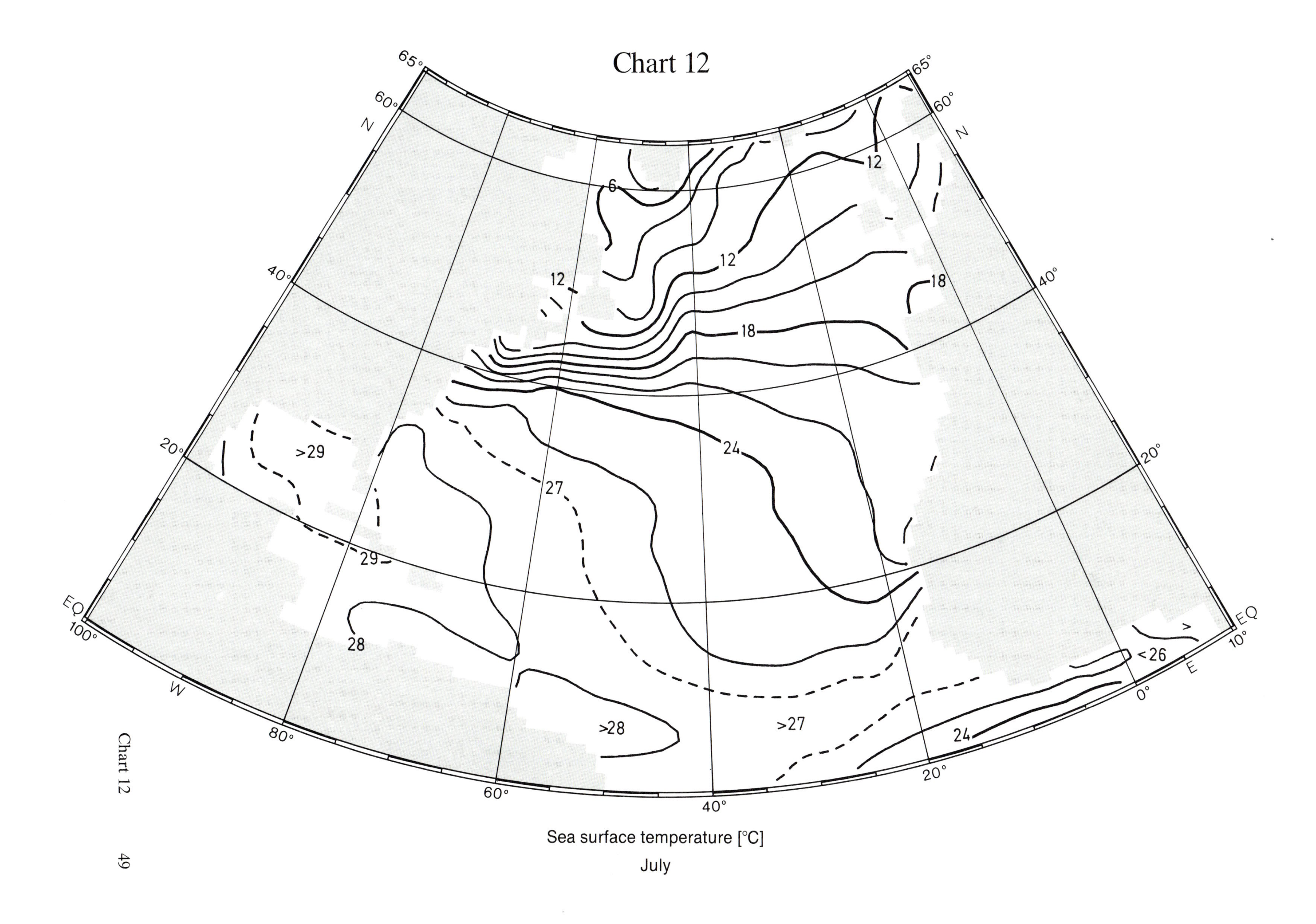

Sea surface temperature [°C]
July

Chart 13

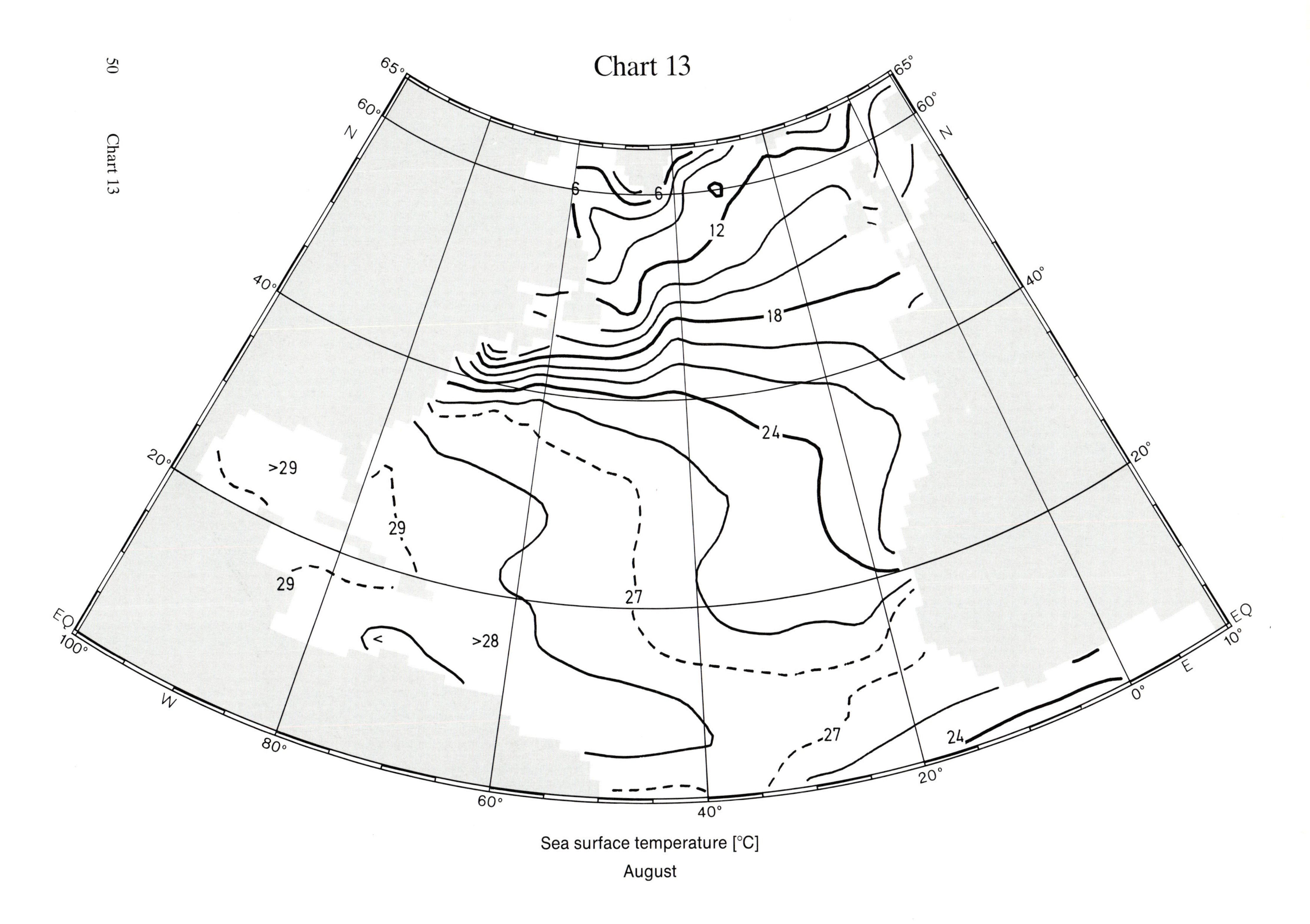

Sea surface temperature [°C]
August

Chart 14

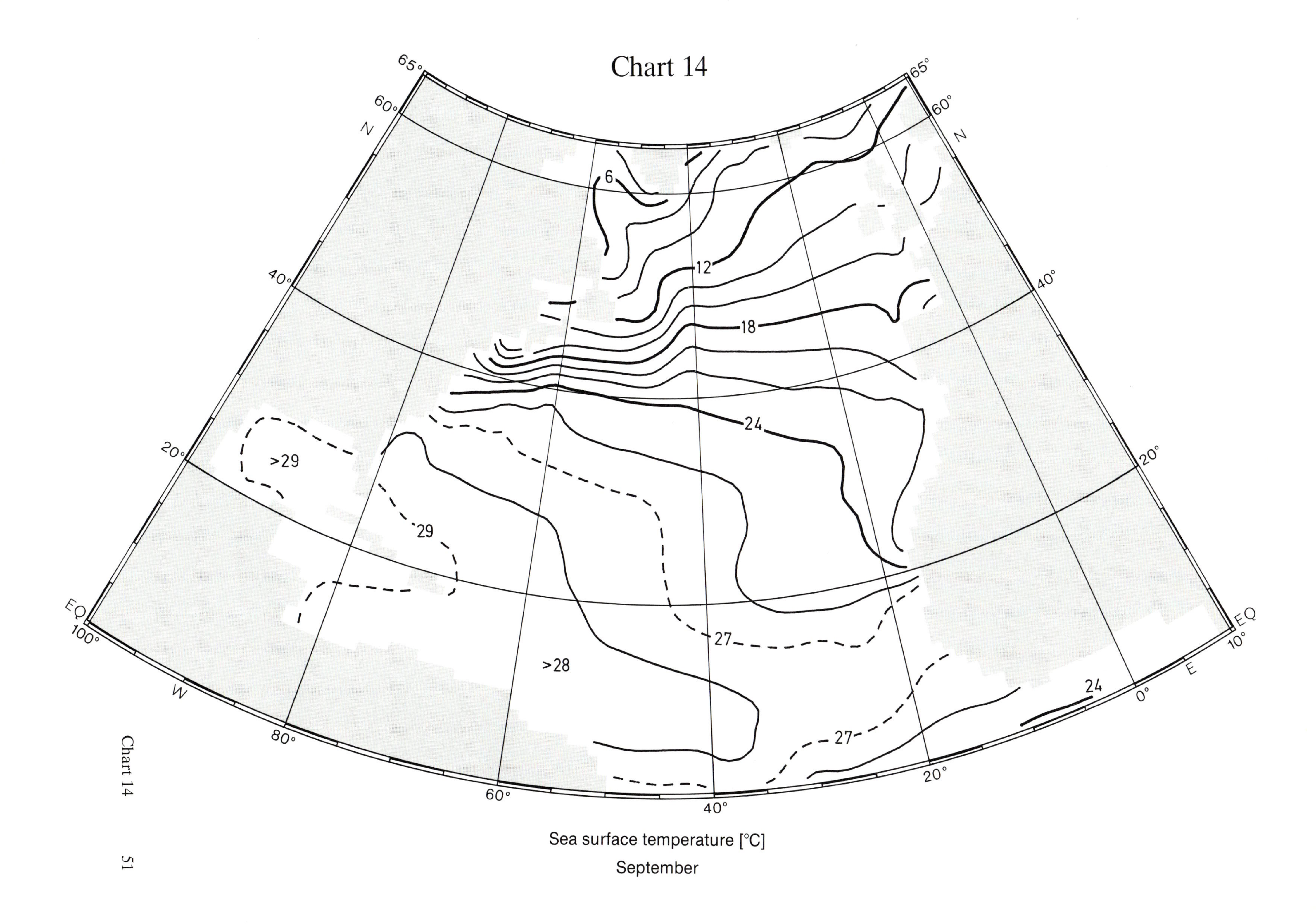

Sea surface temperature [°C]
September

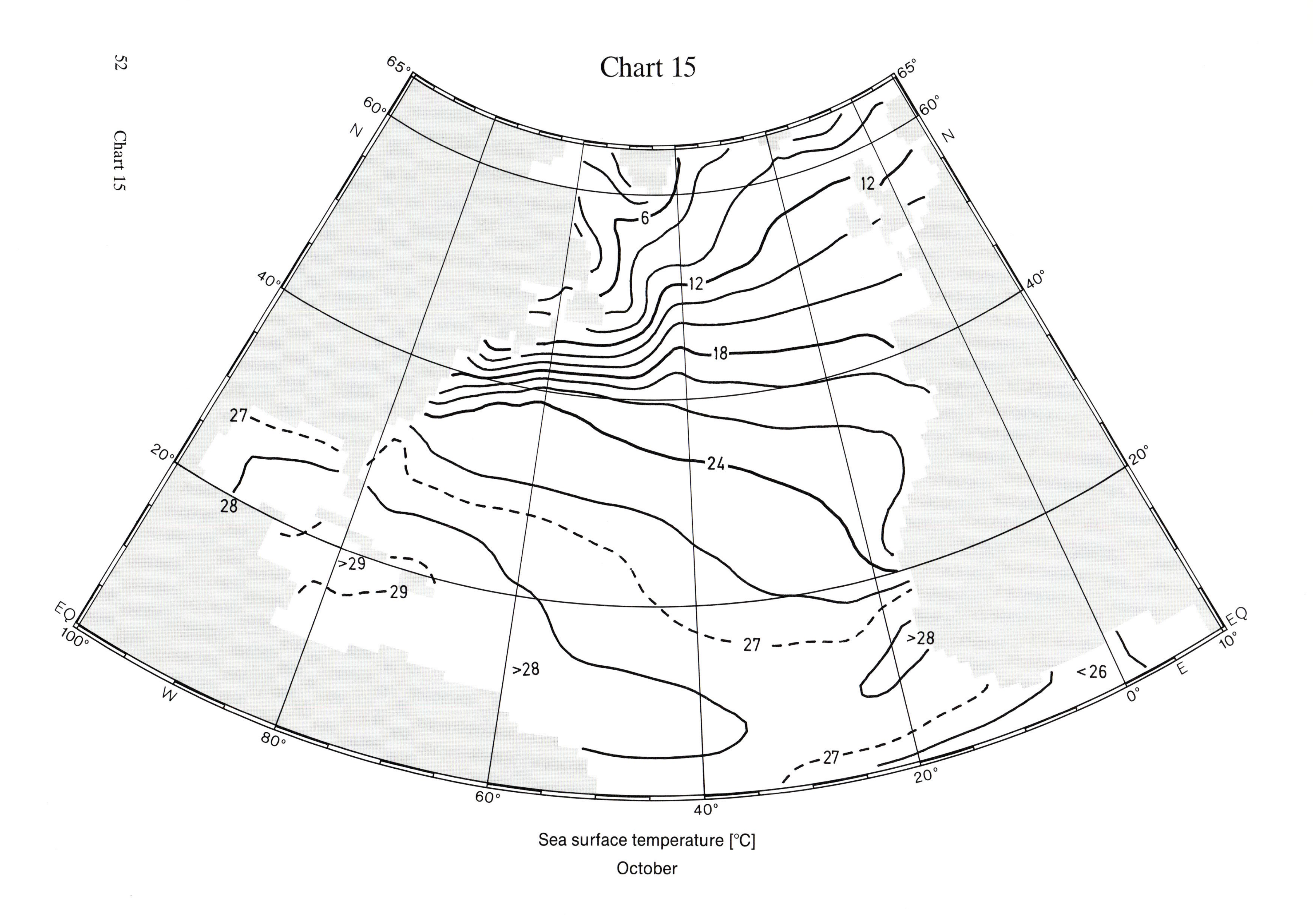

Chart 15
65°
60°
N
40°
20°
EQ
100°
W
80°
60°
40°
20°
0°
E
10°
6
12
12
18
24
27
27
27
27
28
>29
29
>28
>28
<26
Sea surface temperature [°C]
October

Chart 16

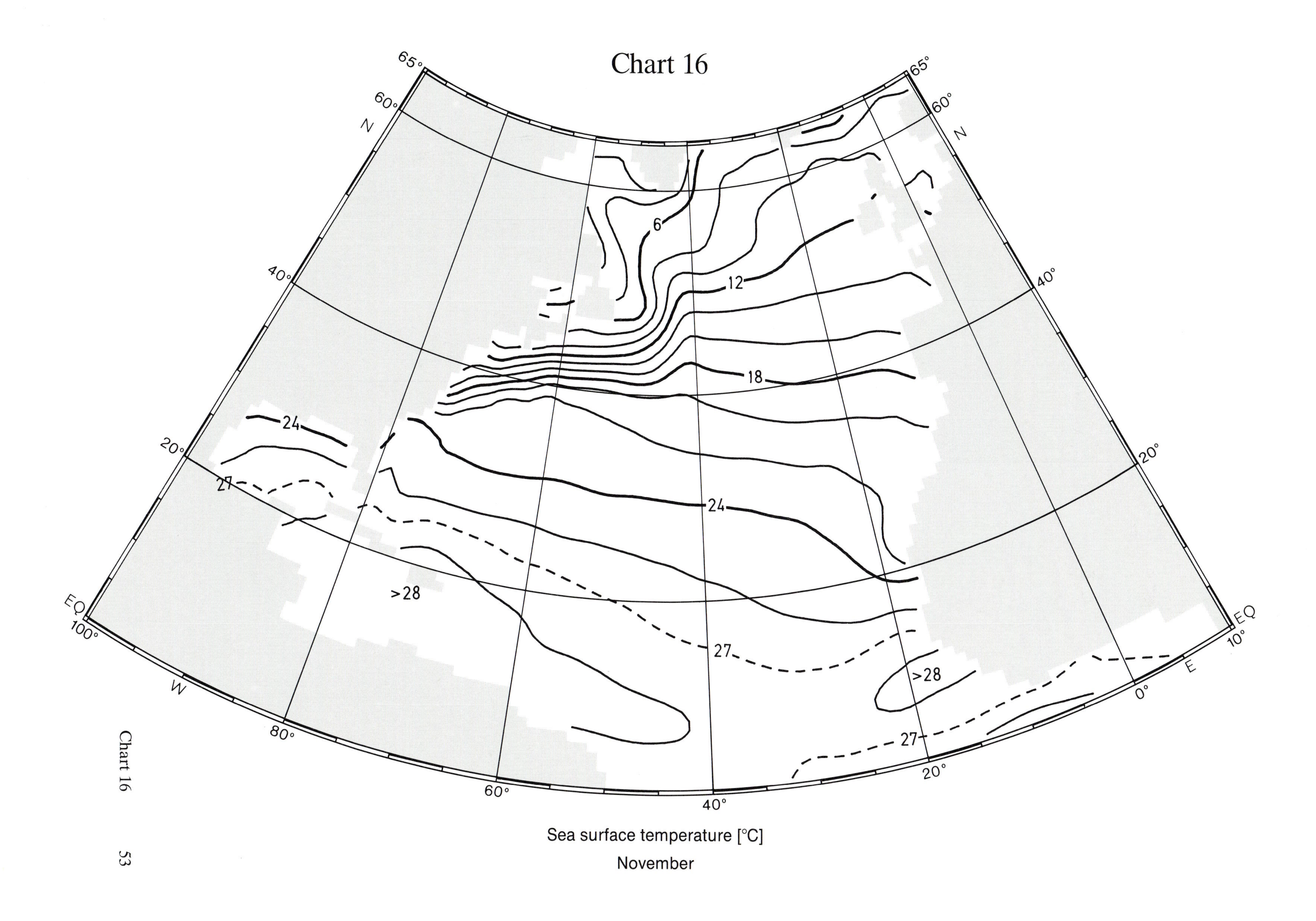

Sea surface temperature [°C]
November

Chart 17

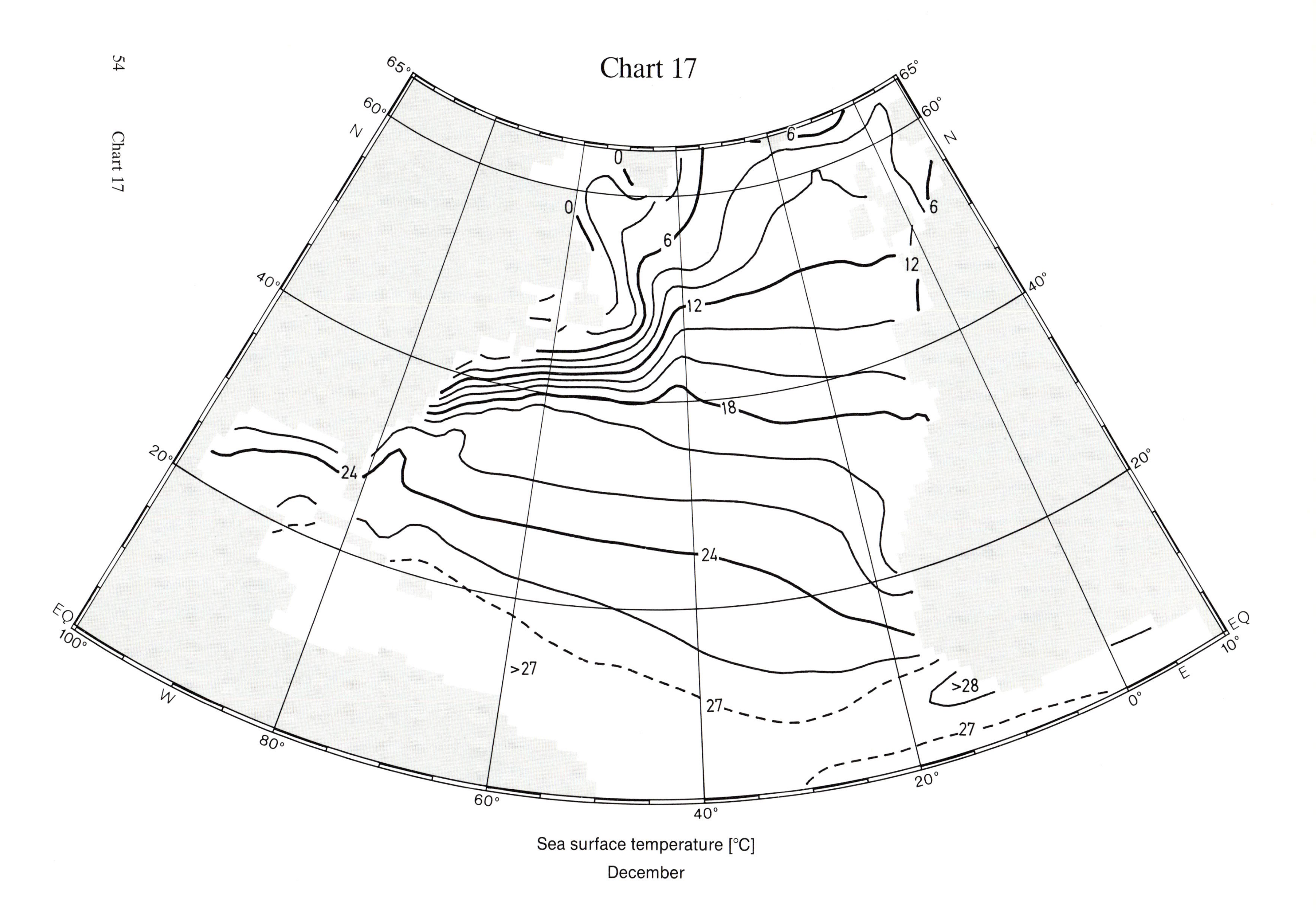

Sea surface temperature [°C]
December

Chart 18

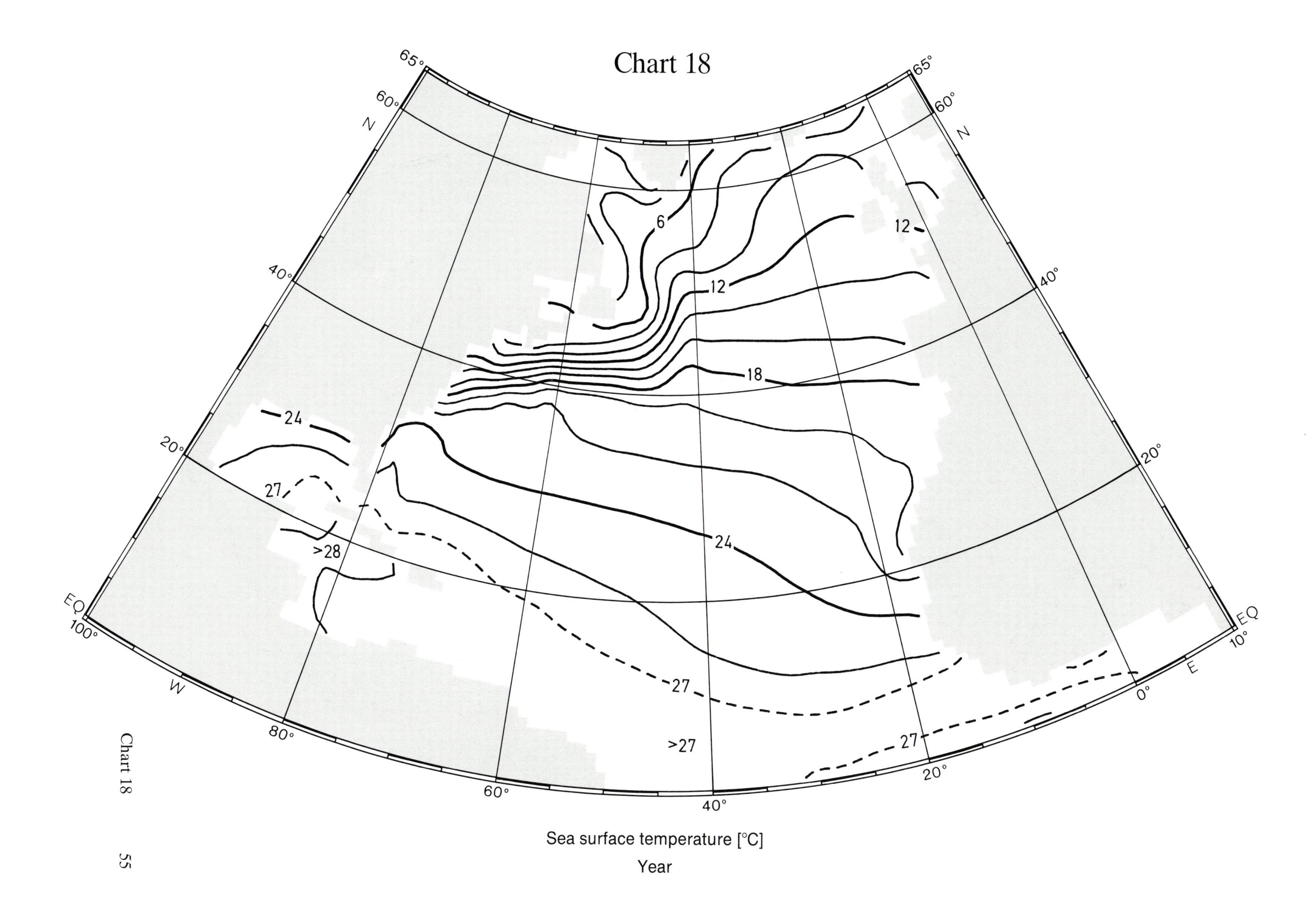

65°
60°
N
40°
20°
EQ
100°
W
80°
60°
40°
20°
0°
E
10°
6
12
12
18
24
24
27
27
27
>27
>28
Sea surface temperature [°C]
Year

Chart 19

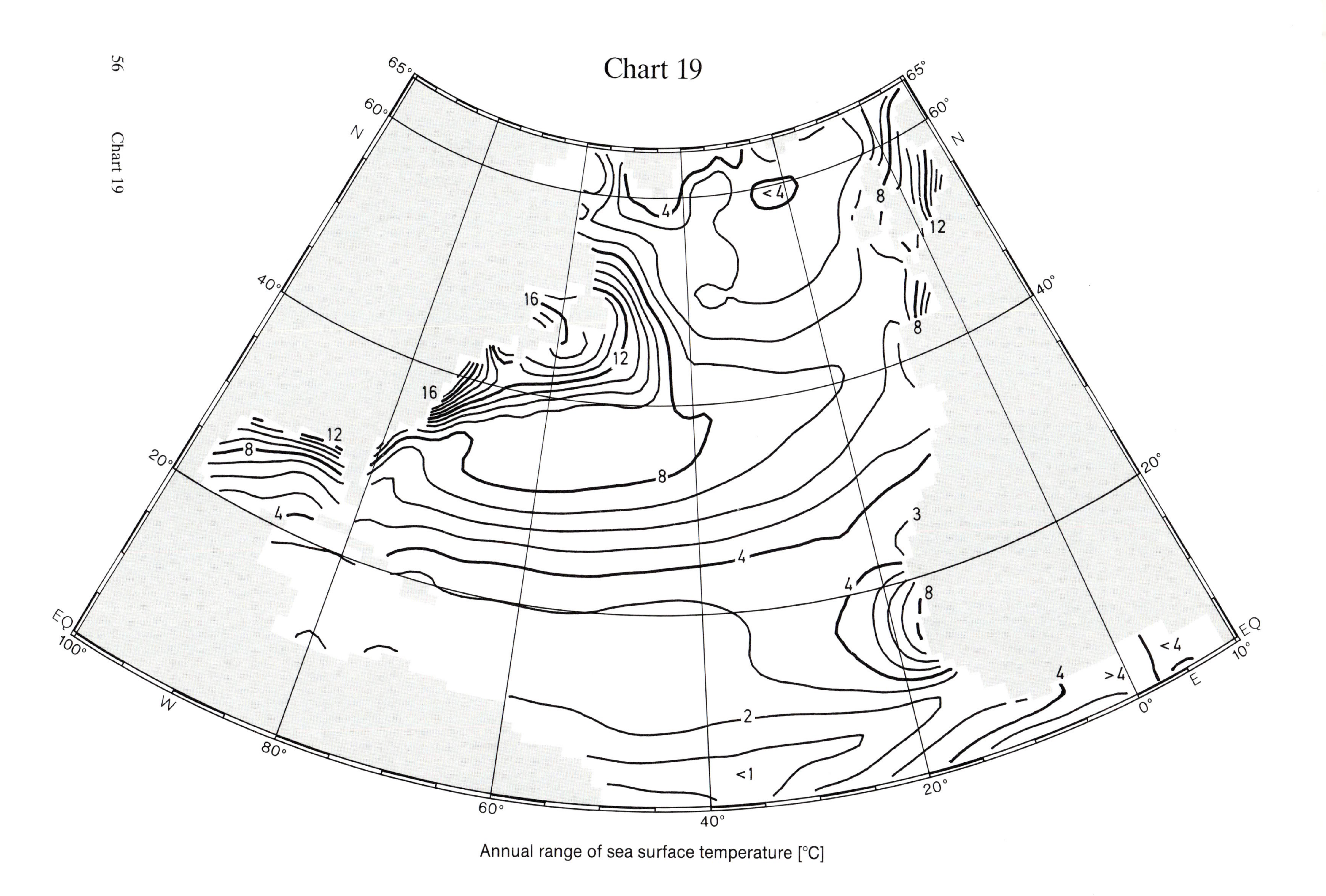

Annual range of sea surface temperature [°C]

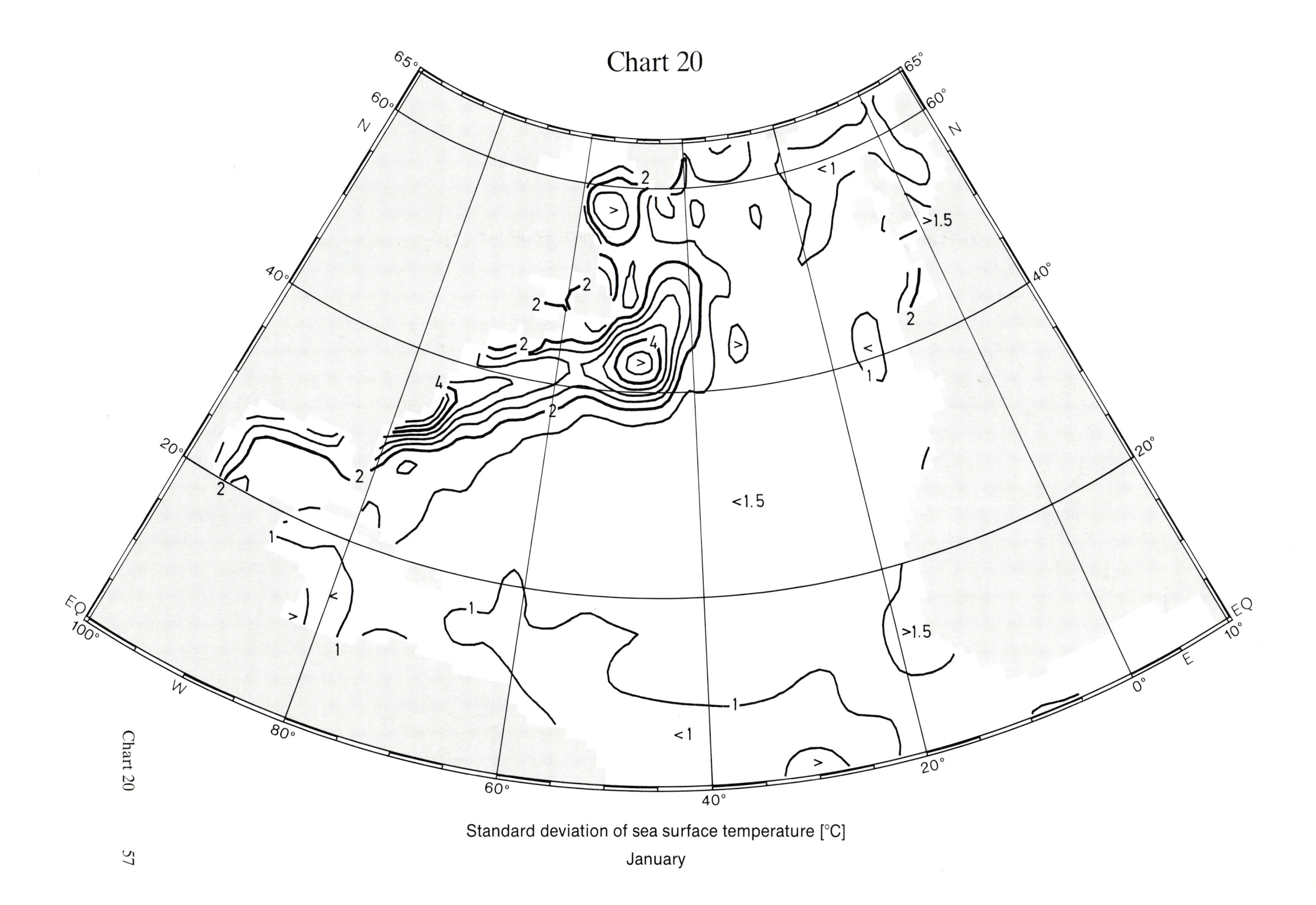
Chart 20
65°
60°
N
40°
20°
EQ
100°
W
80°
60°
40°
20°
0°
E
10°
2
4
>
<
1
<1
>1.5
<1.5
Standard deviation of sea surface temperature [°C]
January

Chart 21

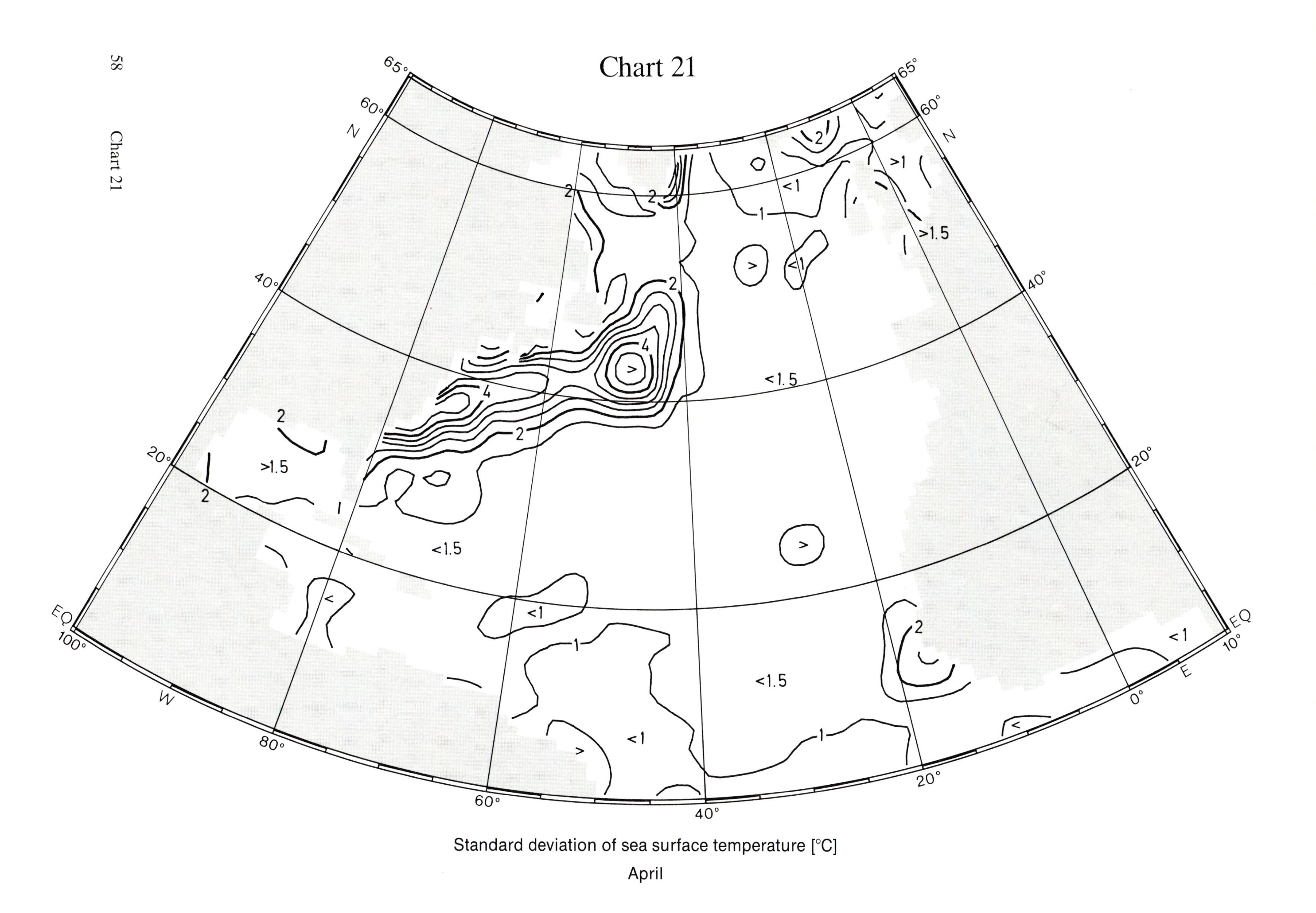

Standard deviation of sea surface temperature [°C]
April

Chart 22

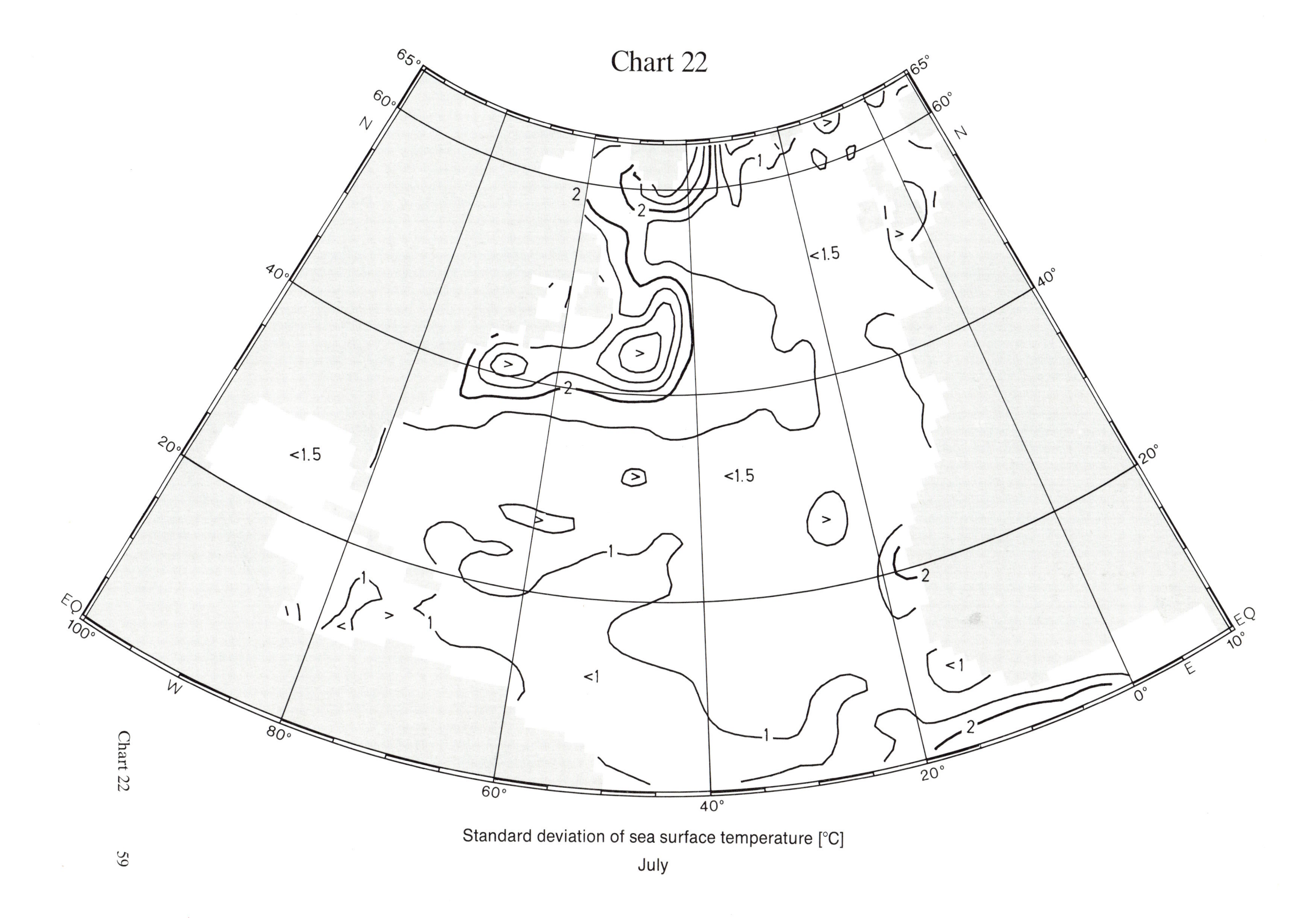

Standard deviation of sea surface temperature [°C]

July

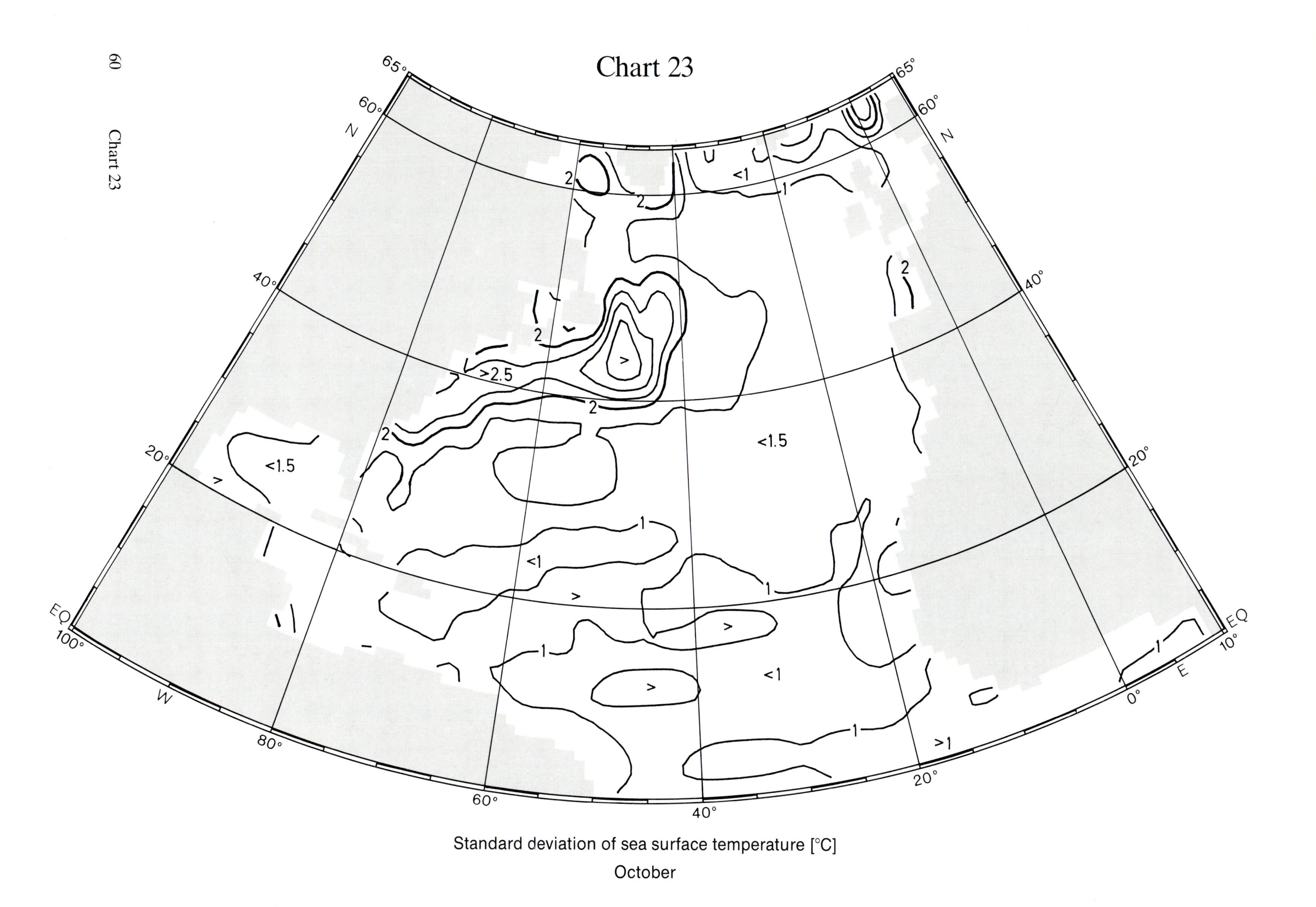

Standard deviation of sea surface temperature [°C]
October

Chart 24

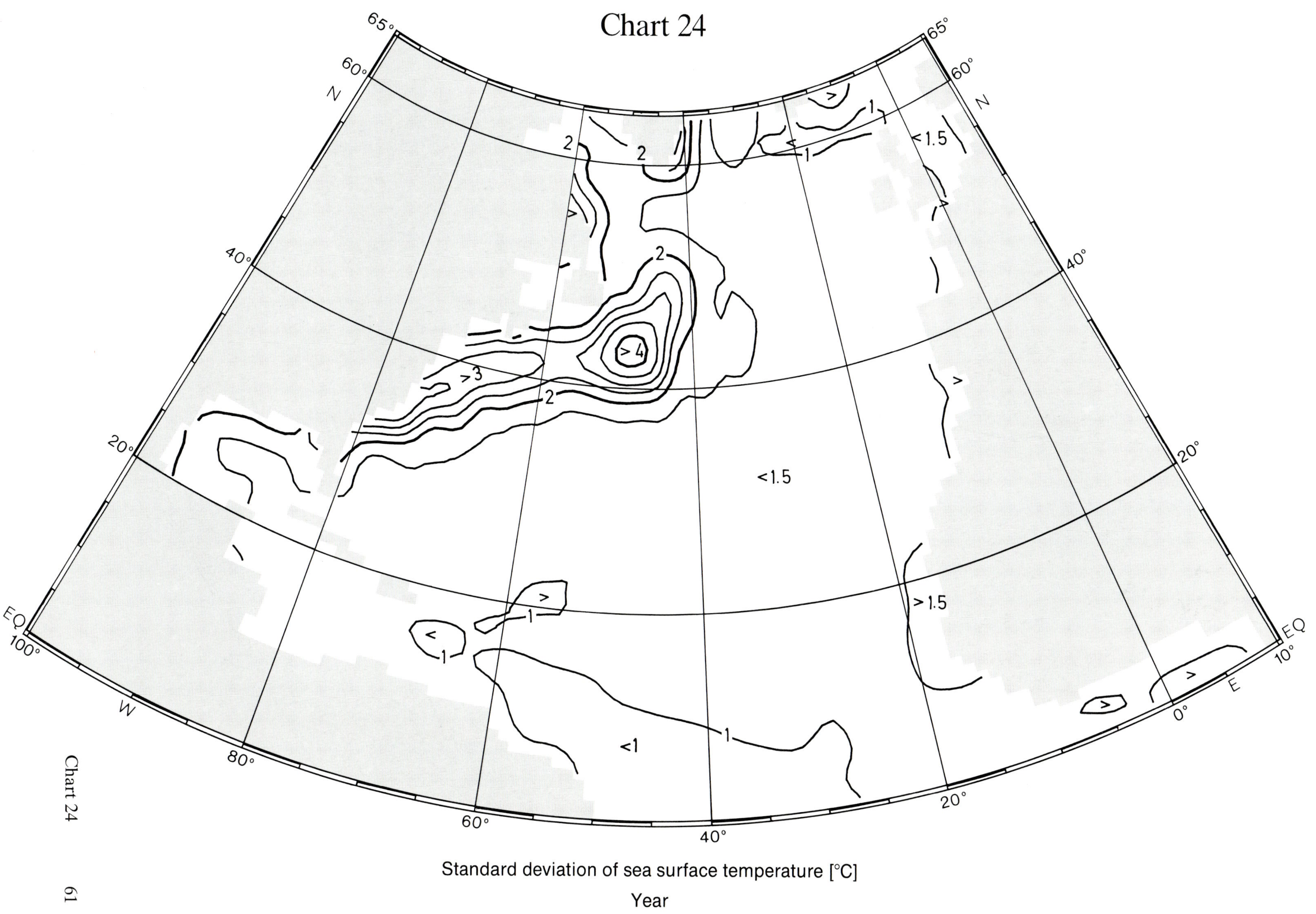

Standard deviation of sea surface temperature [°C]
Year

Chart 25

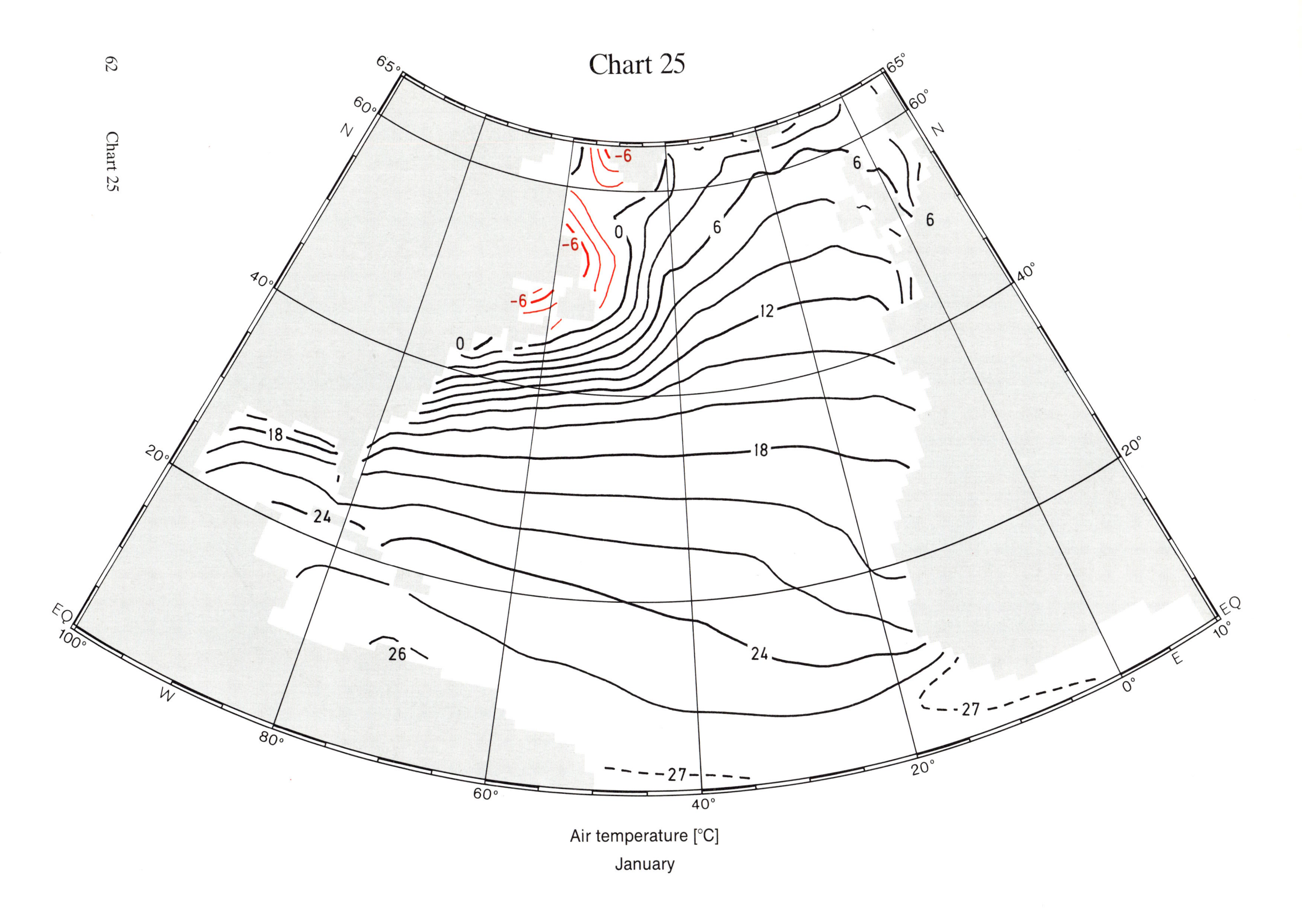

Air temperature [°C]
January

Chart 26

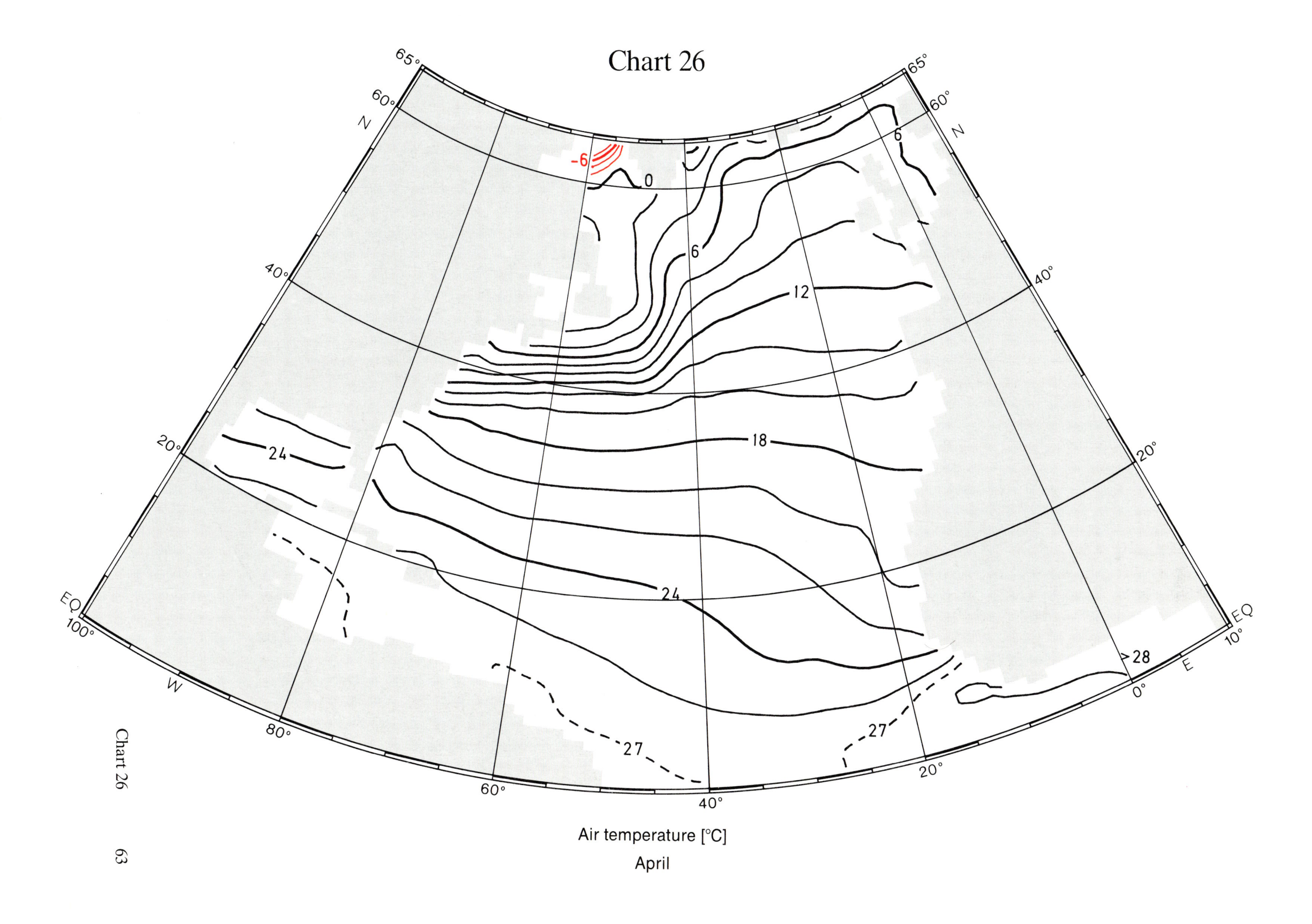
65°
60°
N
40°
20°
EQ
100°
W
80°
60°
40°
20°
0°
E
10°
-6
0
6
12
18
24
27
>28
Air temperature [°C]
April

Chart 27

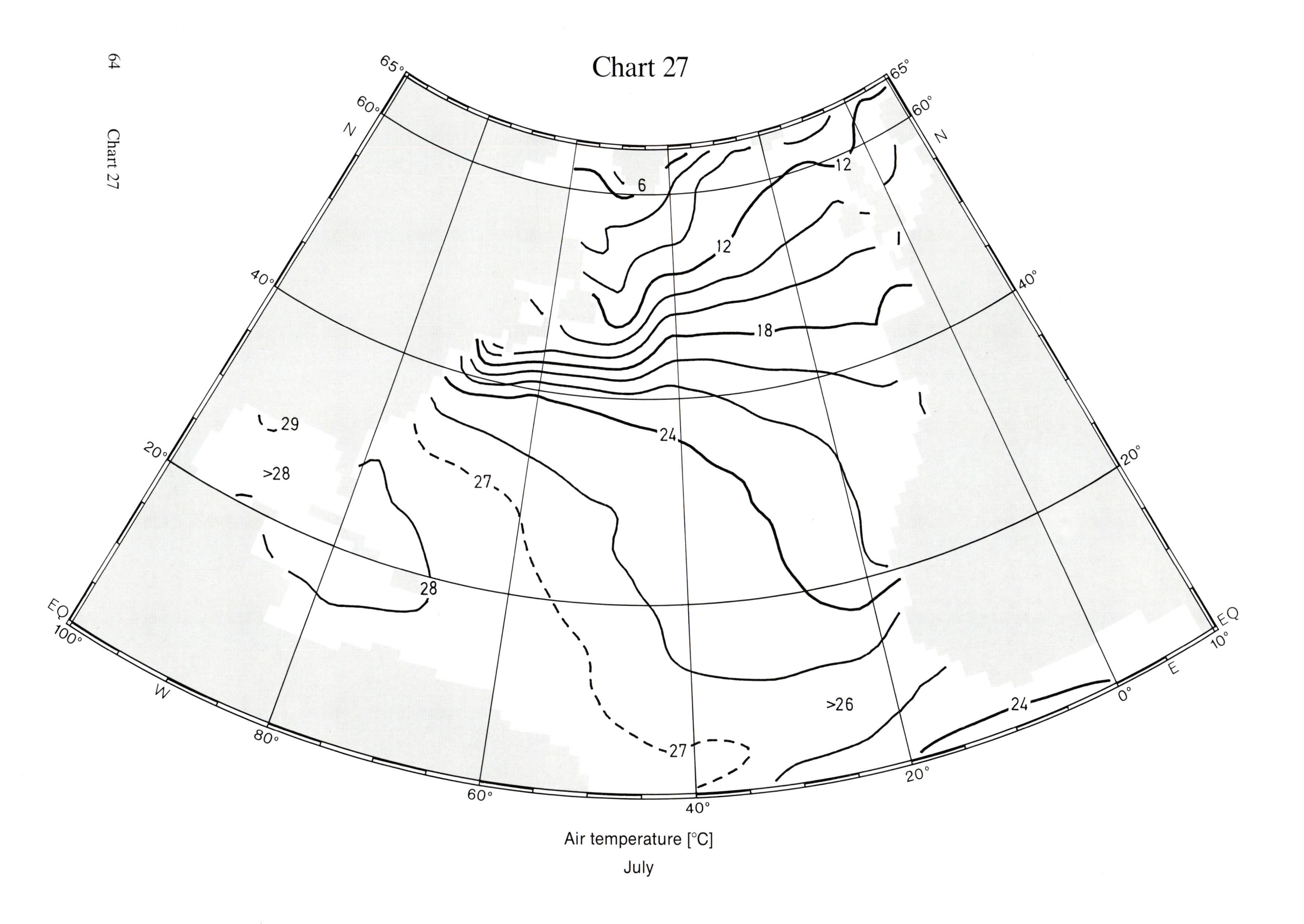

Air temperature [°C]
July

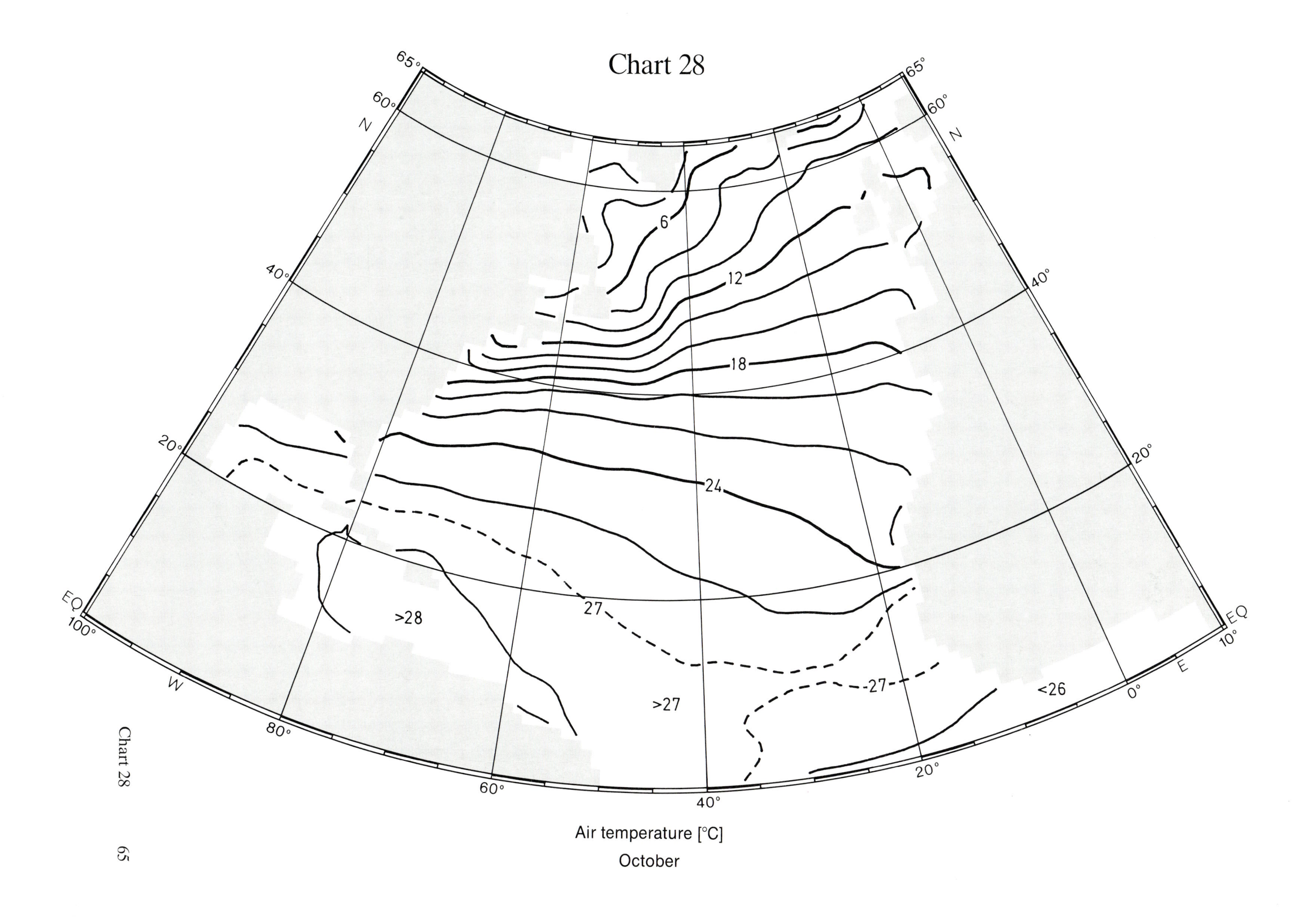

Chart 28
Air temperature [°C]
October
6
12
18
24
27
-27
>27
<26
>28
65°
60°
N
40°
20°
EQ
10°
E
0°
20°
40°
60°
80°
W
100°

Chart 29

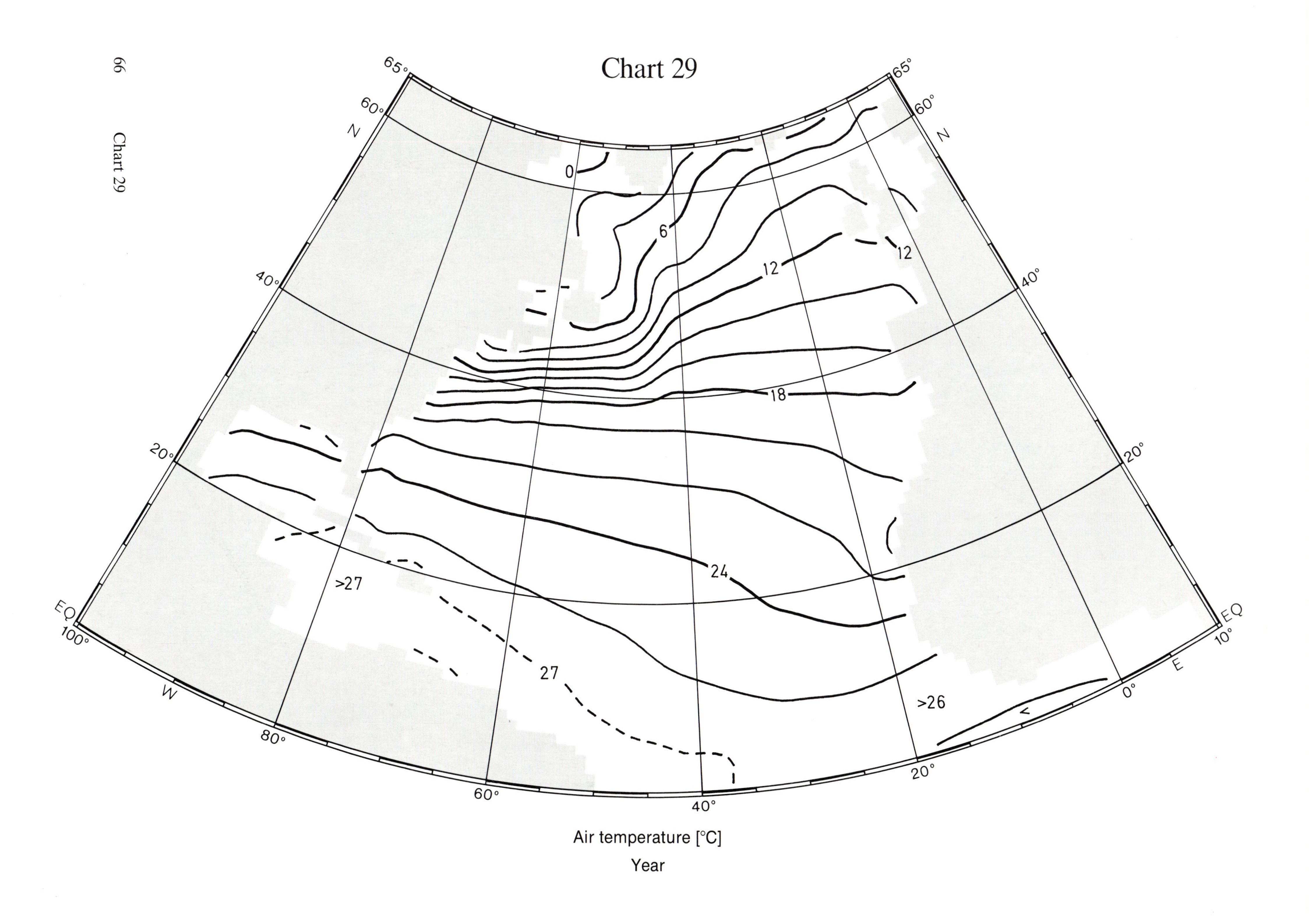

Air temperature [°C]
Year

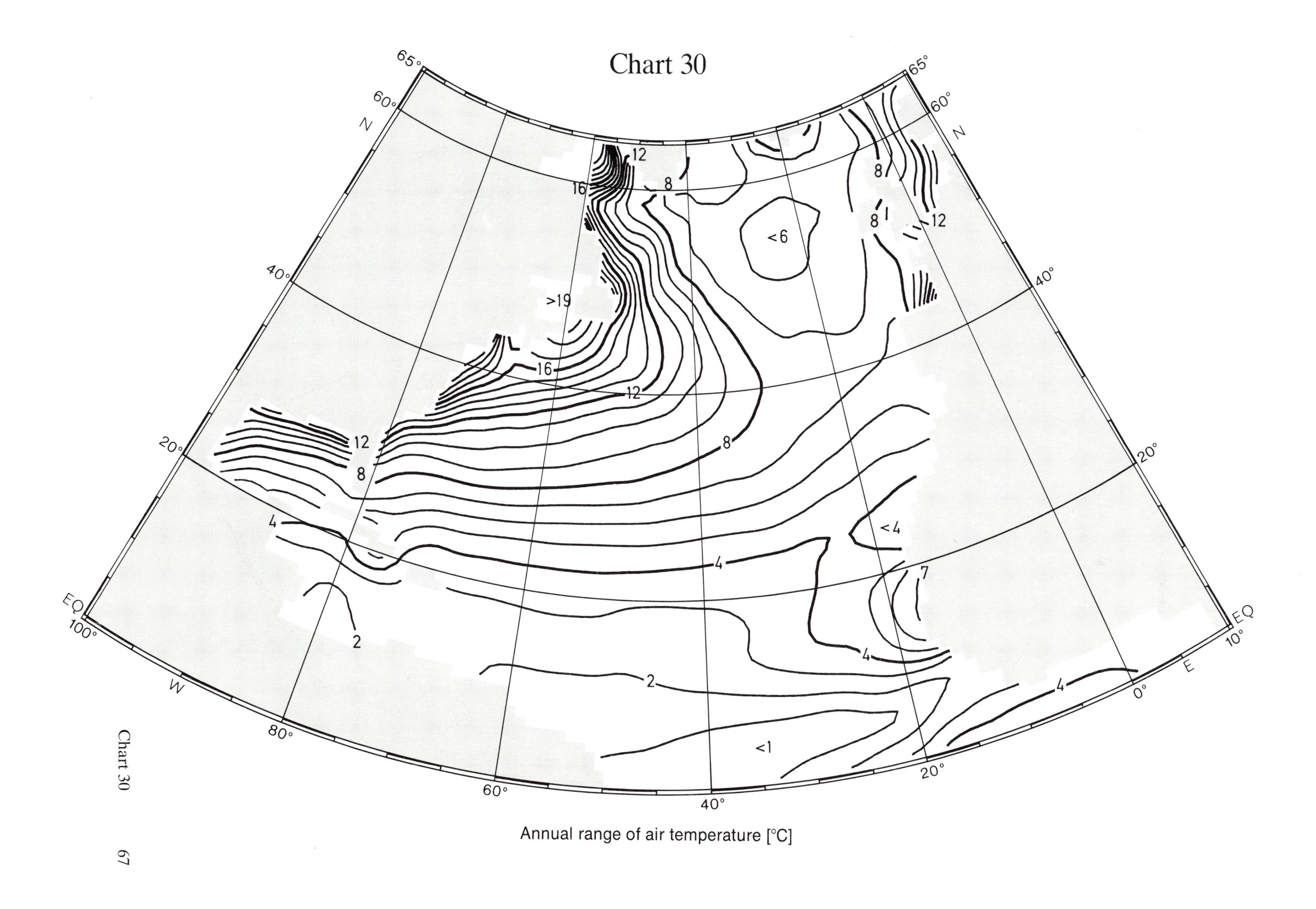
Chart 30
65°
60°
N
40°
20°
EQ
100°
W
80°
60°
40°
20°
0°
E
10°
12
16
8
<6
>19
12
8
4
<4
7
2
<1
Annual range of air temperature [°C]

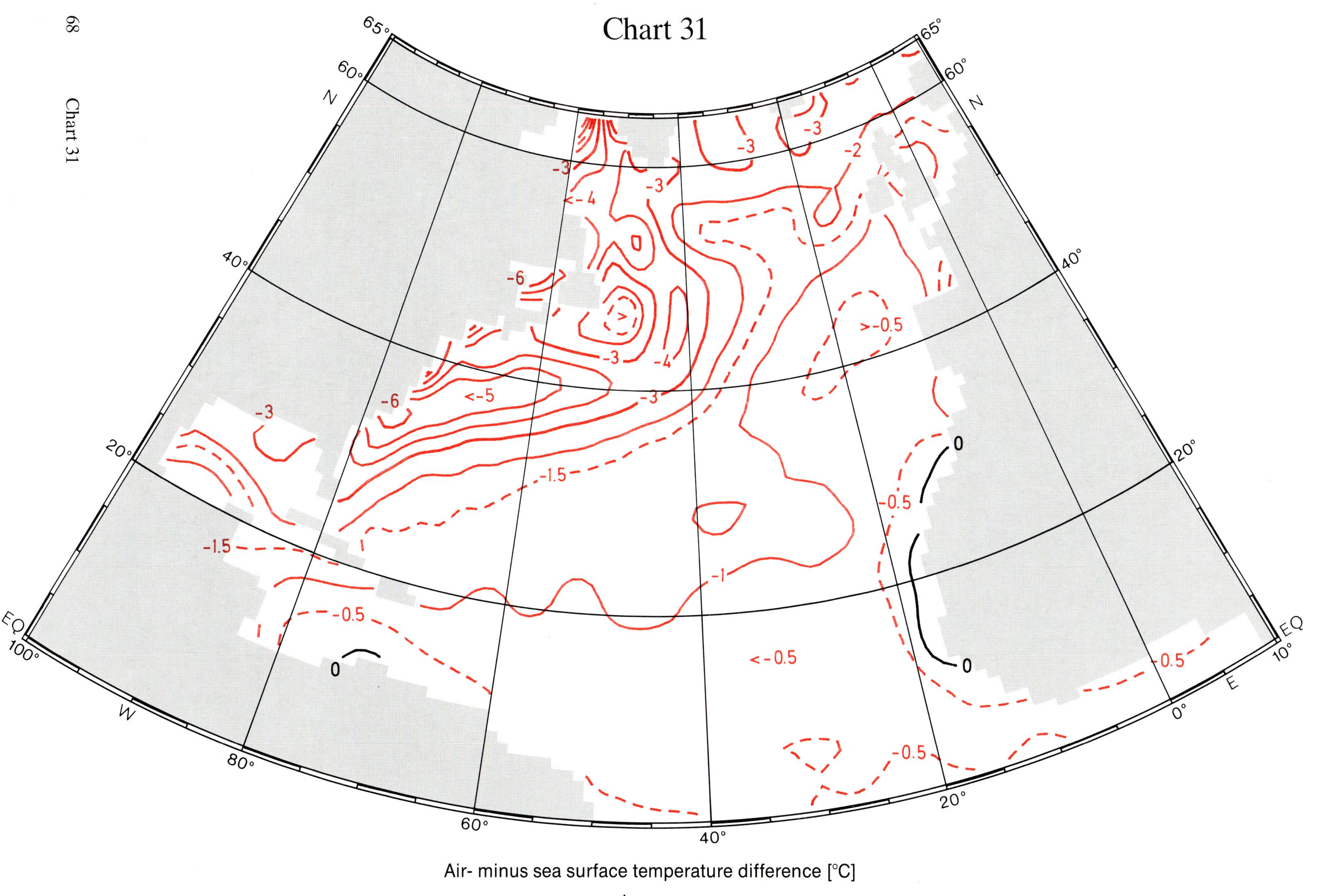

Air- minus sea surface temperature difference [°C]
January

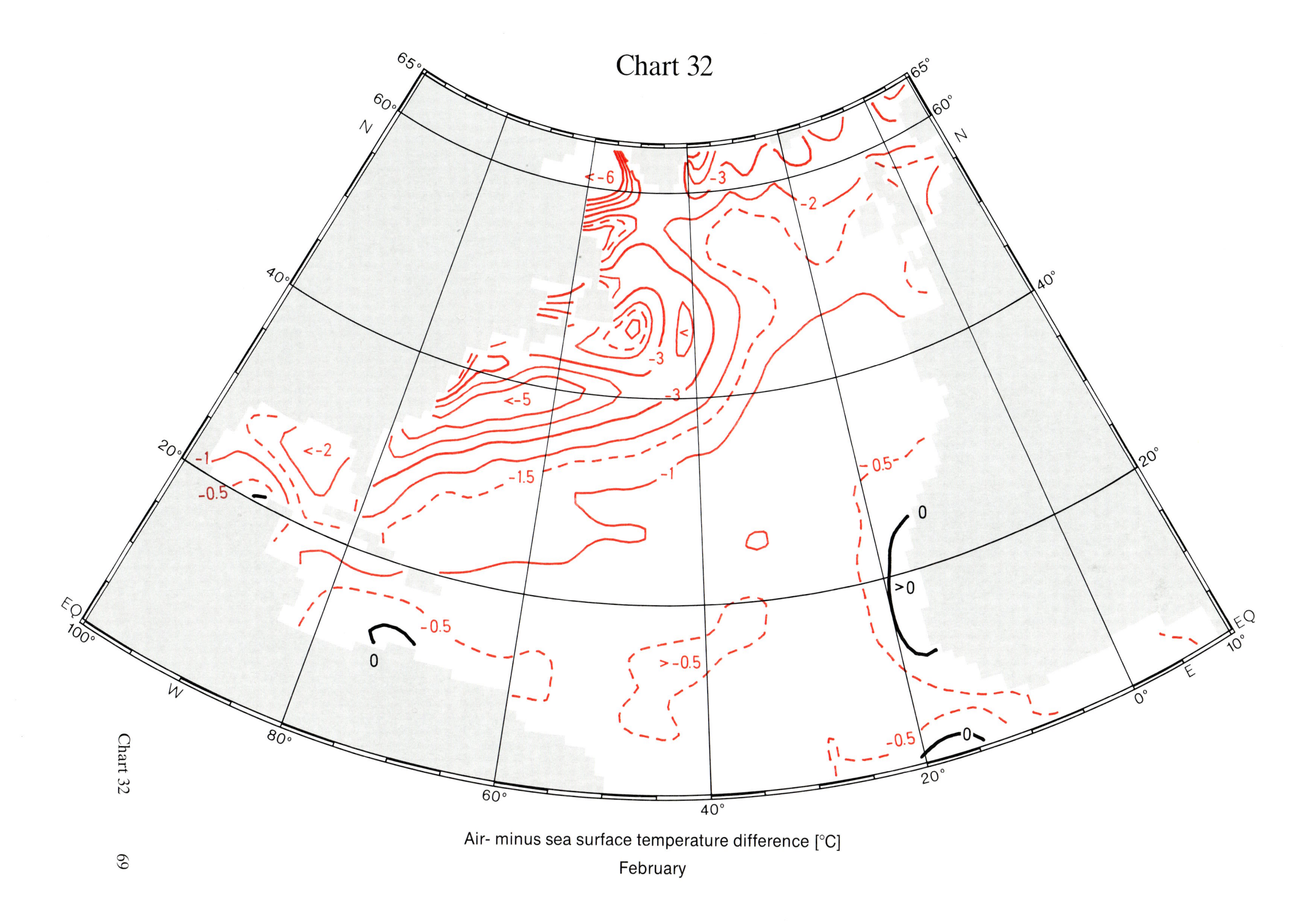
Chart 32
65°
60°
N
40°
20°
EQ
100°
W
80°
60°
40°
20°
0°
E
10°
<-6
-3
-2
<
-3
-3
<-5
<-2
-1
-0.5
-1.5
-1
-0.5
0
>0
-0.5
0
>-0.5
-0.5
0
Air- minus sea surface temperature difference [°C]
February

Chart 33

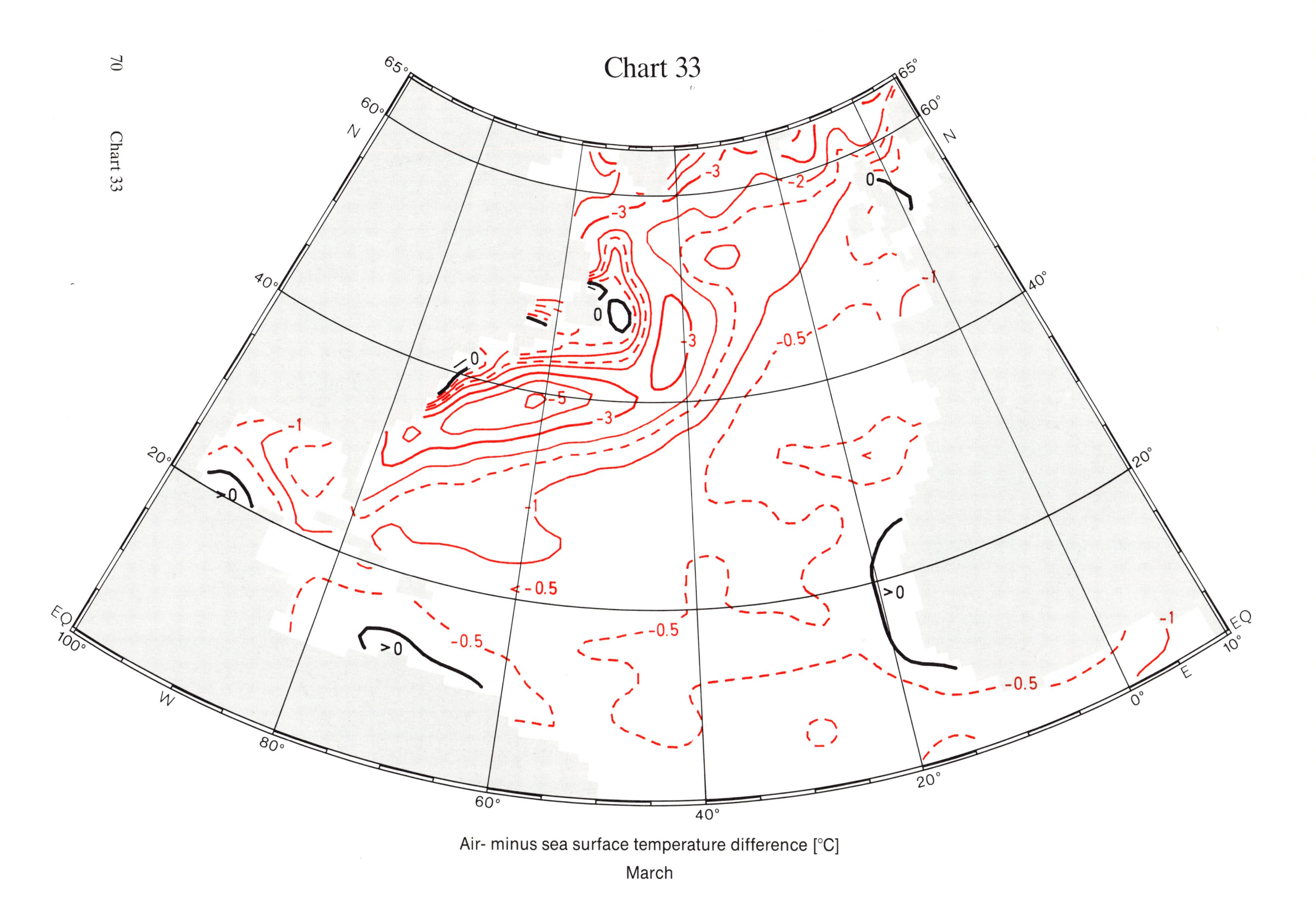

Air- minus sea surface temperature difference [°C]
March

Chart 34

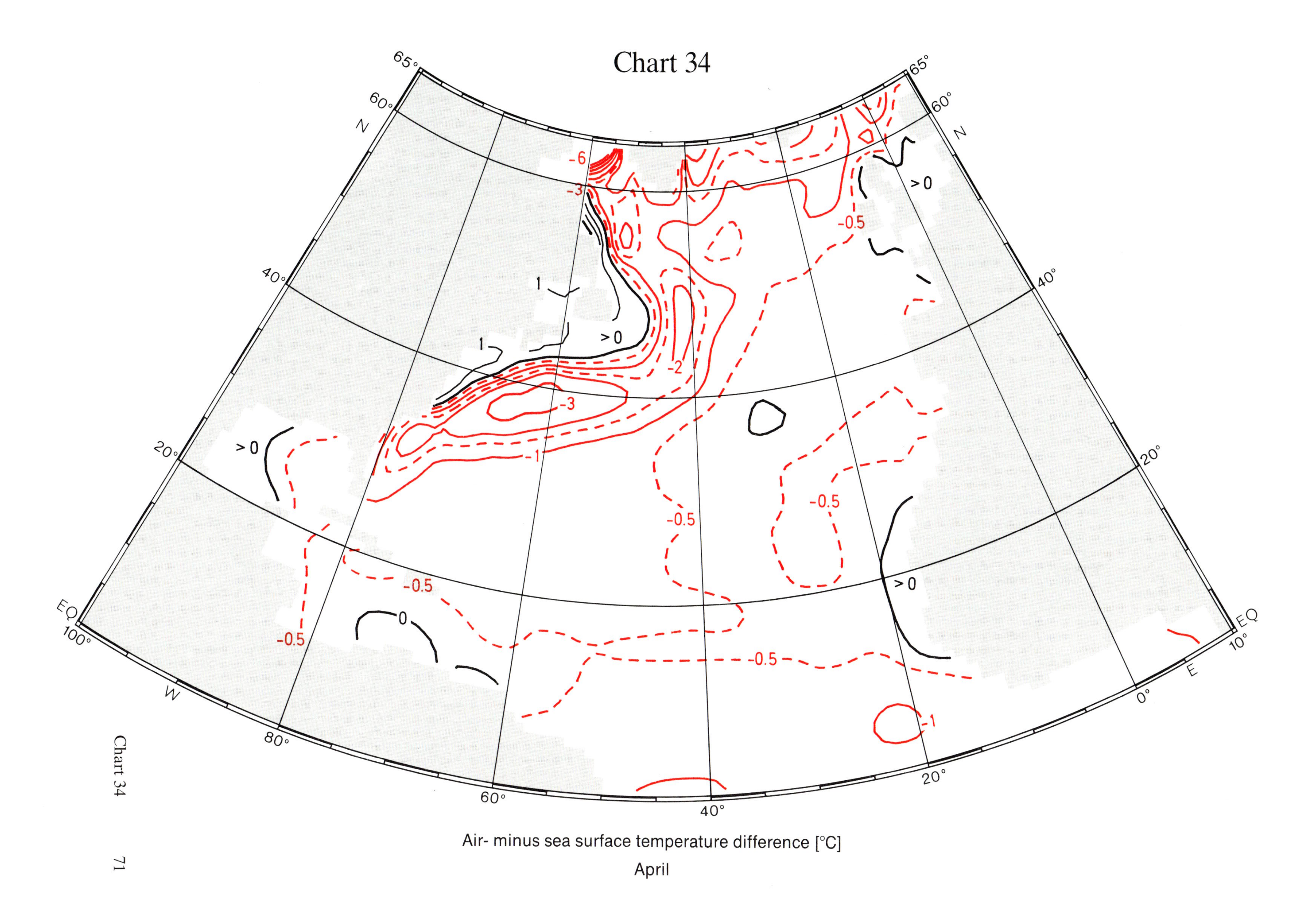

Air- minus sea surface temperature difference [°C]
April

Chart 35

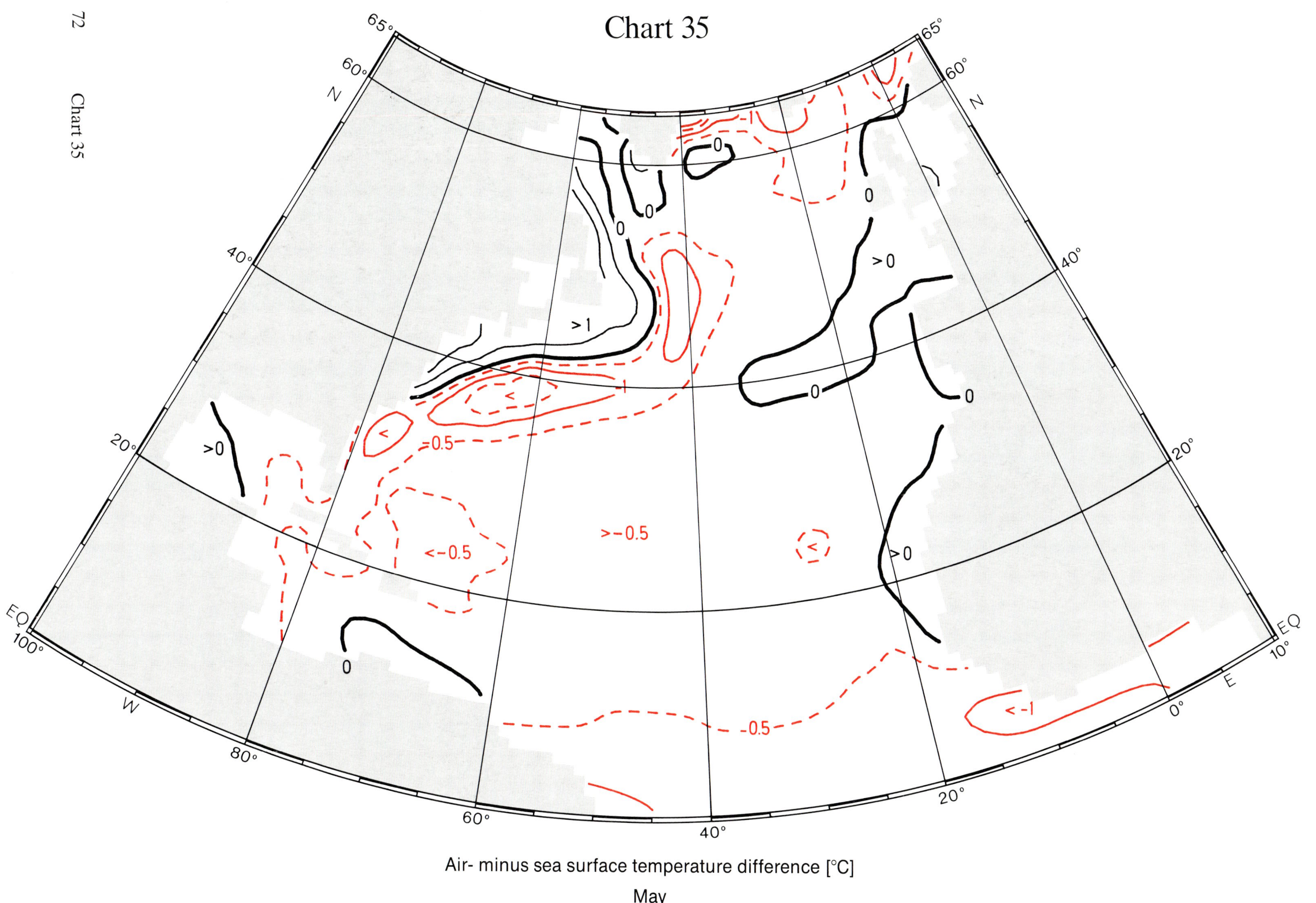

Air- minus sea surface temperature difference [°C]

May

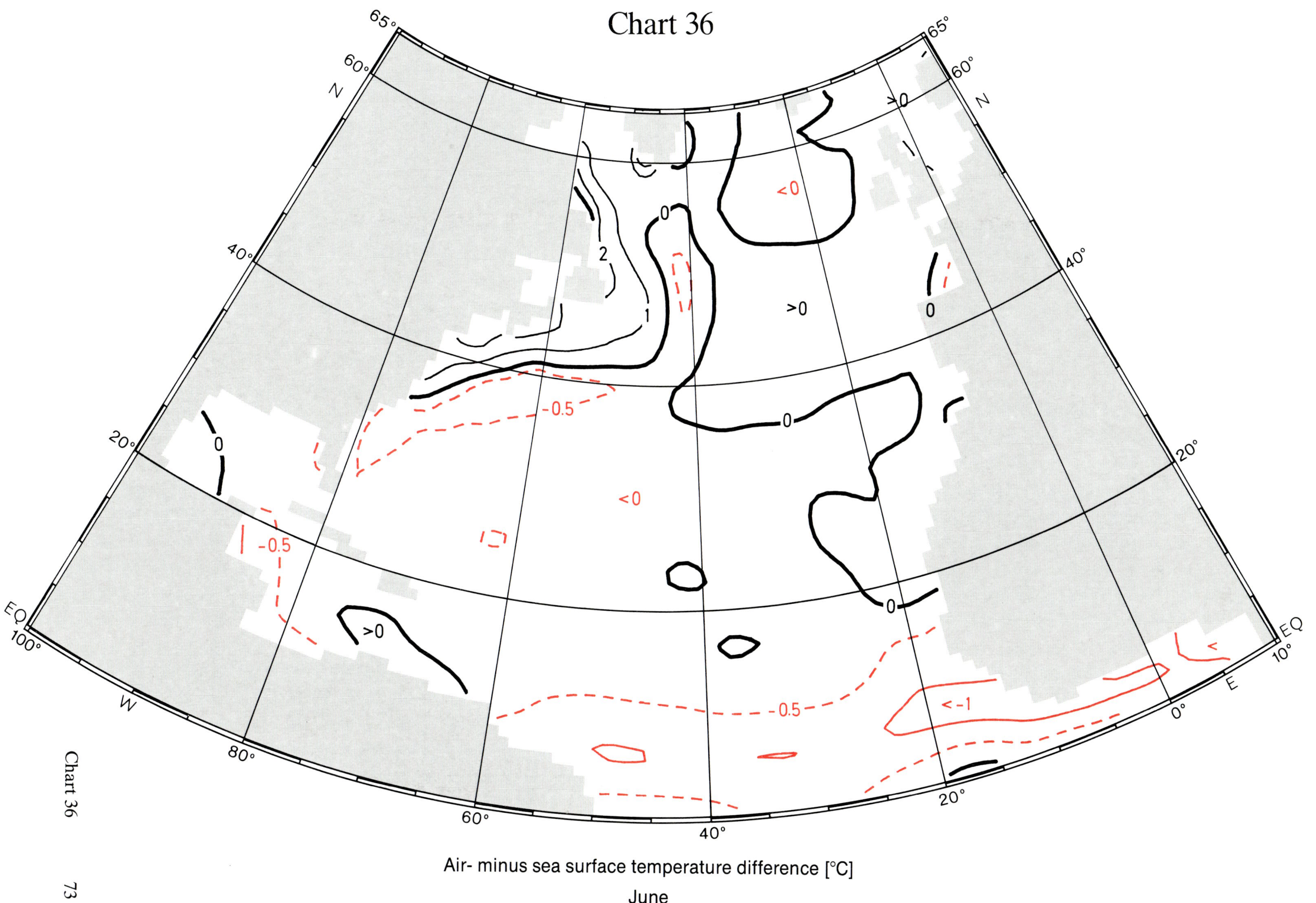
Chart 36
65°
60°
N
40°
20°
EQ
100°
W
80°
60°
40°
20°
0°
E
10°
0
1
2
>0
<0
-0.5
<-1
Air- minus sea surface temperature difference [°C]
June

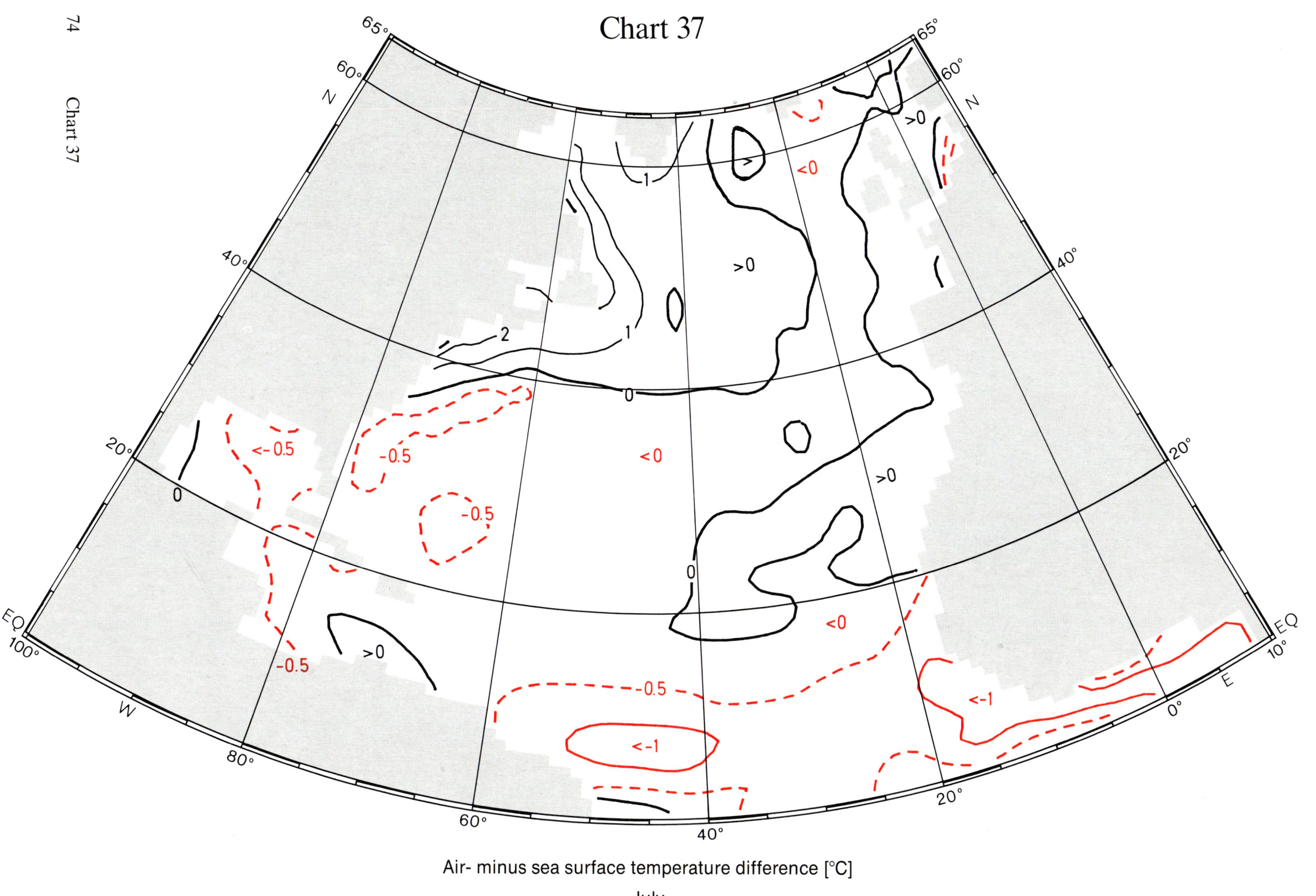
Chart 37
65°
60°
40°
20°
EQ
100°
80°
60°
40°
20°
0°
10°
N
W
E
>0
<0
0
1
2
-0.5
<-0.5
<-1
Air- minus sea surface temperature difference [°C]
July

Chart 38

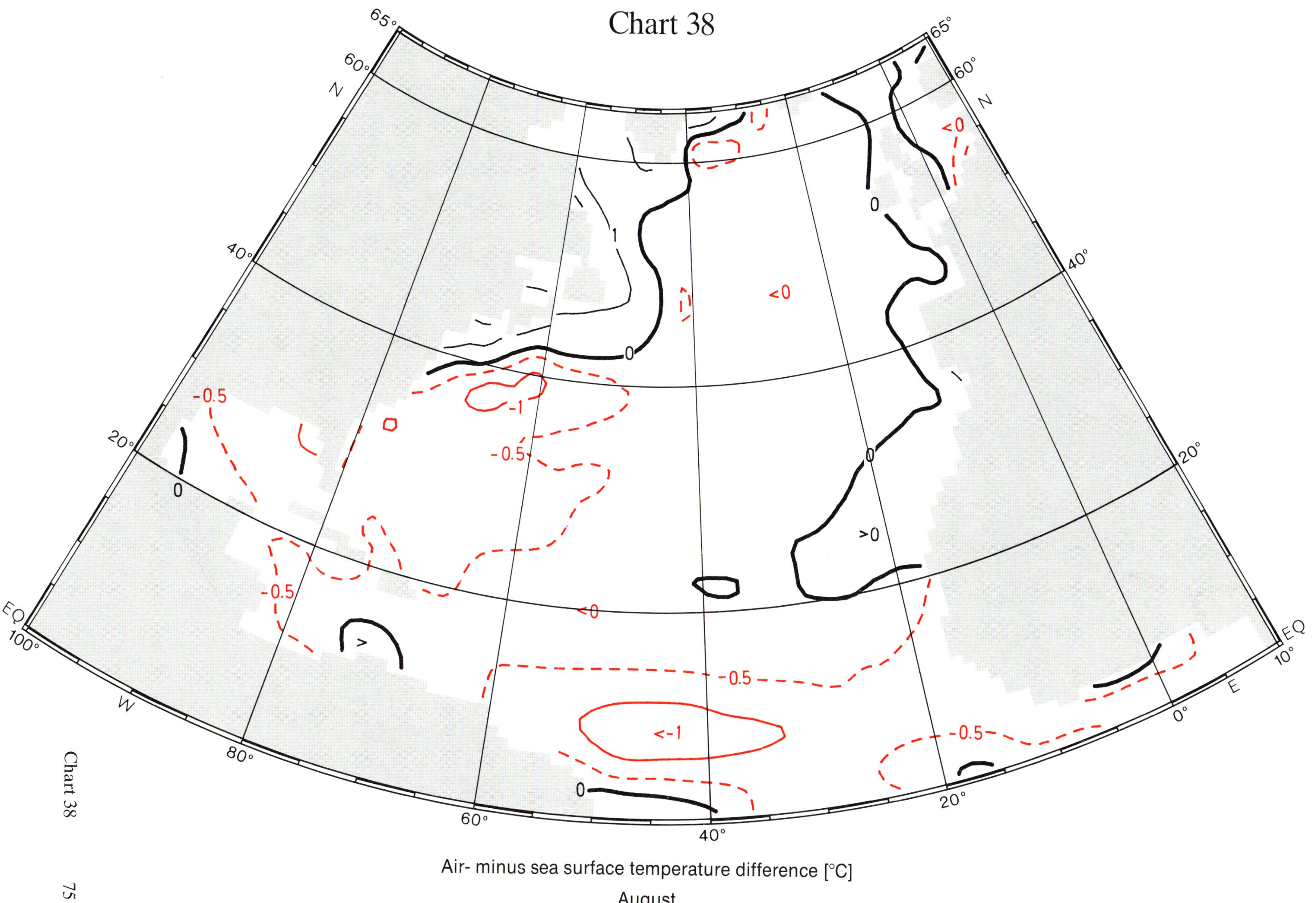

Air- minus sea surface temperature difference [°C]
August

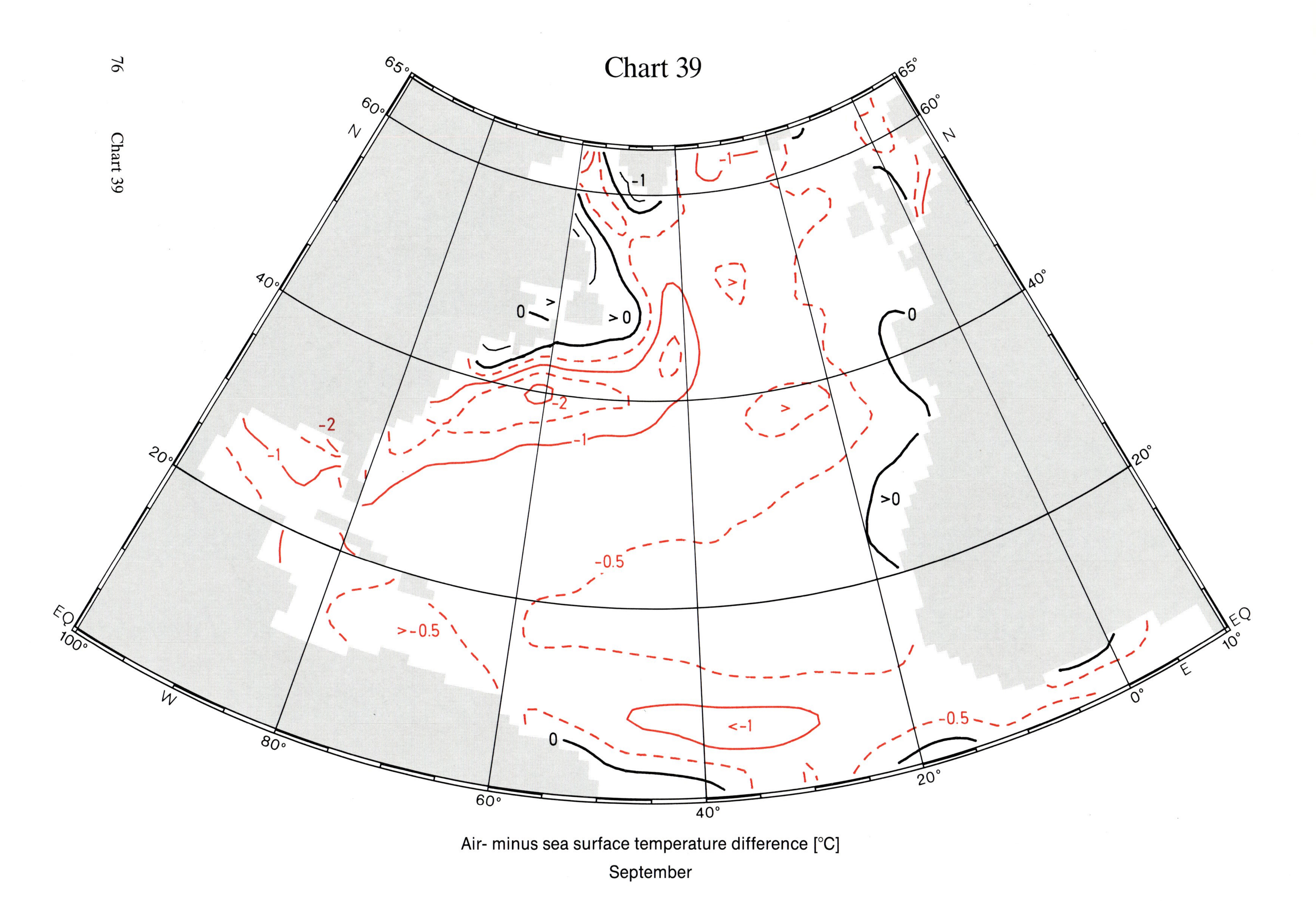

Chart 39
Air- minus sea surface temperature difference [°C]
September

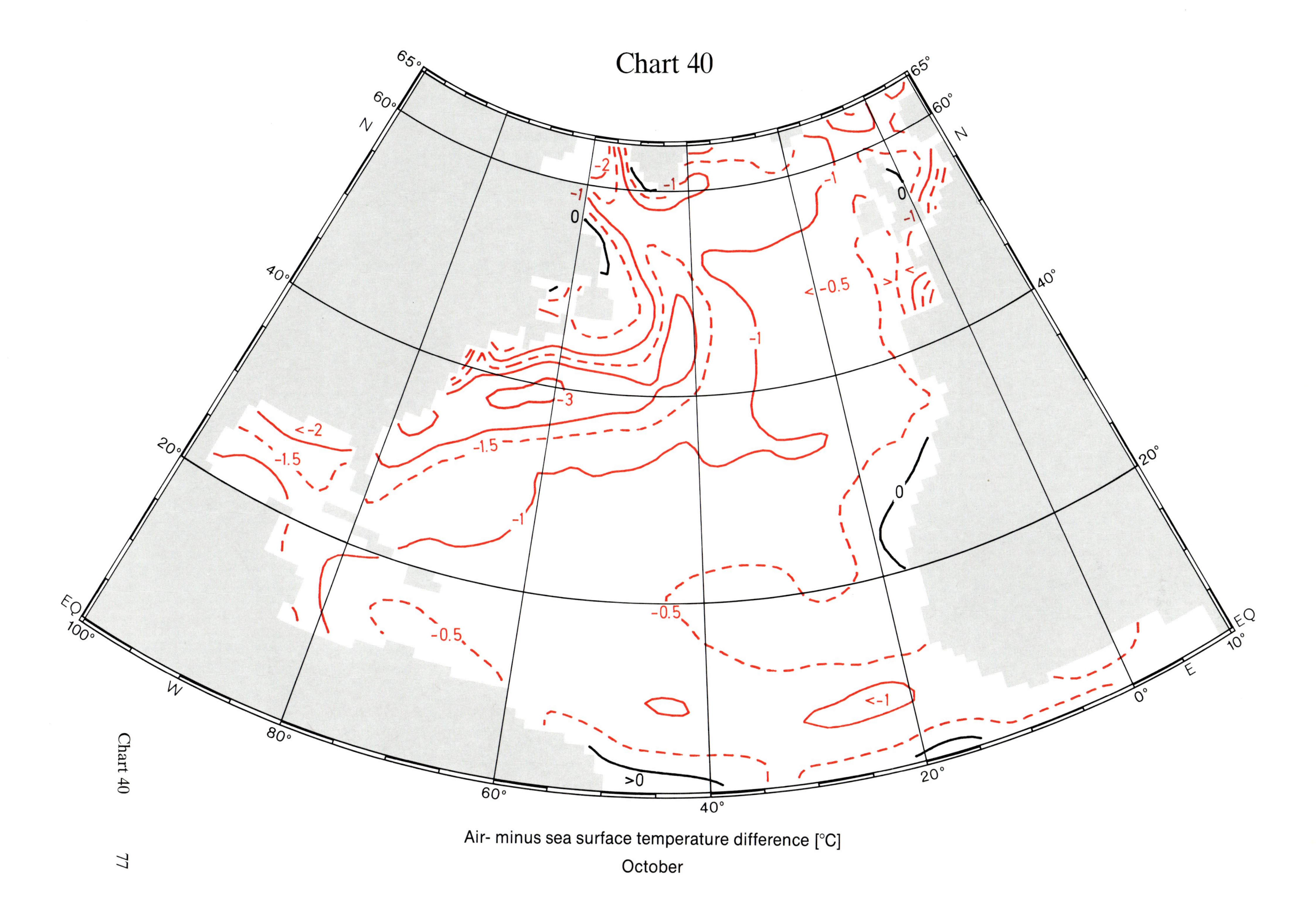
Chart 40
65°
60°
N
40°
20°
EQ
100°
W
80°
60°
40°
20°
0°
E
10°
0
>0
-0.5
<-0.5
-1
<-1
-1.5
-2
<-2
-3
Air- minus sea surface temperature difference [°C]
October

Chart 41

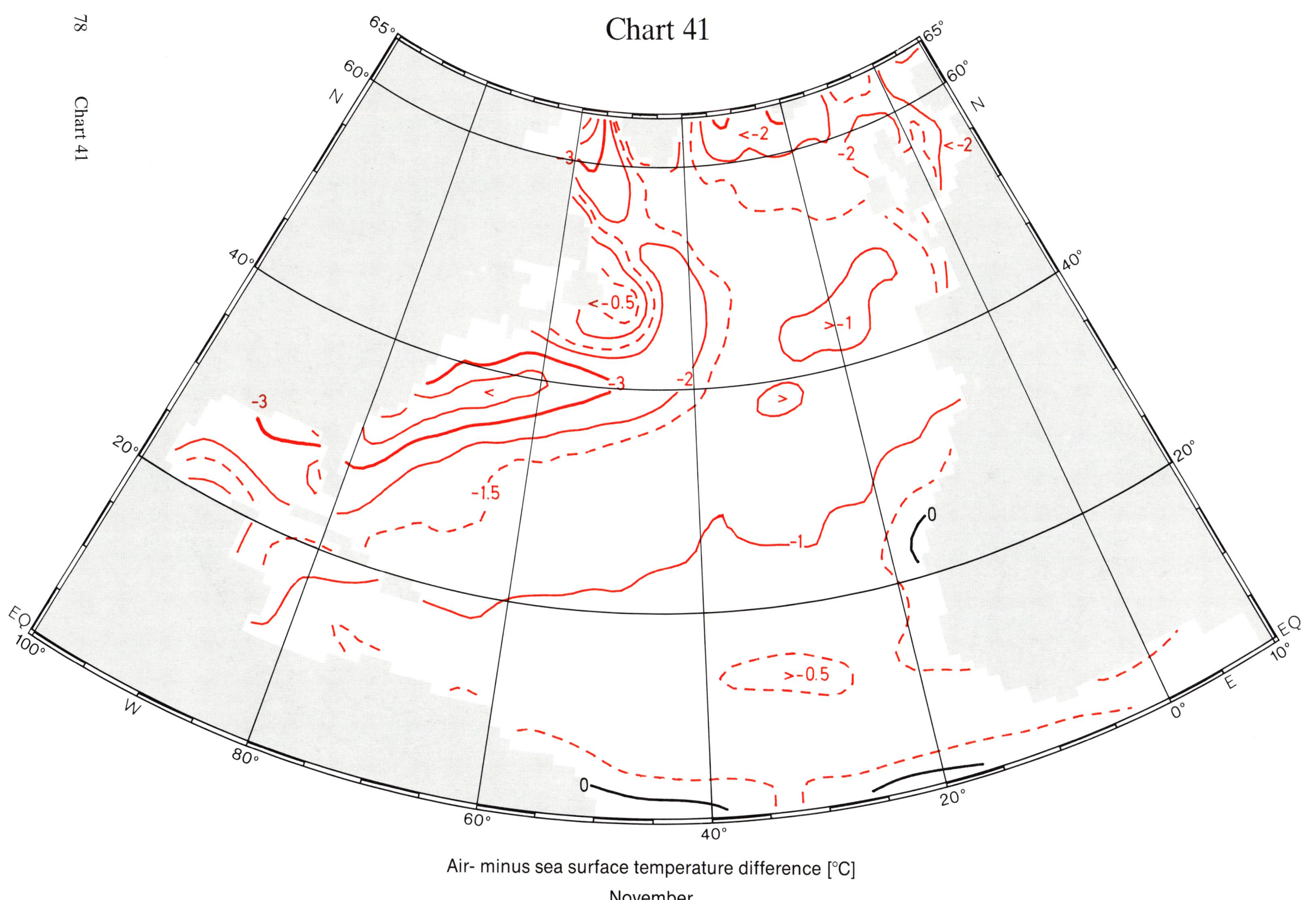

Air- minus sea surface temperature difference [°C]

November

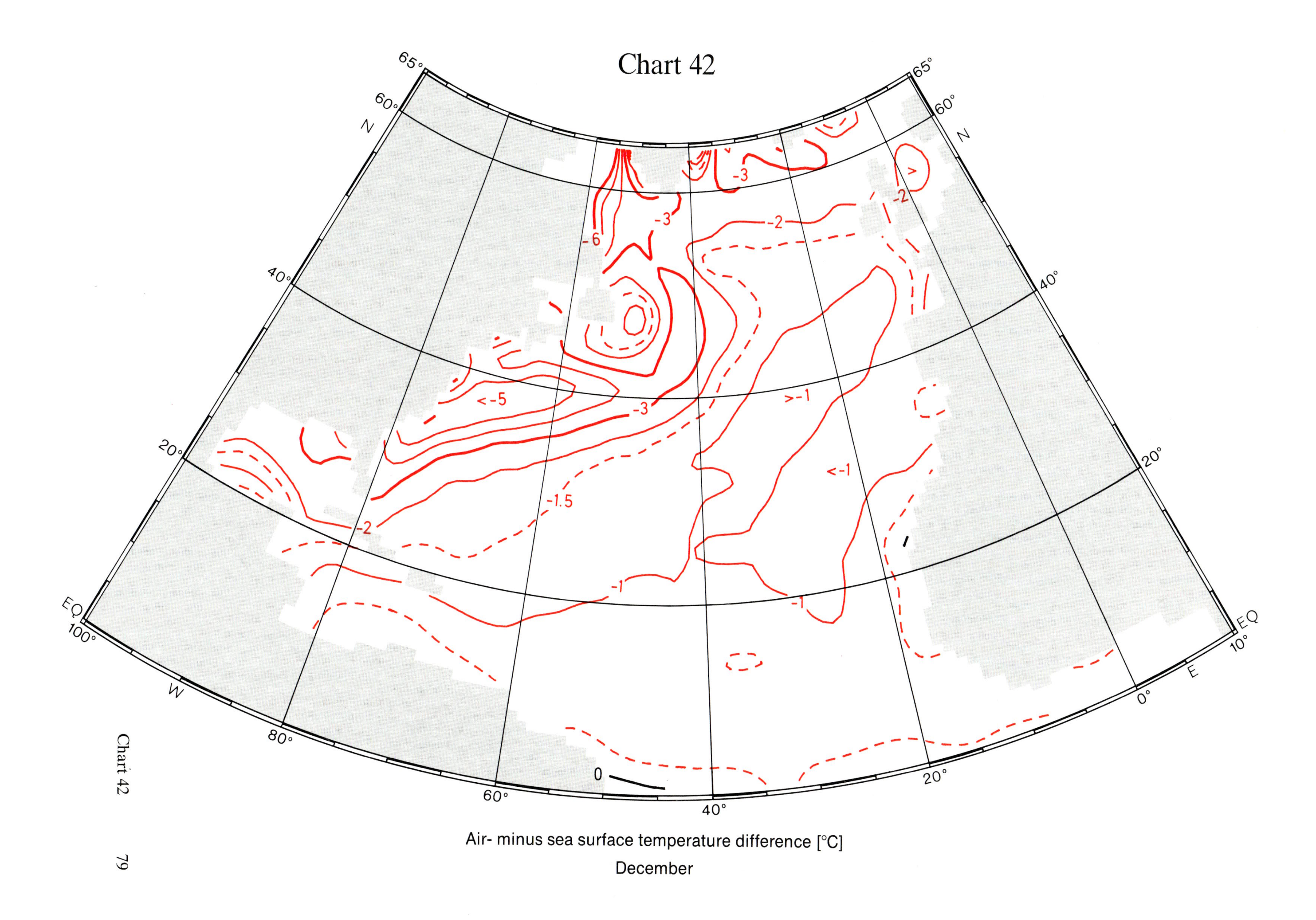
Chart 42
65°
60°
N
40°
20°
EQ
100°
W
80°
60°
40°
20°
0°
E
10°
-6
-3
-2
<-5
>-1
<-1
-1.5
-1
0
Air- minus sea surface temperature difference [°C]
December

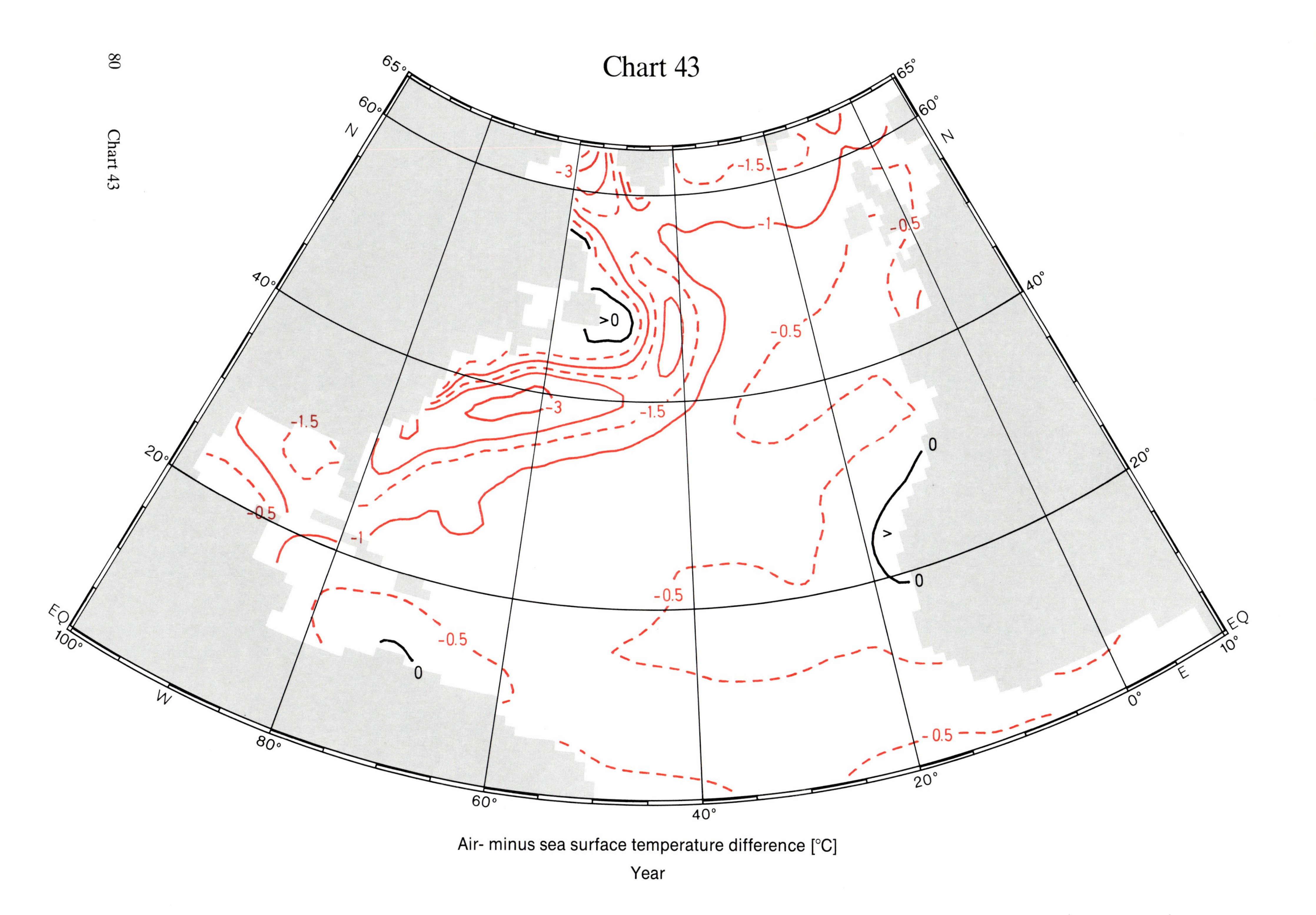
Chart 43
Air- minus sea surface temperature difference [°C]
Year
65°
60°
N
40°
20°
EQ
100°
W
80°
60°
40°
20°
0°
E
10°
-0.5
-1
-1.5
-3
0
>0

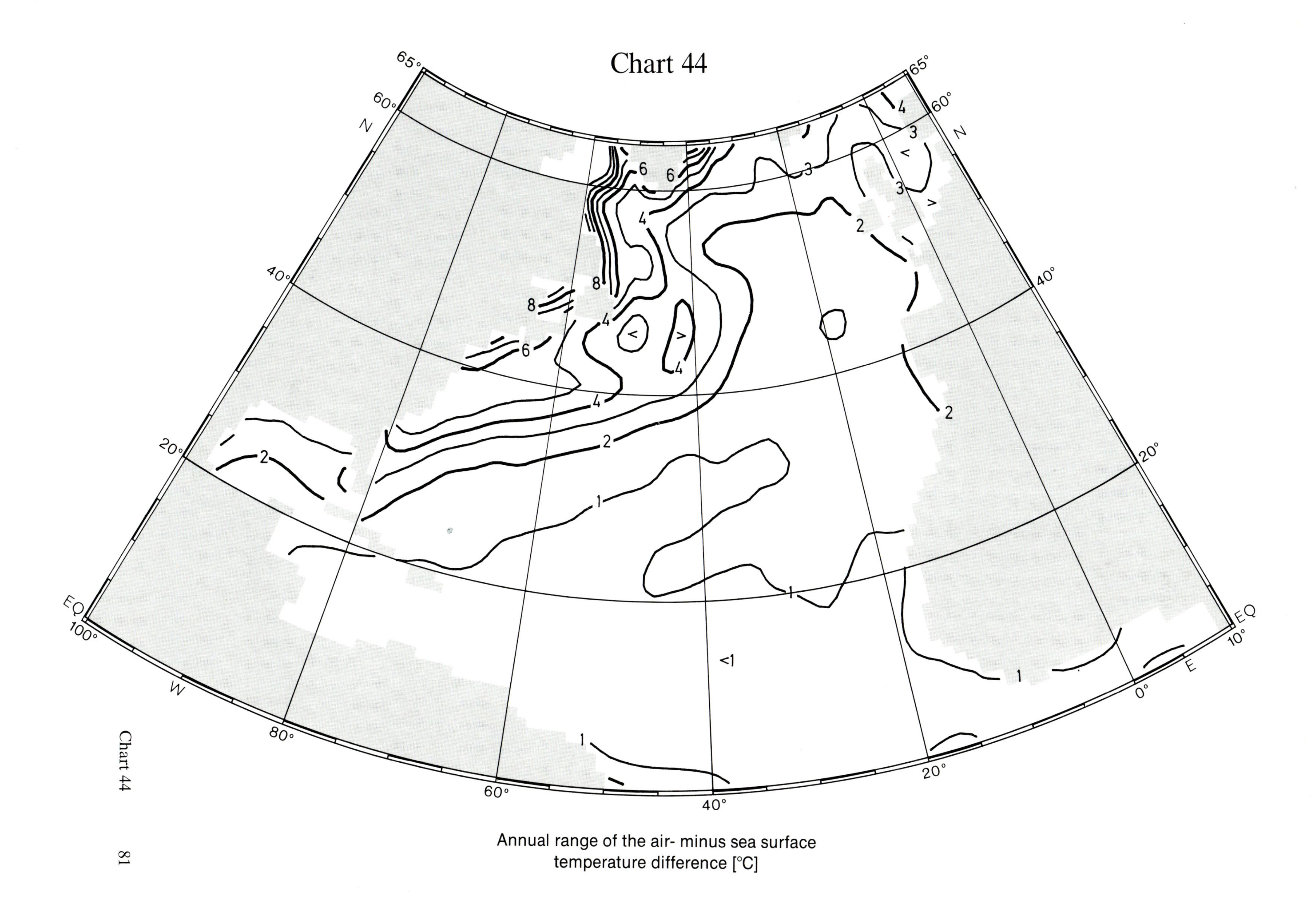

Annual range of the air- minus sea surface temperature difference [°C]

Chart 45

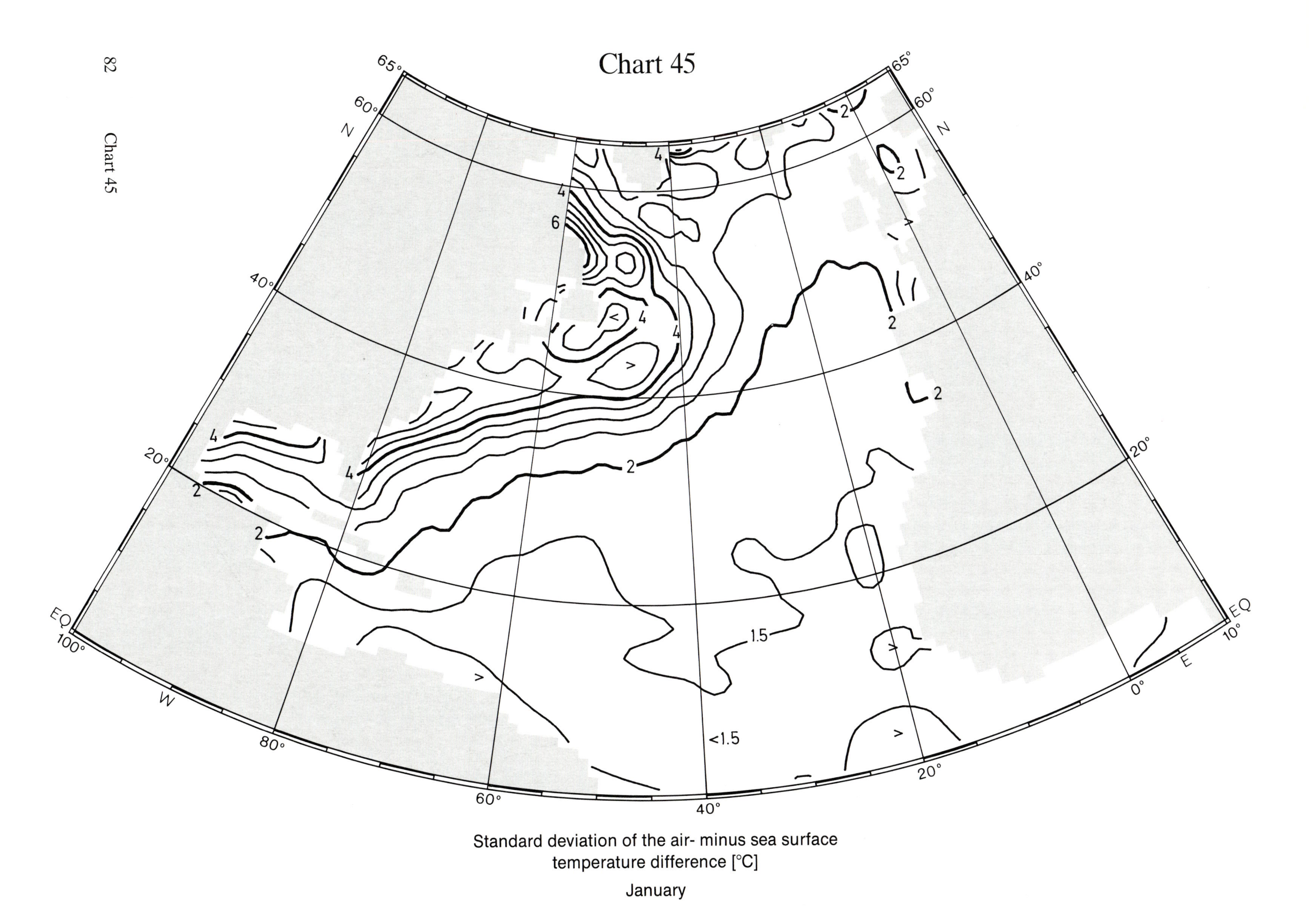

Standard deviation of the air- minus sea surface temperature difference [°C]

January

Chart 46

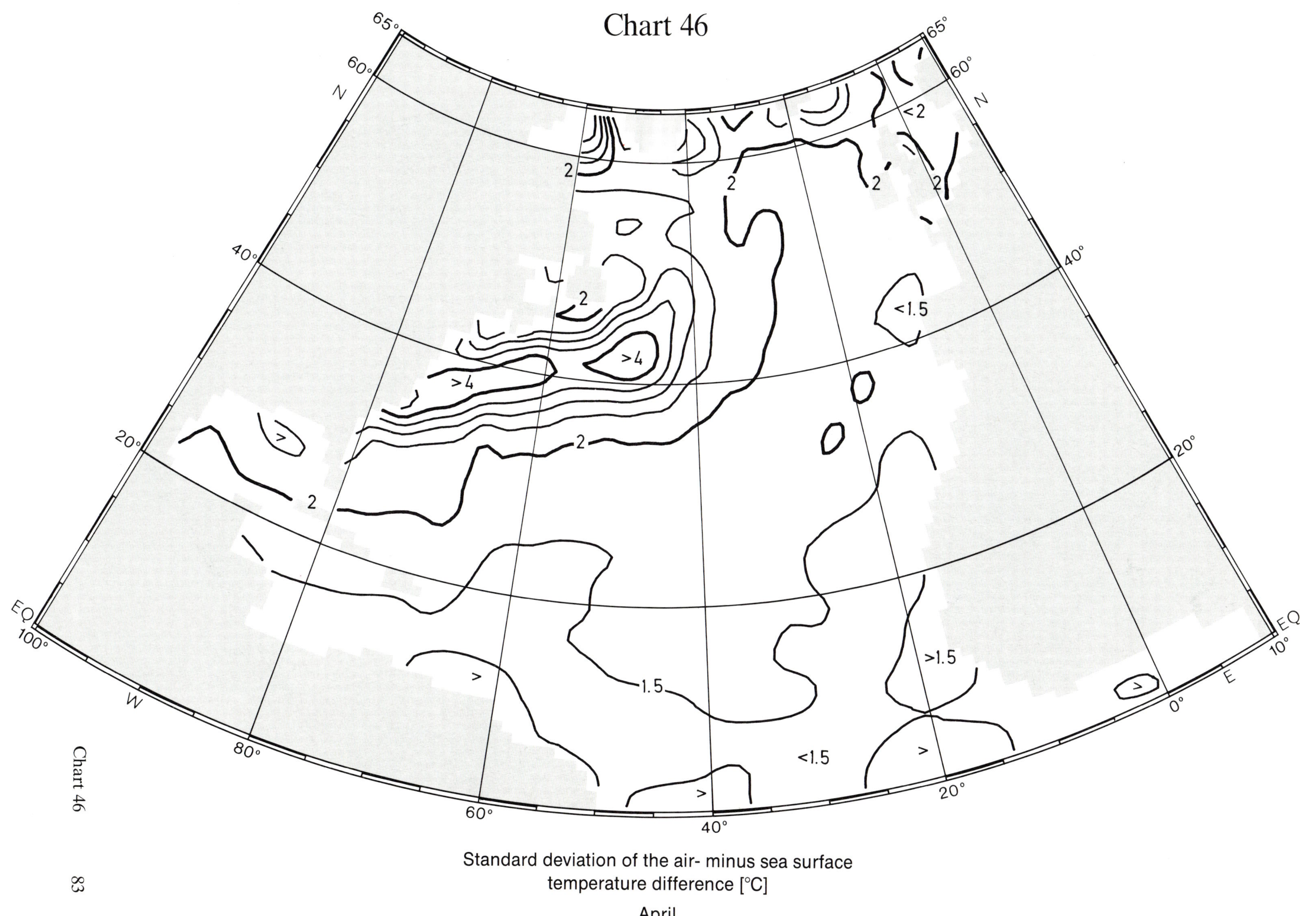

Standard deviation of the air- minus sea surface temperature difference [°C]

April

Chart 47

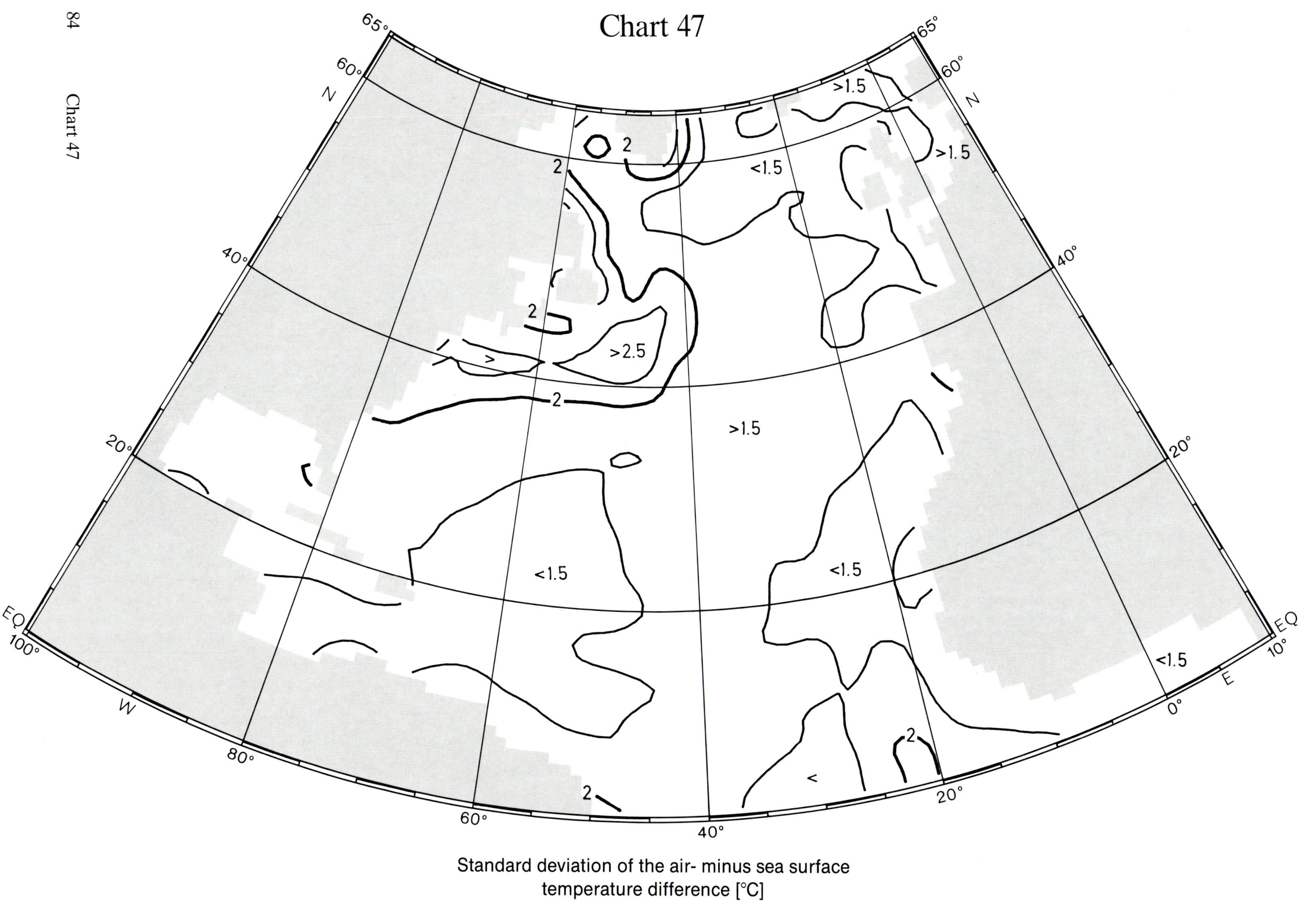

Standard deviation of the air- minus sea surface temperature difference [°C]
July

Chart 48

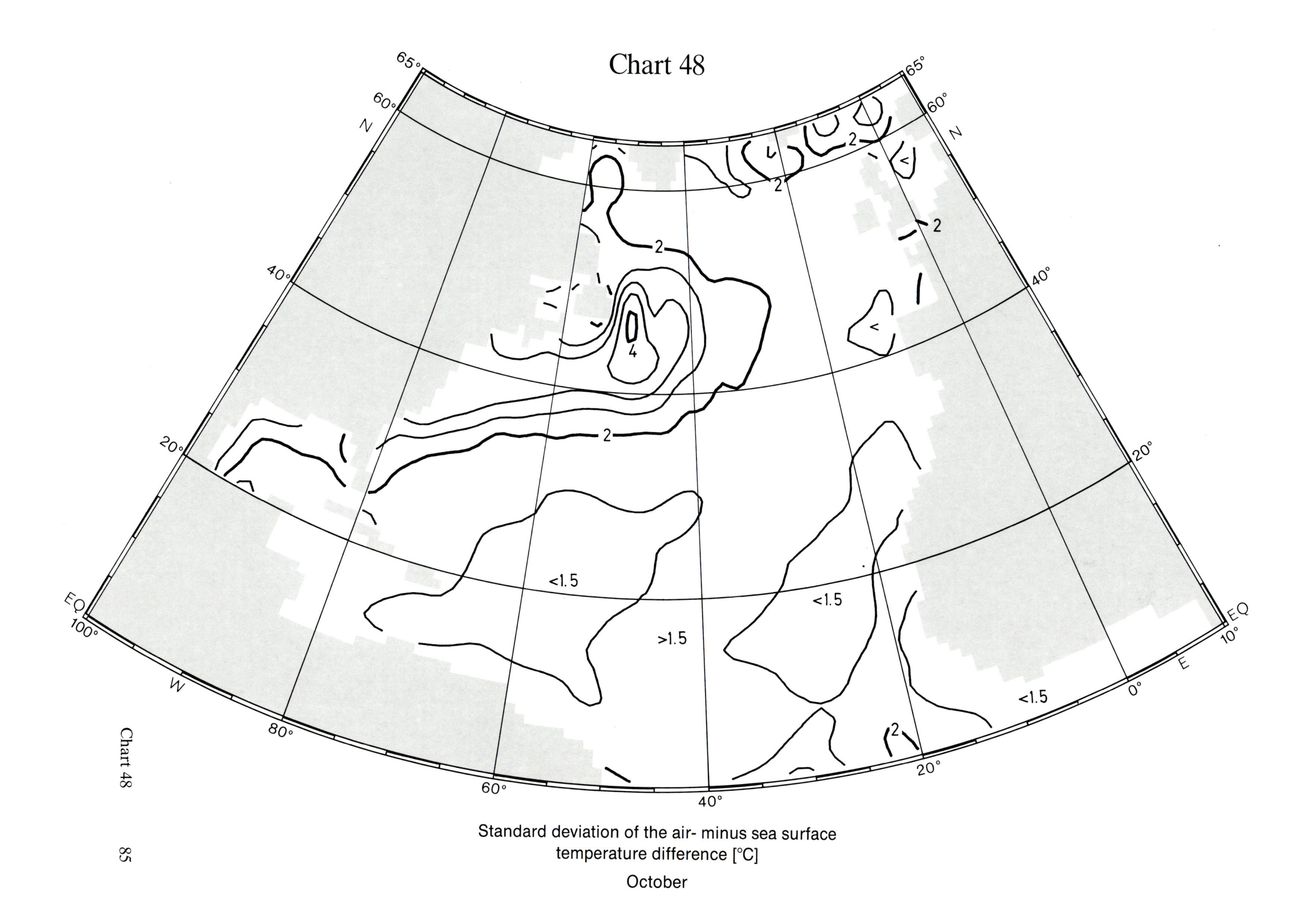

Standard deviation of the air- minus sea surface temperature difference [°C]

October

Chart 49

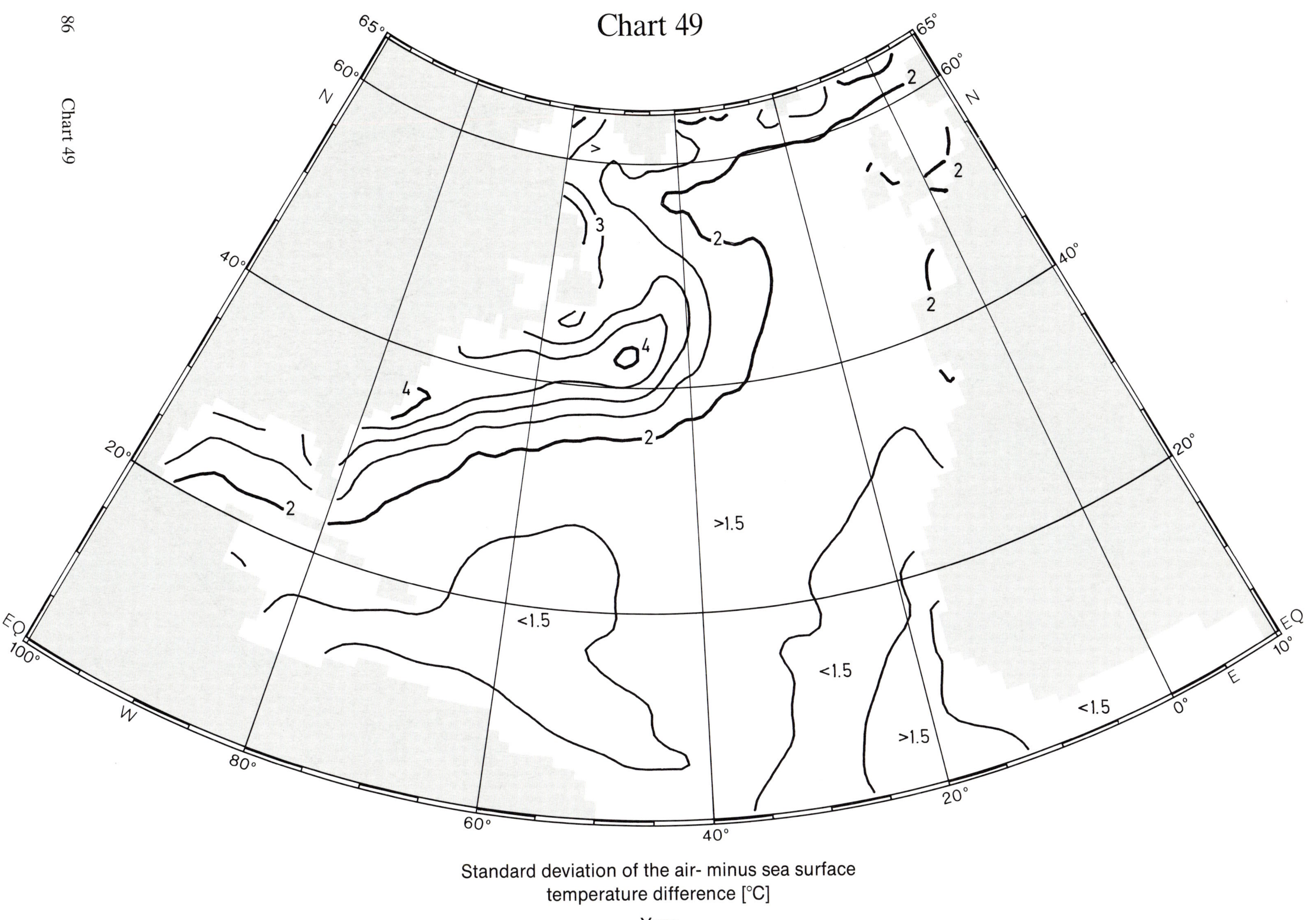

Standard deviation of the air- minus sea surface temperature difference [°C]

Year

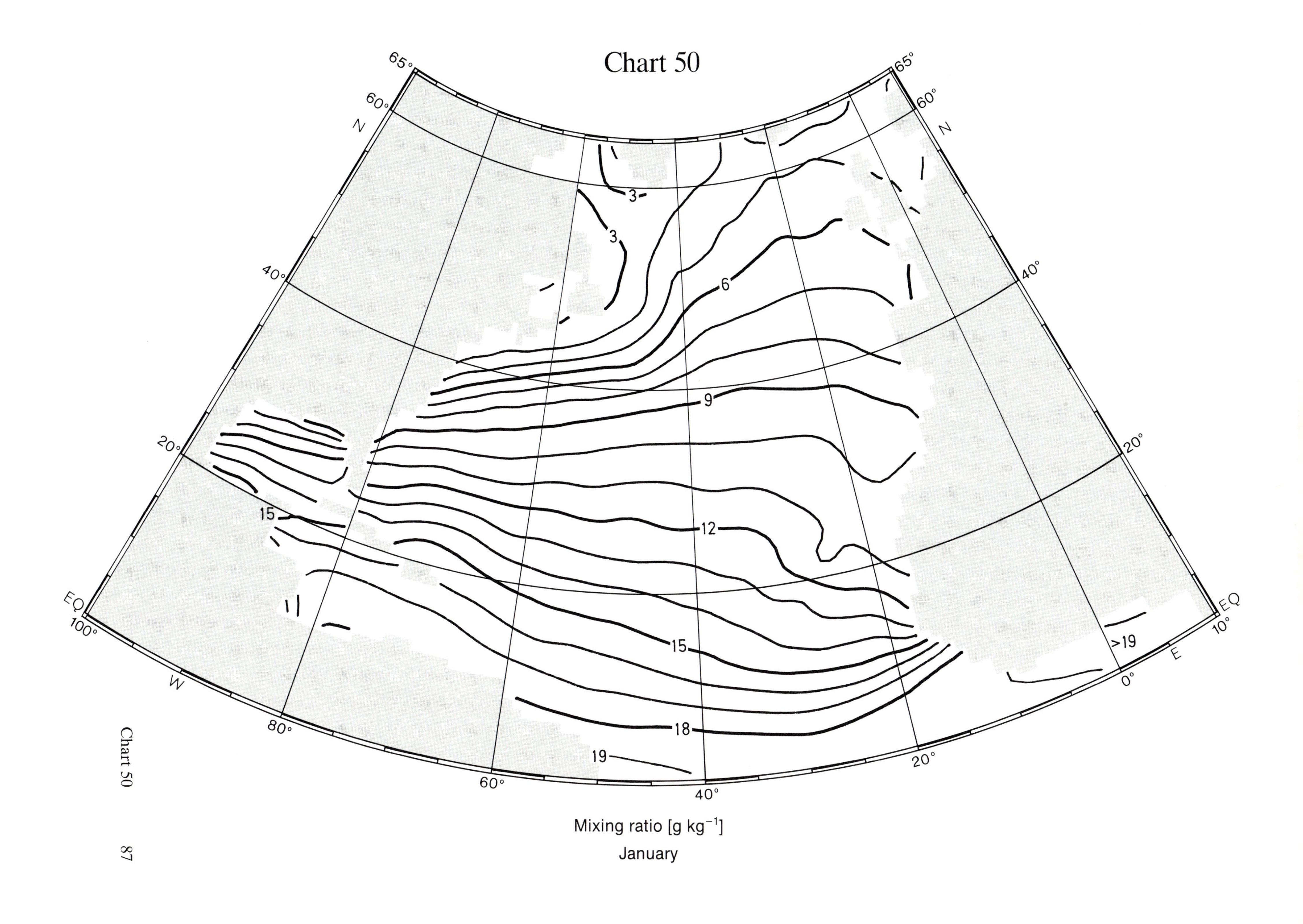
Chart 50
3
3
6
9
12
15
15
18
19
>19
65°
60°
40°
20°
EQ
100°
80°
60°
40°
20°
0°
10°
N
W
E
Mixing ratio [g kg^{-1}]
January

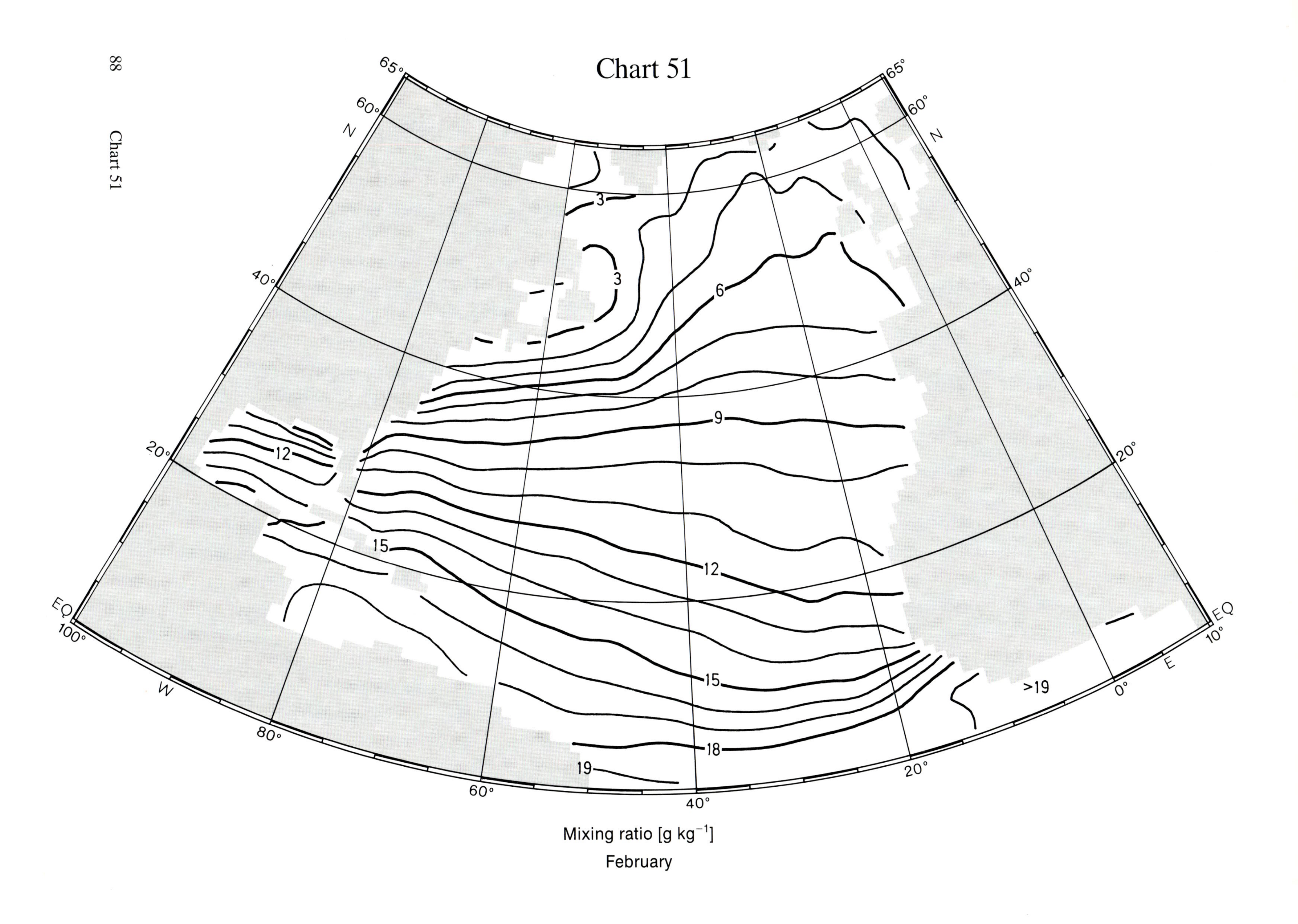
Chart 51
65°
60°
N
40°
20°
EQ
100°
W
80°
60°
40°
20°
0°
E
10°
3
3
6
9
12
12
15
15
18
19
>19
Mixing ratio [g kg⁻¹]
February

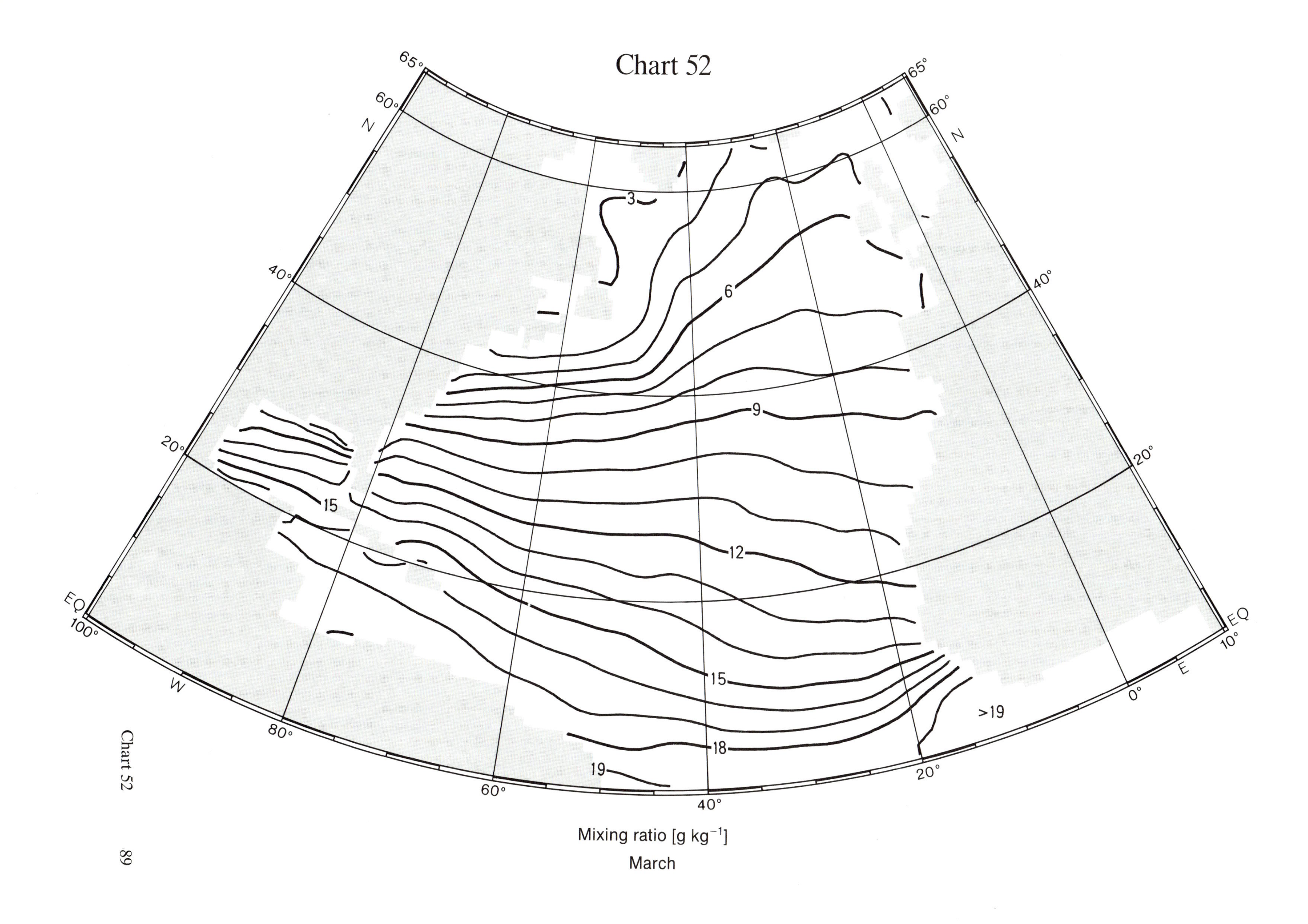
Chart 52
65°
60°
N
40°
20°
EQ
100°
W
80°
60°
40°
20°
0°
E
10°
3
6
9
12
15
15
18
19
>19
Mixing ratio [g kg^{-1}]
March

Chart 53

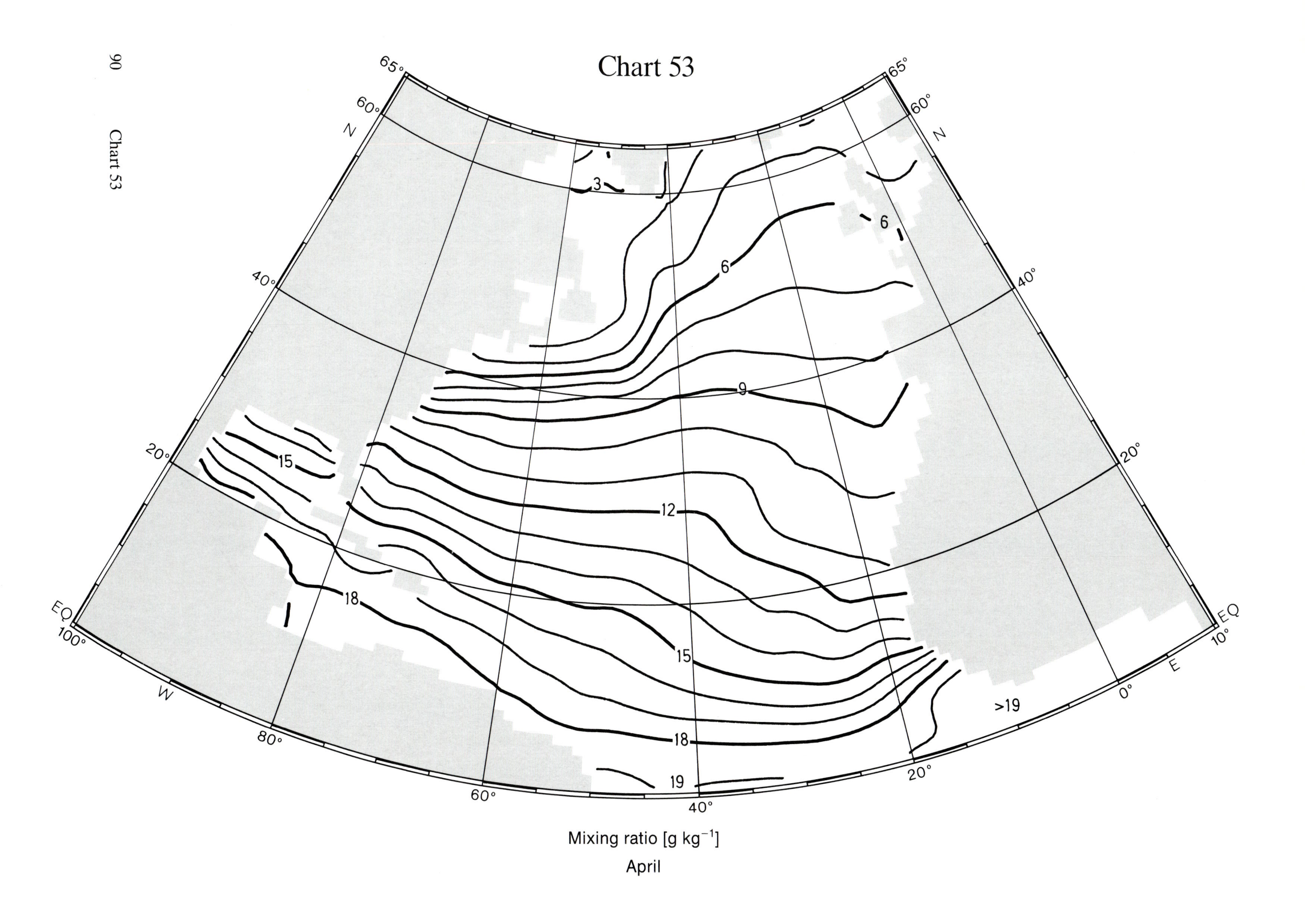

Mixing ratio [g kg^{-1}]
April

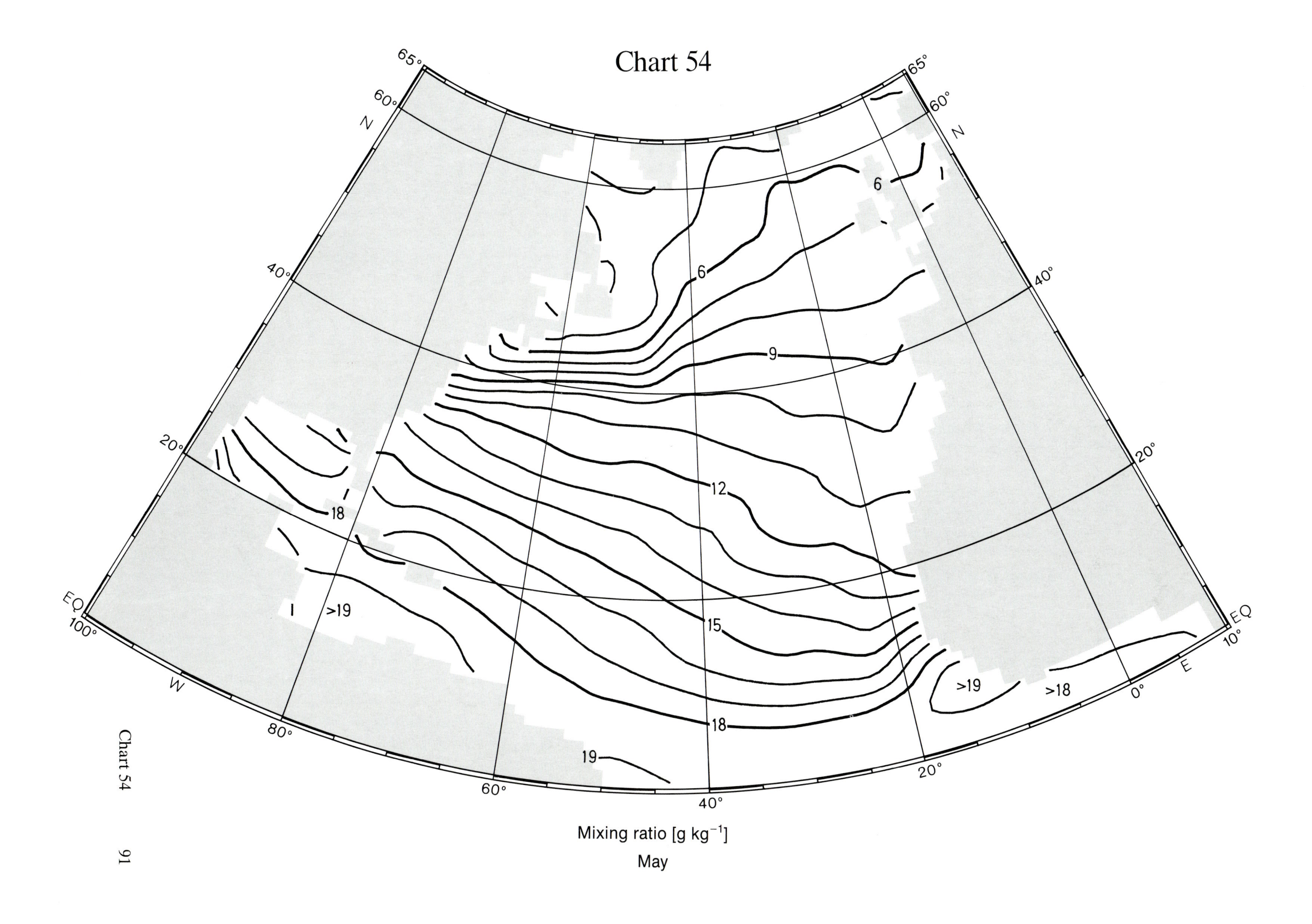

Chart 54
Mixing ratio [g kg⁻¹]
May
6
9
12
15
18
19
>18
>19
65°
60°
40°
20°
10°
0°
20°
40°
60°
80°
100°
N
E
W
EQ

Chart 55

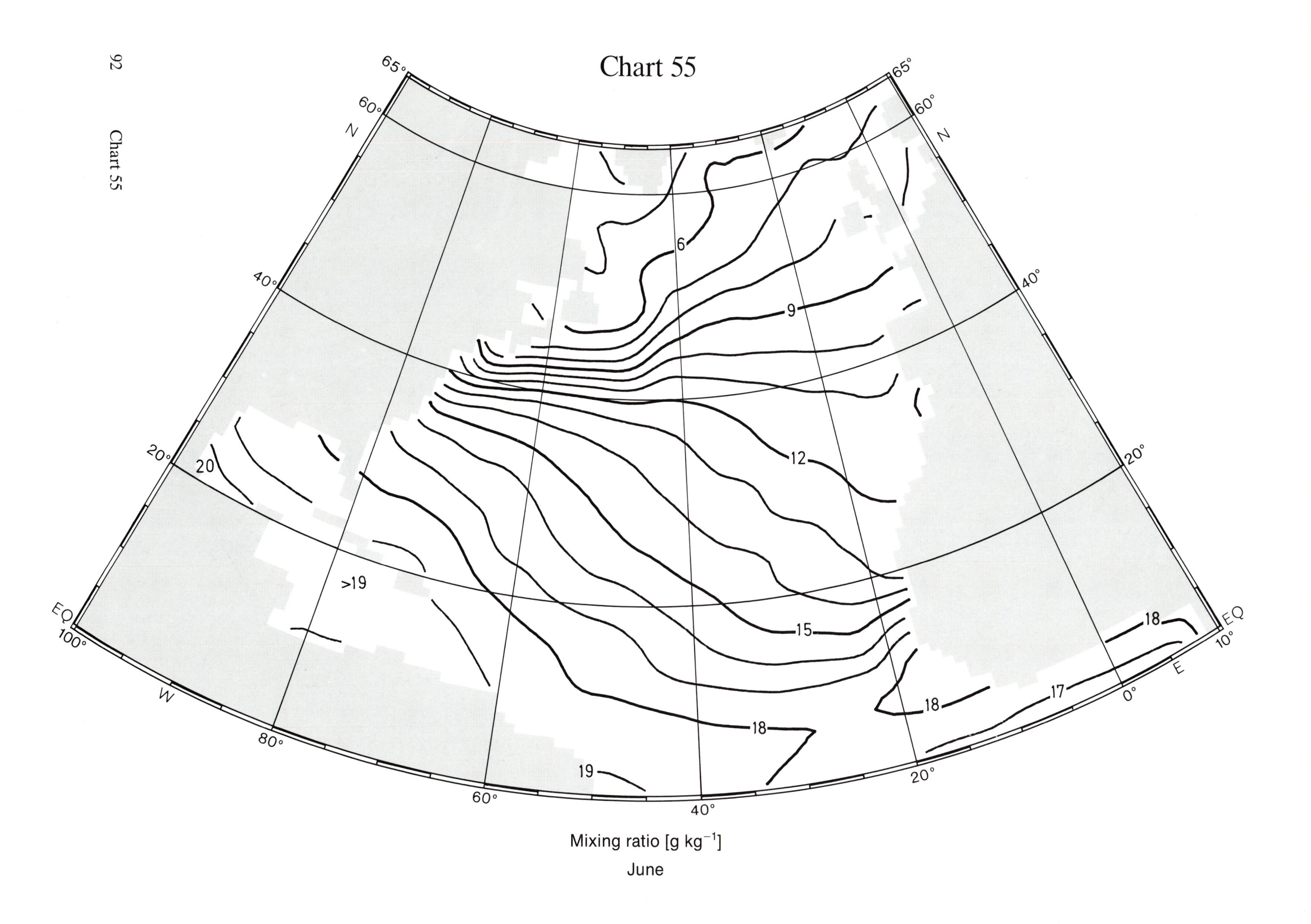

Mixing ratio [g kg^{-1}]

June

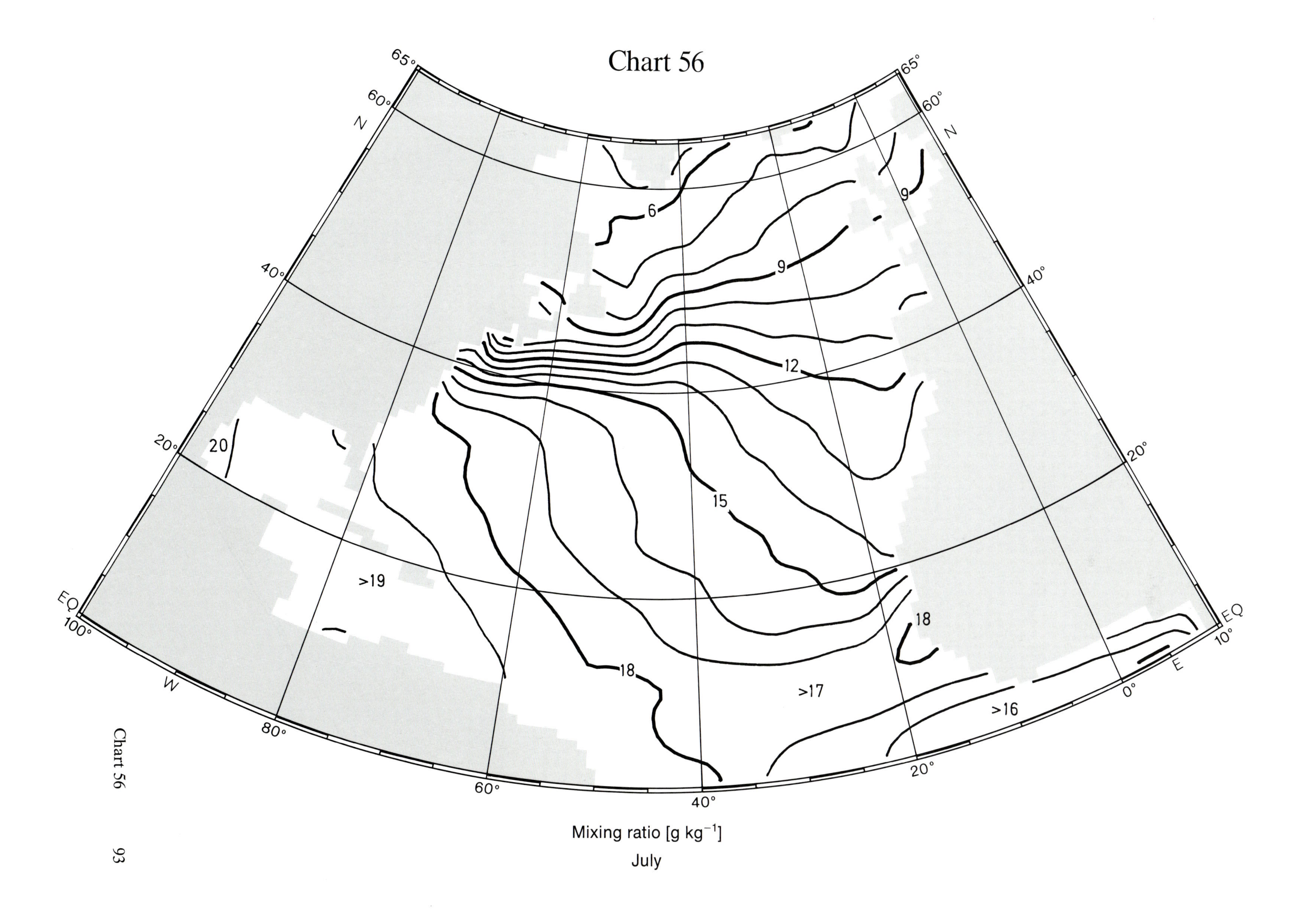

Chart 56
65°
60°
N
40°
20°
EQ
100°
W
80°
60°
40°
20°
0°
E
10°
6
9
9
12
15
18
18
20
>19
>17
>16
Mixing ratio [g kg^{-1}]
July

Chart 57

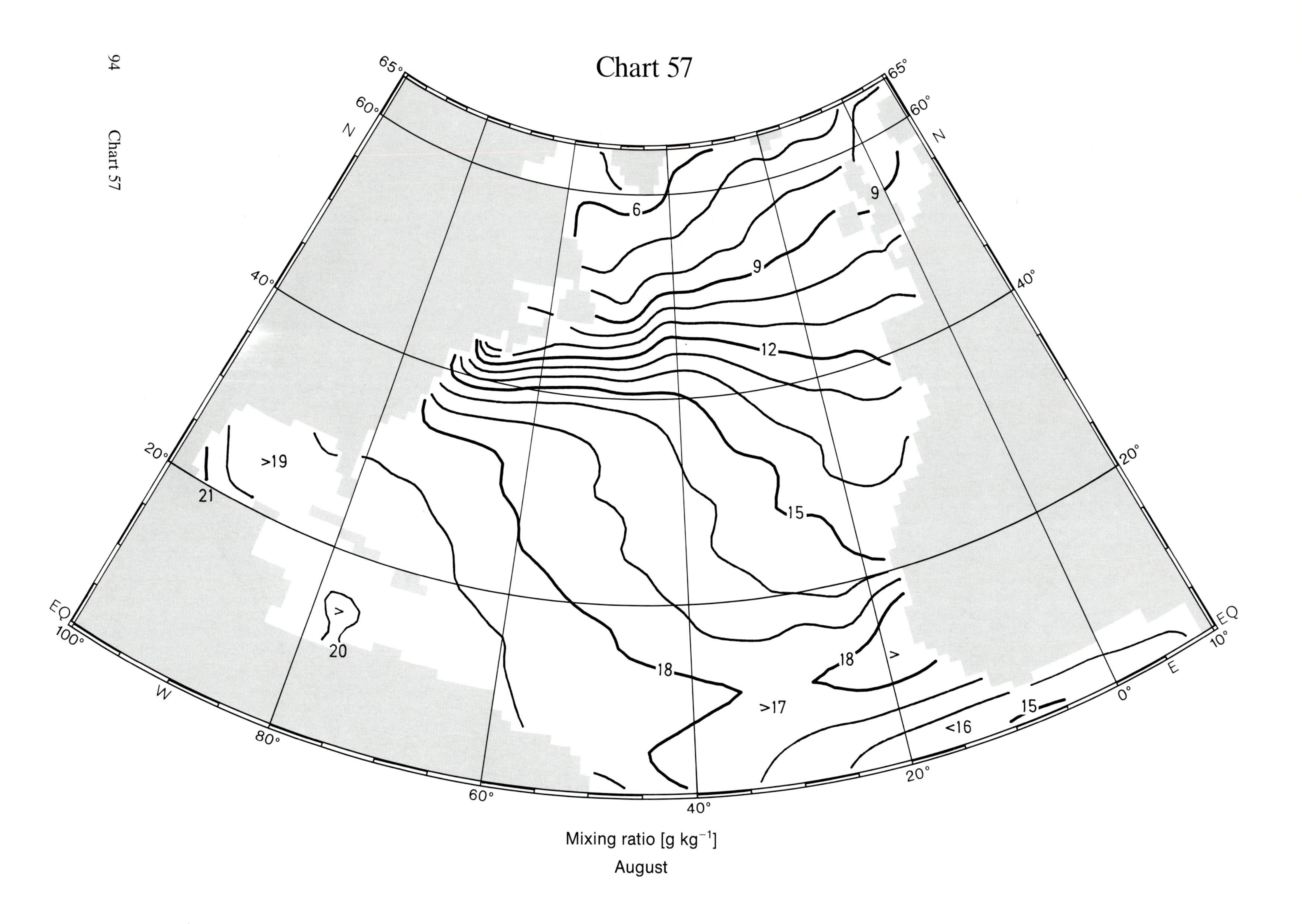

Mixing ratio [g kg^{-1}]
August

Chart 58

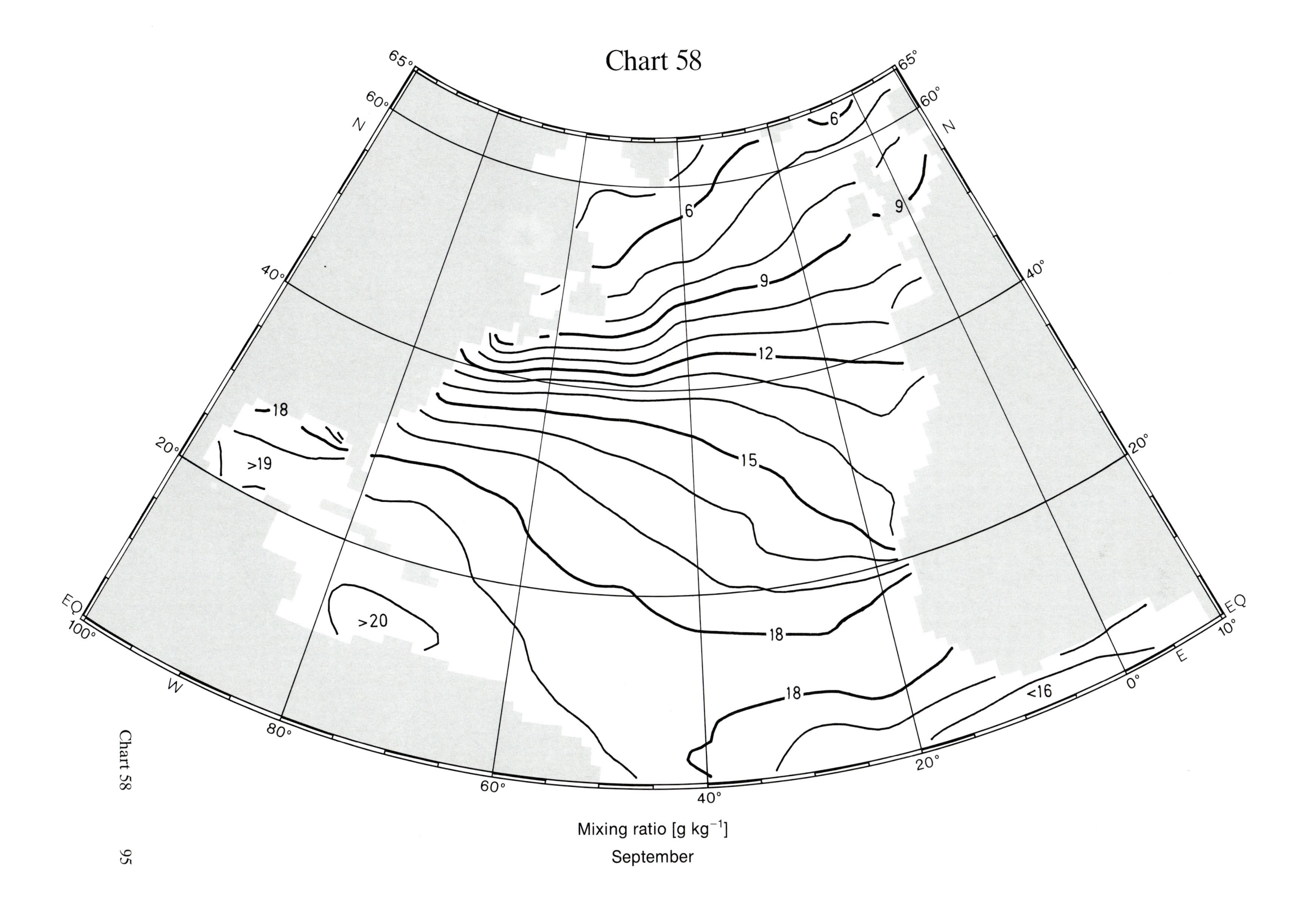

Mixing ratio [g kg^{-1}]
September

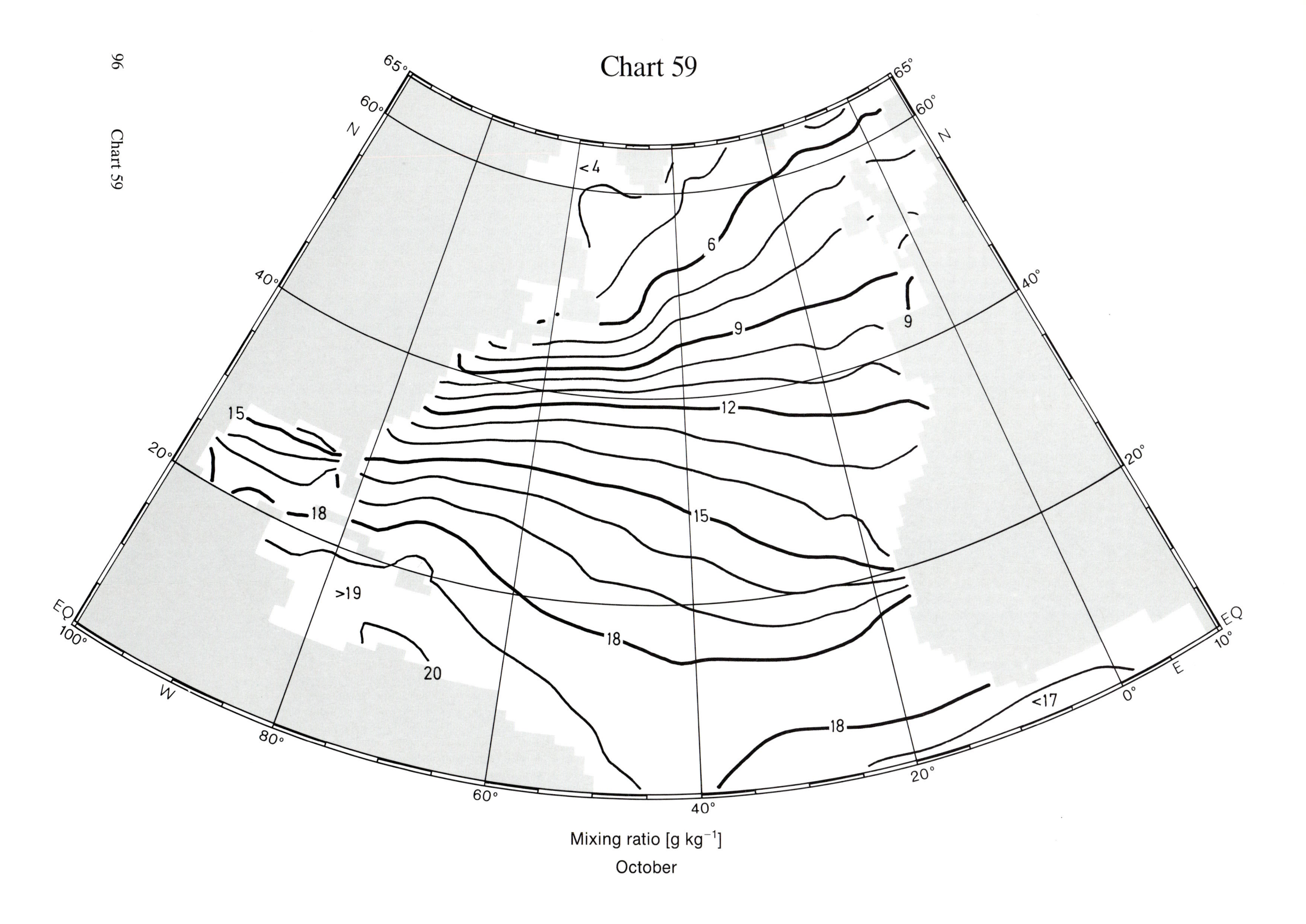

Chart 59
65°
60°
N
40°
20°
EQ
100°
W
80°
60°
40°
20°
0°
E
10°
<4
6
9
12
15
18
>19
20
<17
Mixing ratio [g kg⁻¹]
October

Chart 60

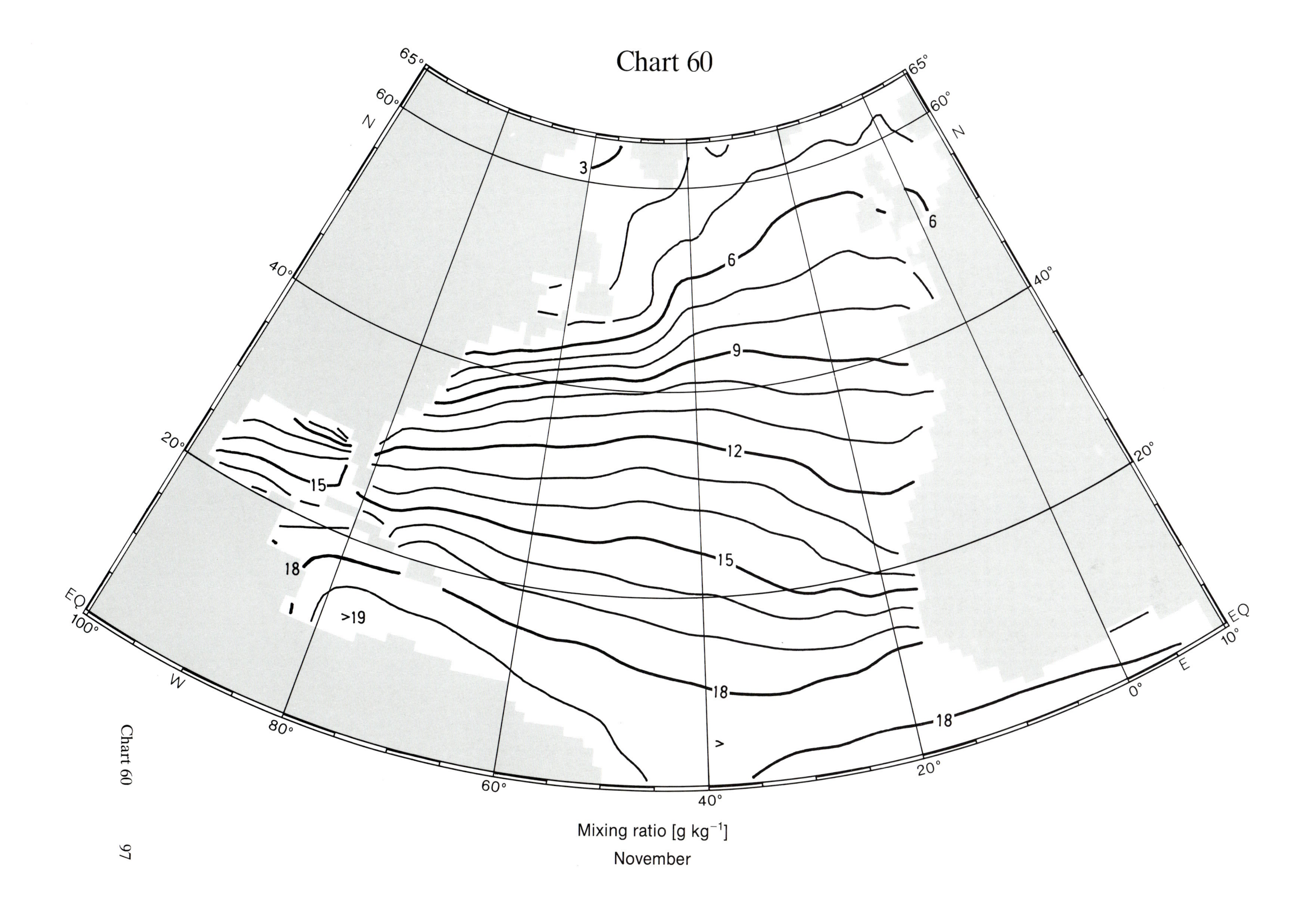

Mixing ratio [g kg^{-1}]
November

Chart 61

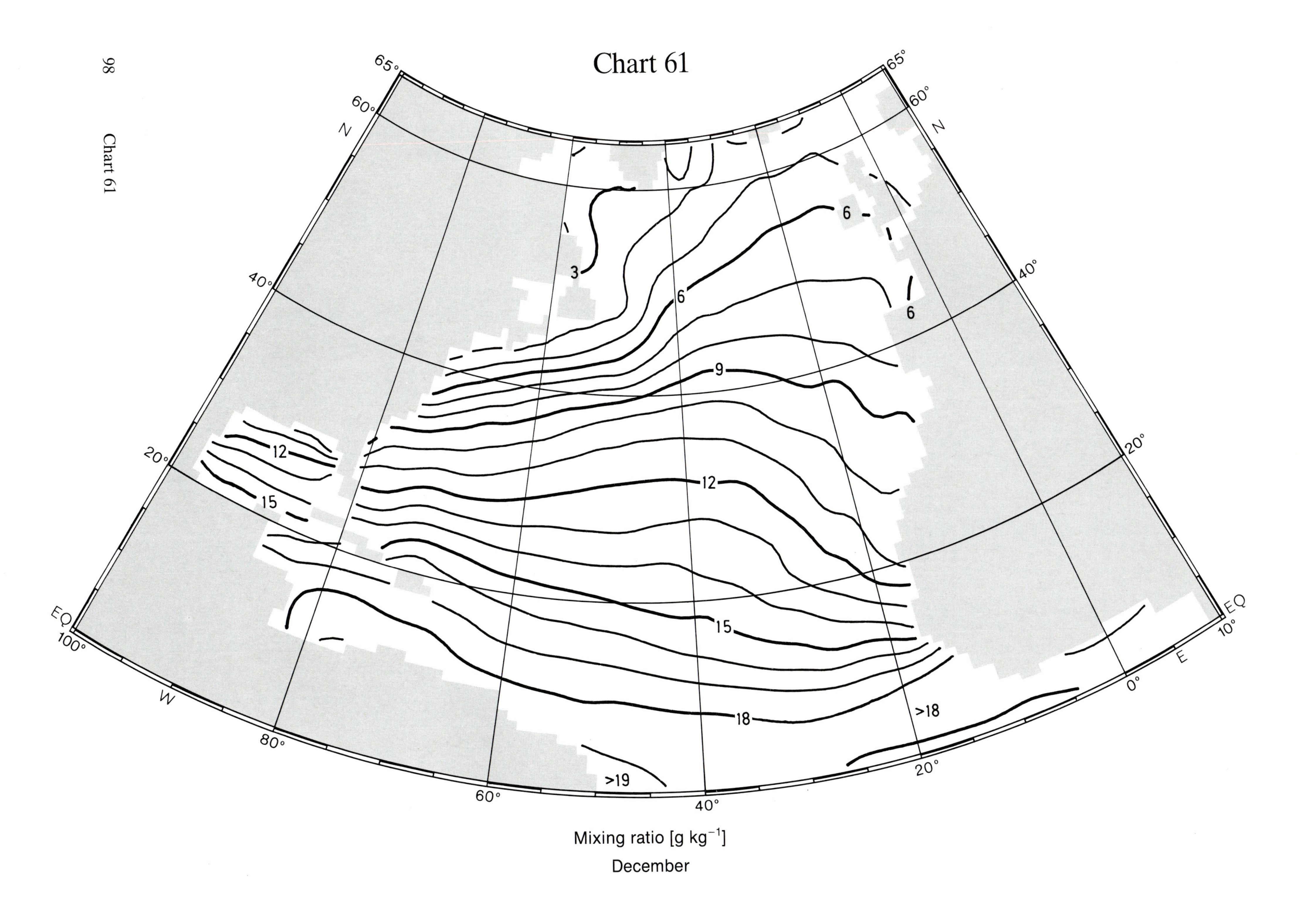

Mixing ratio [g kg^{-1}]
December

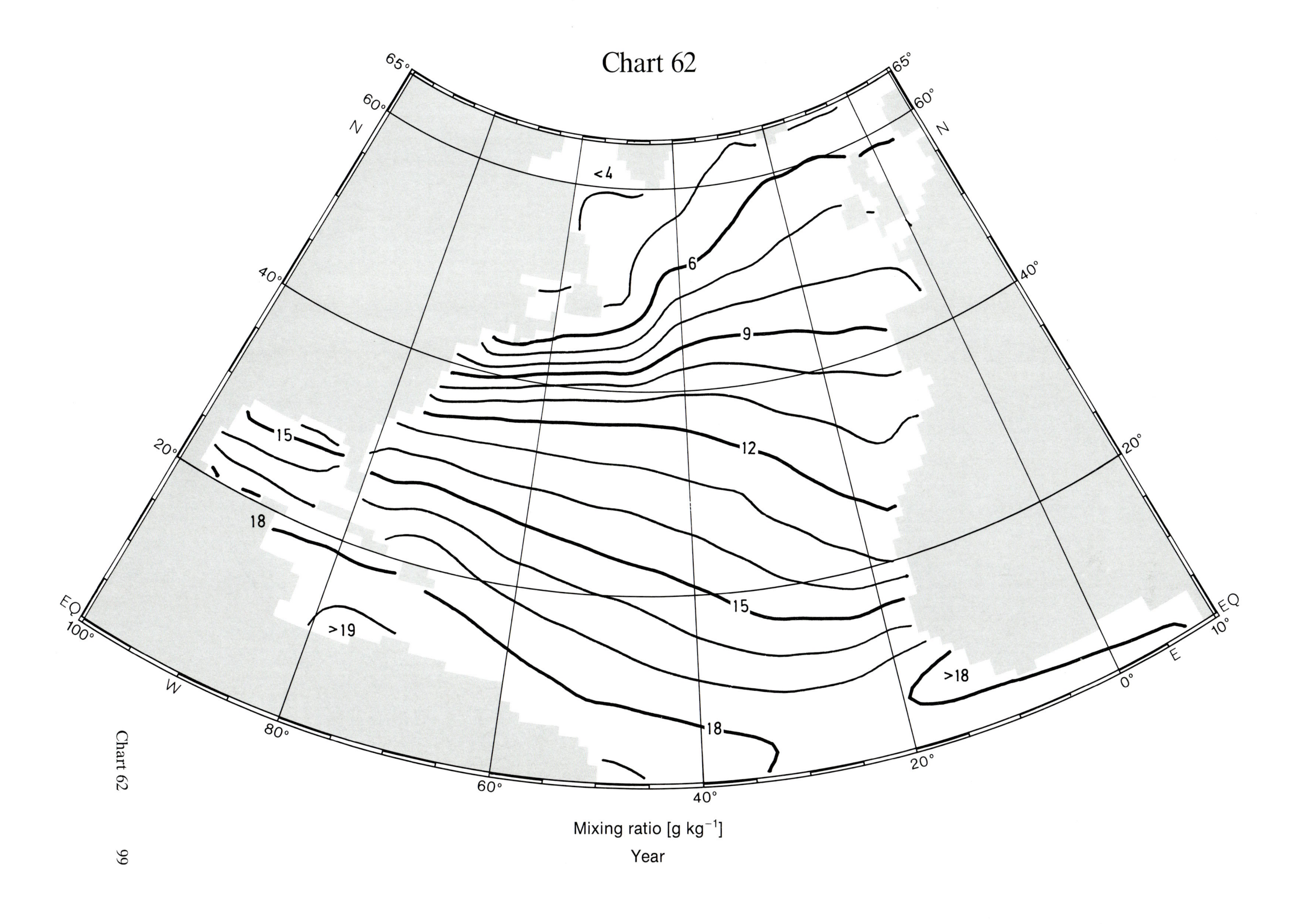
Chart 62
65°
60°
N
40°
20°
EQ
100°
W
80°
60°
40°
20°
0°
E
10°
<4
6
9
12
15
18
>18
>19
Mixing ratio [g kg⁻¹]
Year

Chart 63

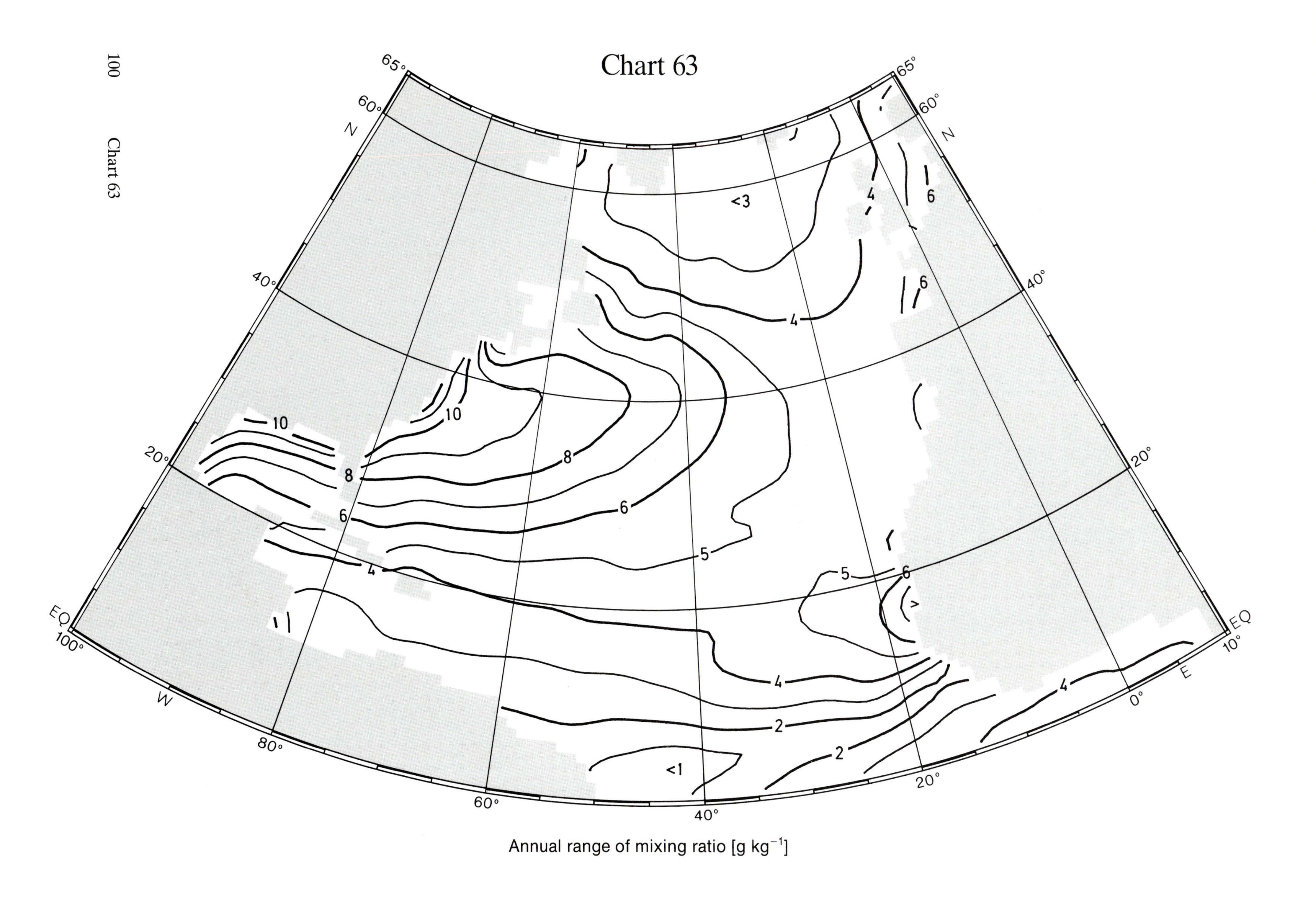

Annual range of mixing ratio [g kg^{-1}]

Chart 64

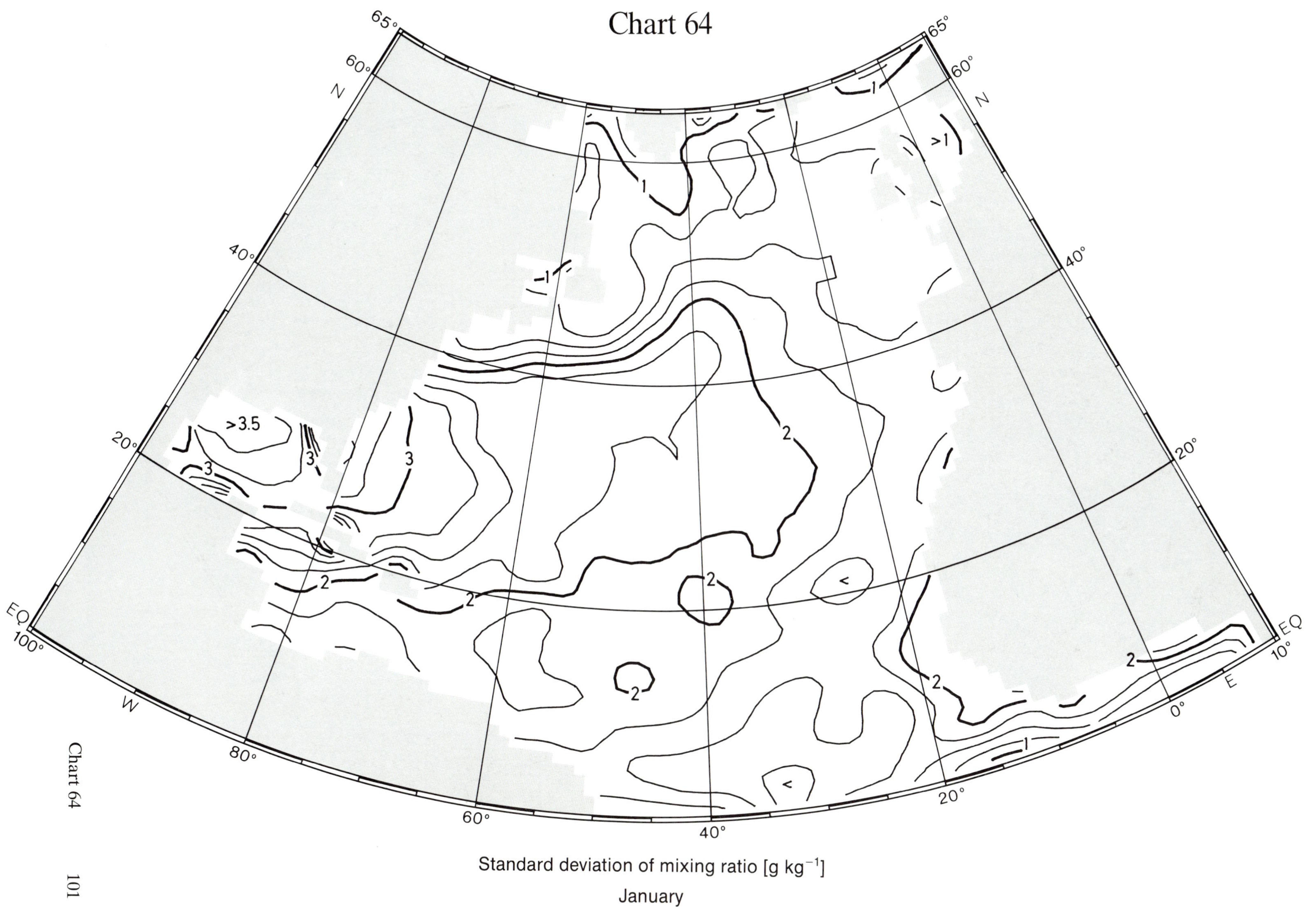

Standard deviation of mixing ratio [g kg^{-1}]

January

Chart 65

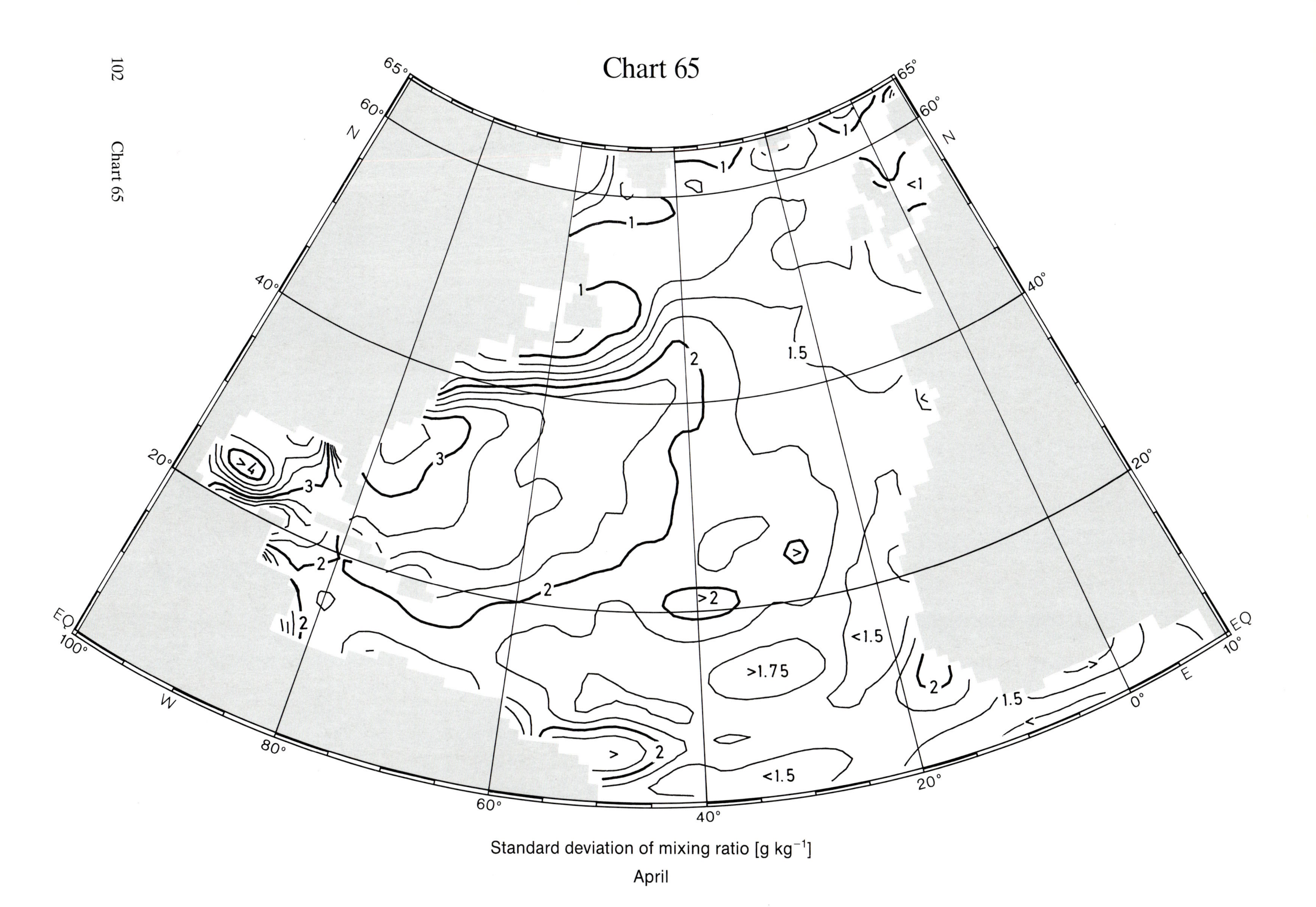

Standard deviation of mixing ratio [g kg^{-1}]
April

Chart 66

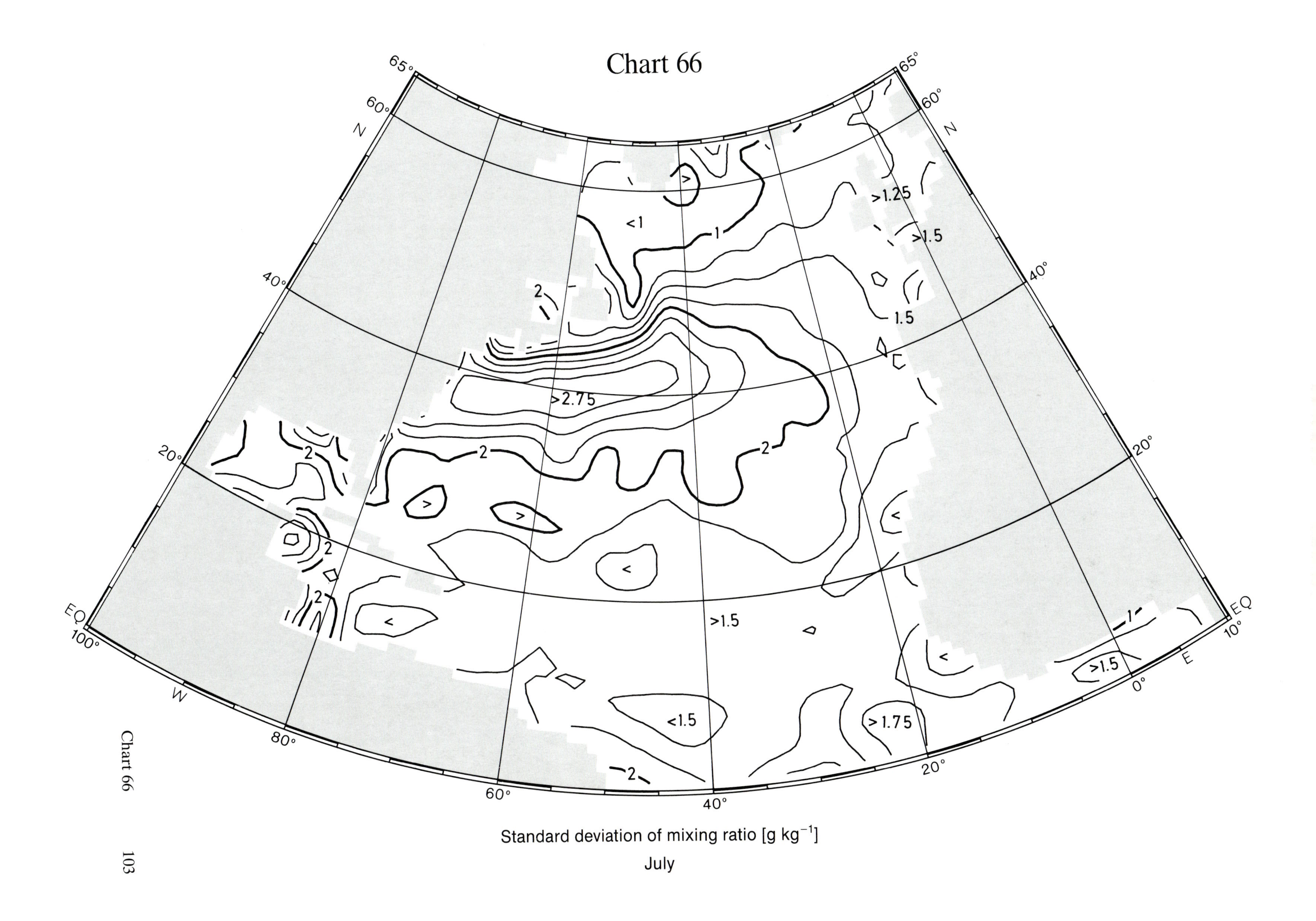

Standard deviation of mixing ratio [g kg^{-1}]

July

Chart 67

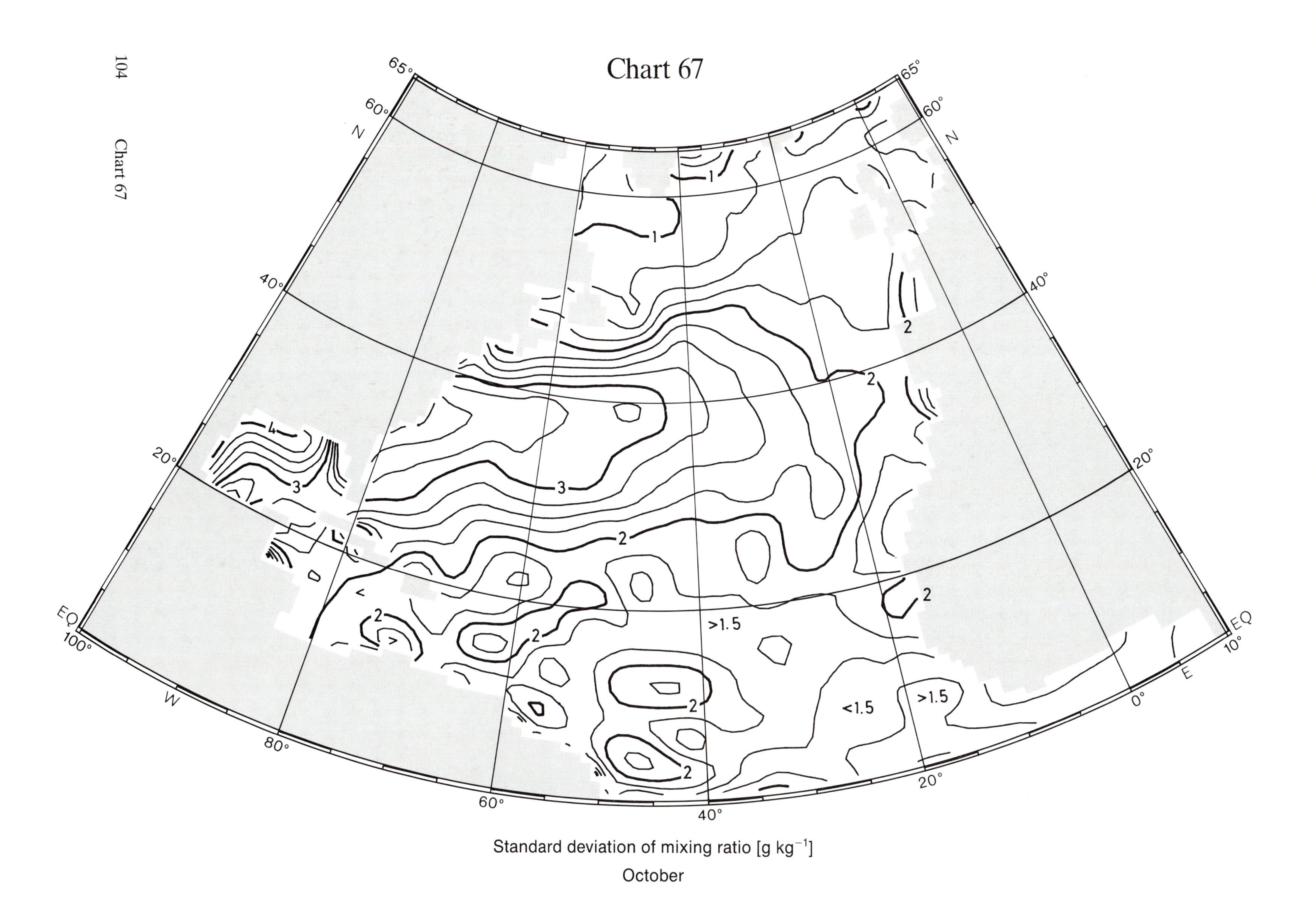

Standard deviation of mixing ratio [g kg^{-1}]

October

Chart 68

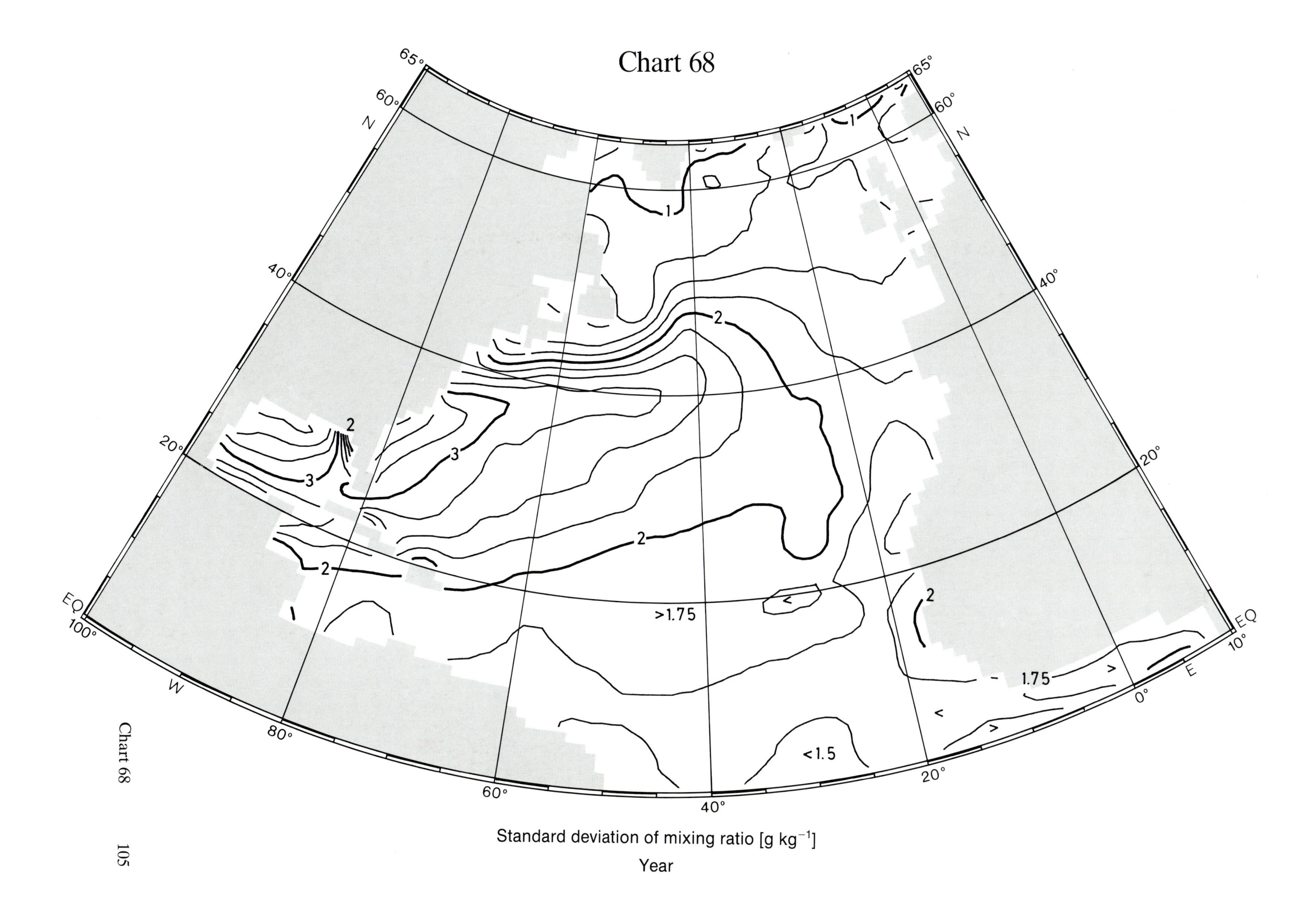

Standard deviation of mixing ratio [g kg^{-1}]
Year

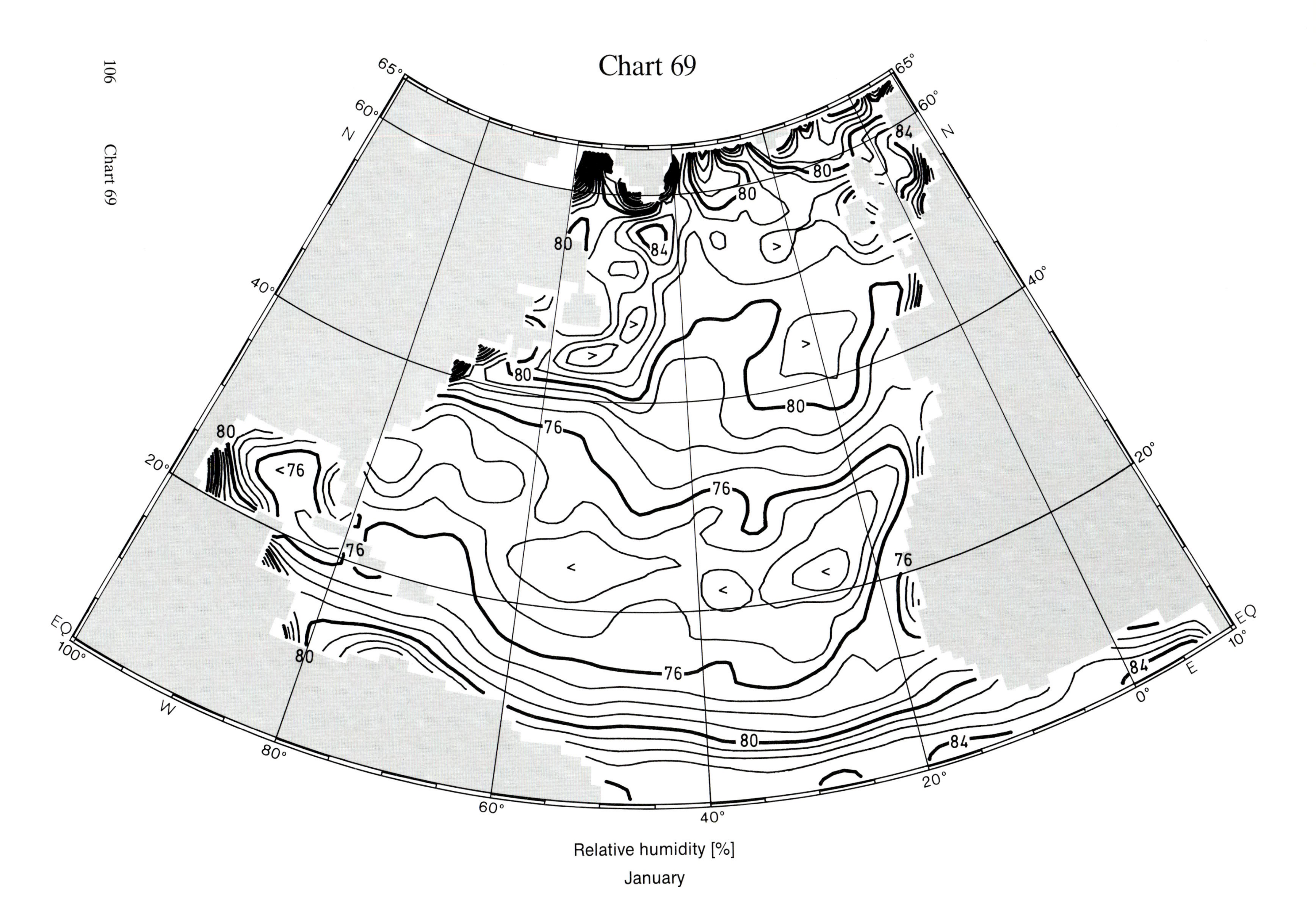
Chart 69
65°
60°
N
40°
20°
EQ
100°
W
80°
60°
40°
20°
0°
E
10°
76
80
84
<76
Relative humidity [%]
January

Chart 70

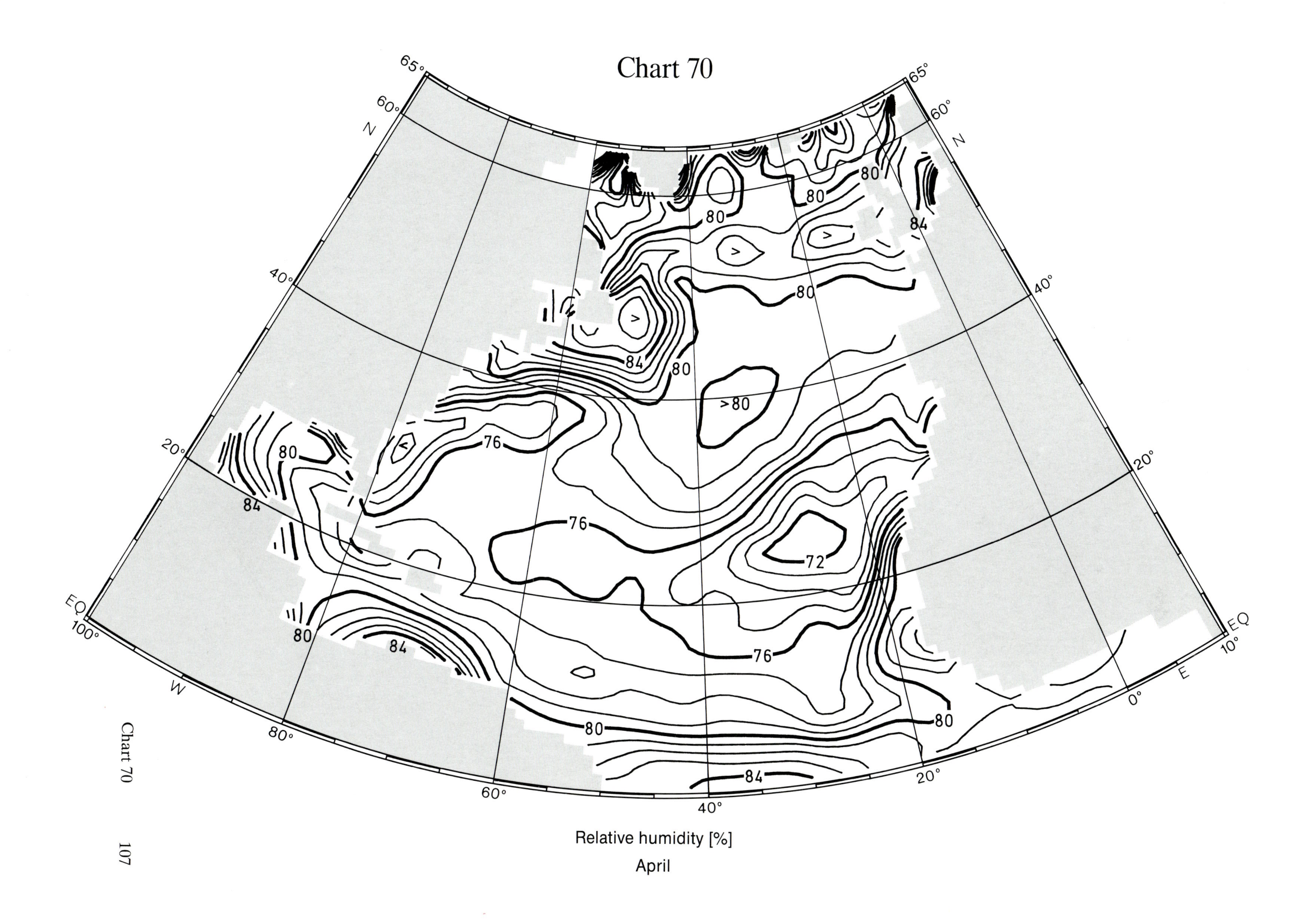

Relative humidity [%]
April

Chart 71

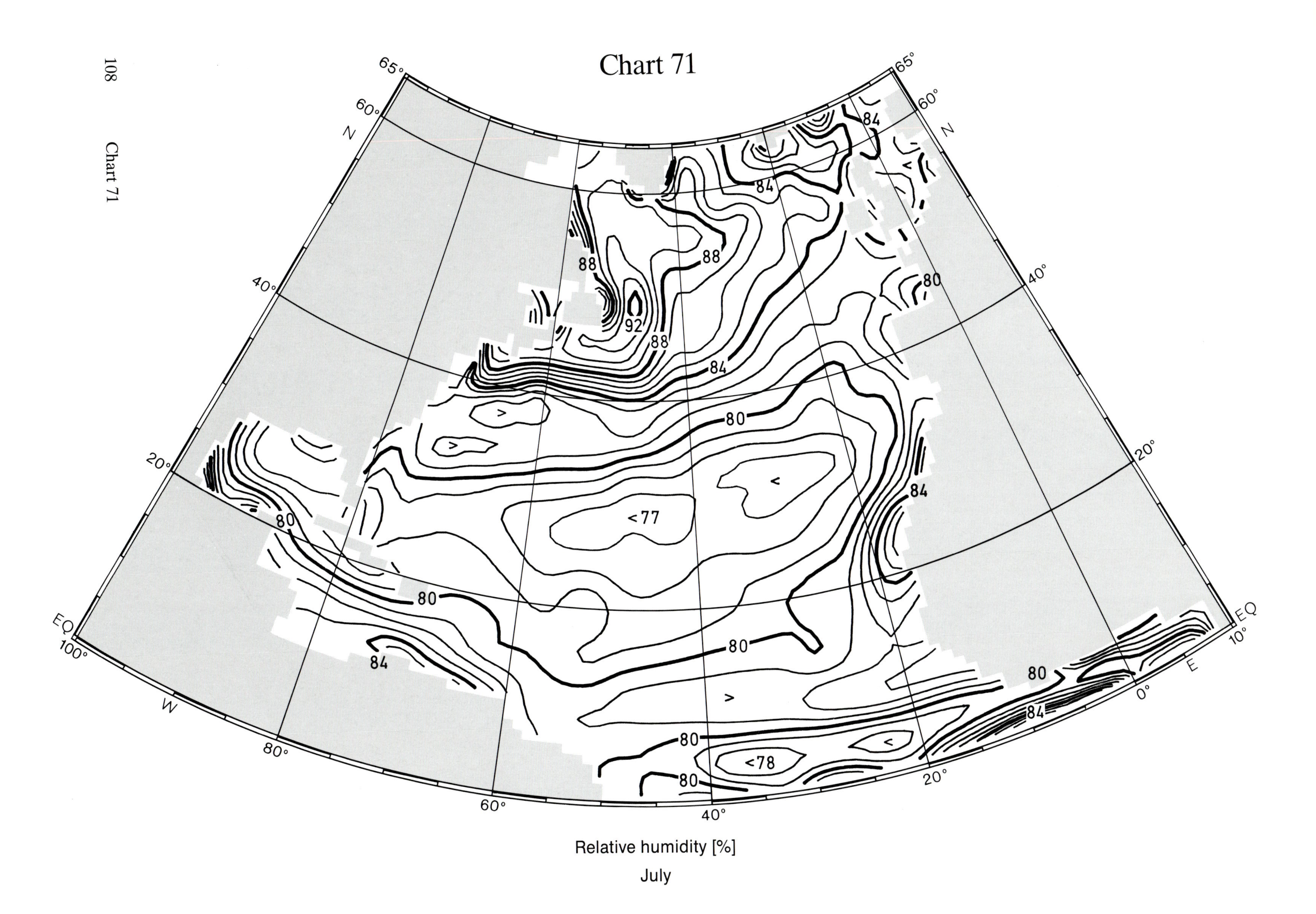

Relative humidity [%]
July

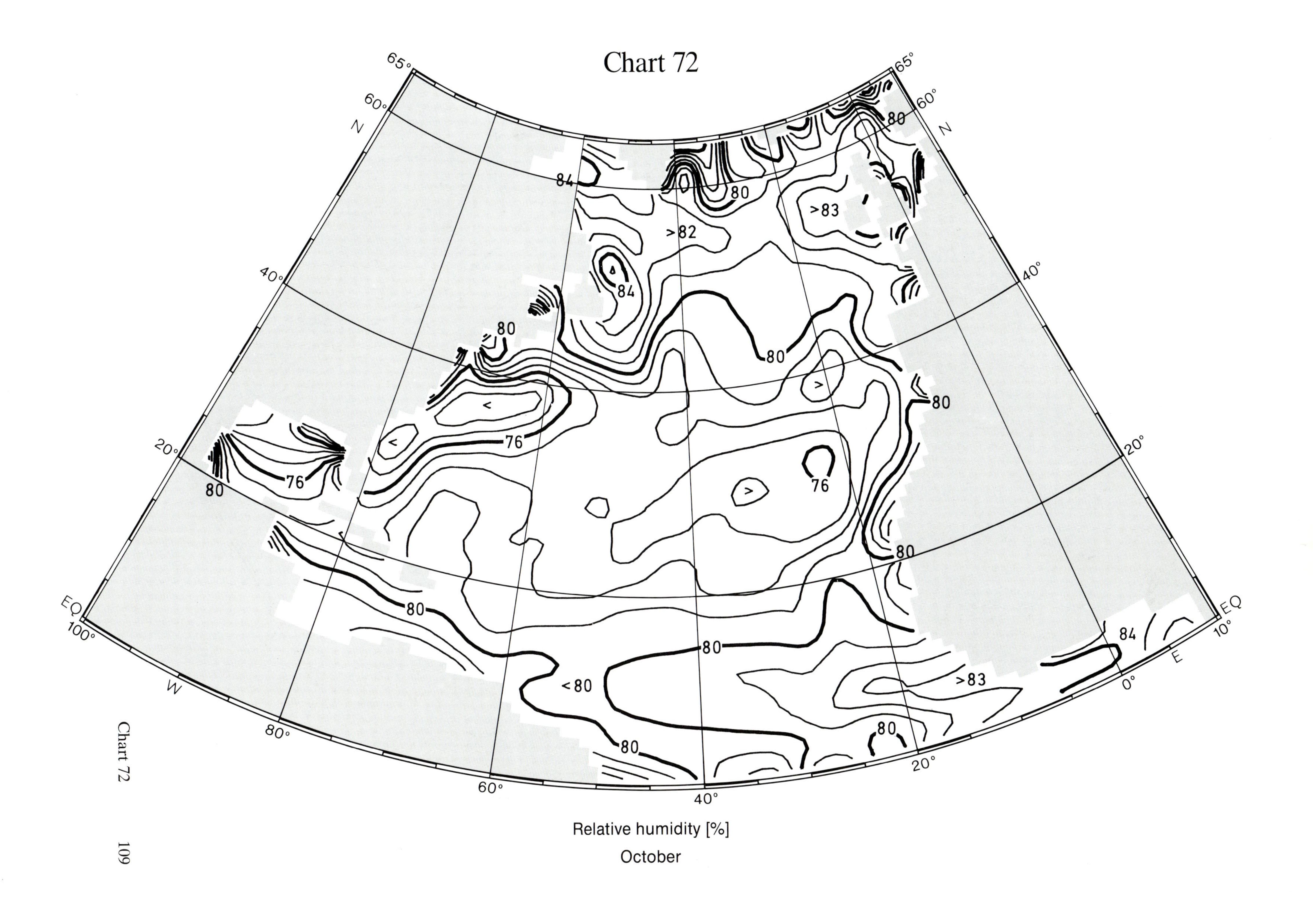
Chart 72
65°
60°
N
40°
20°
EQ
100°
W
80°
60°
40°
20°
0°
E
10°
84
80
>82
>83
80
<
76
>
<80
>83
Relative humidity [%]
October

Chart 73

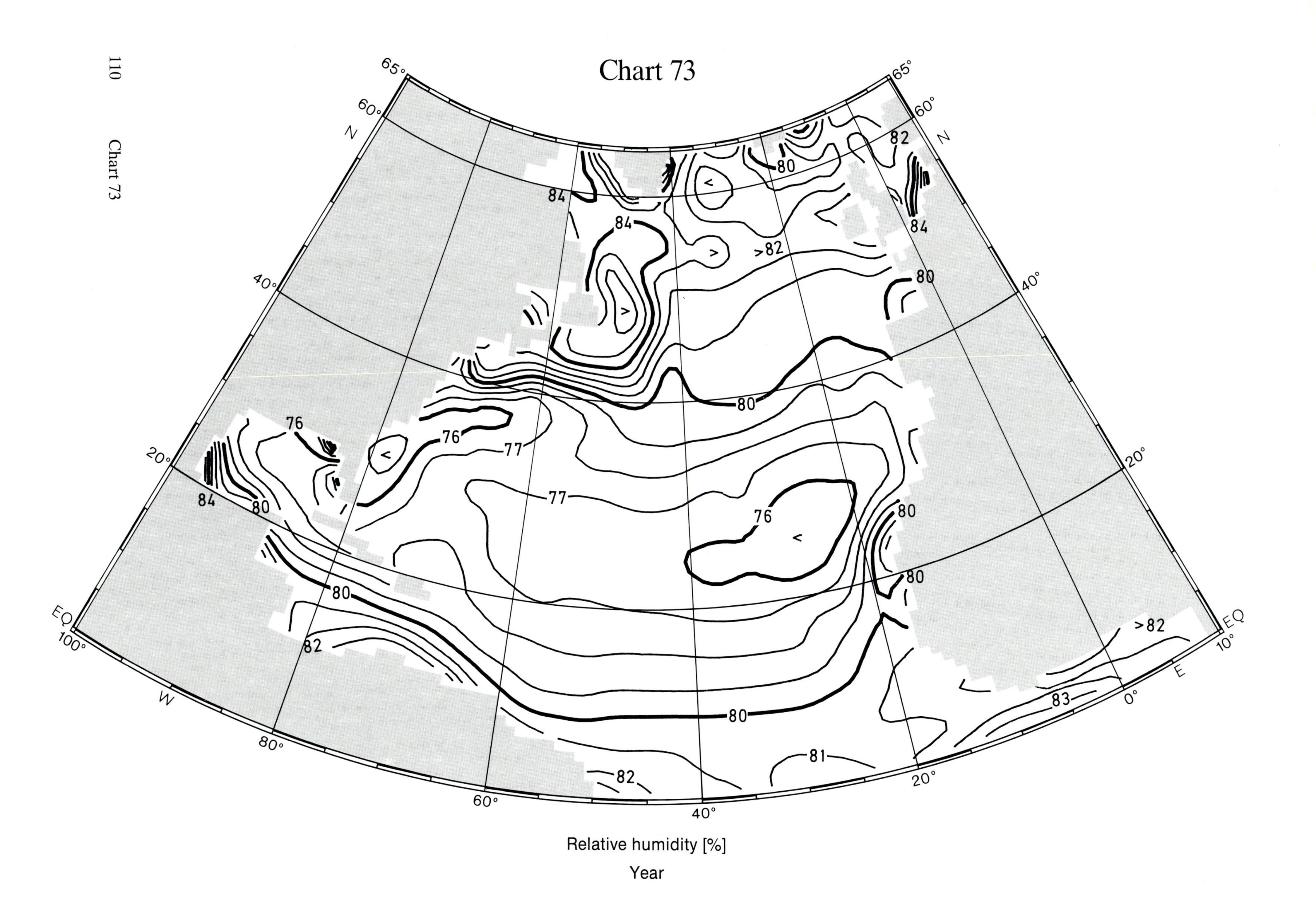

Relative humidity [%]
Year

Chart 74

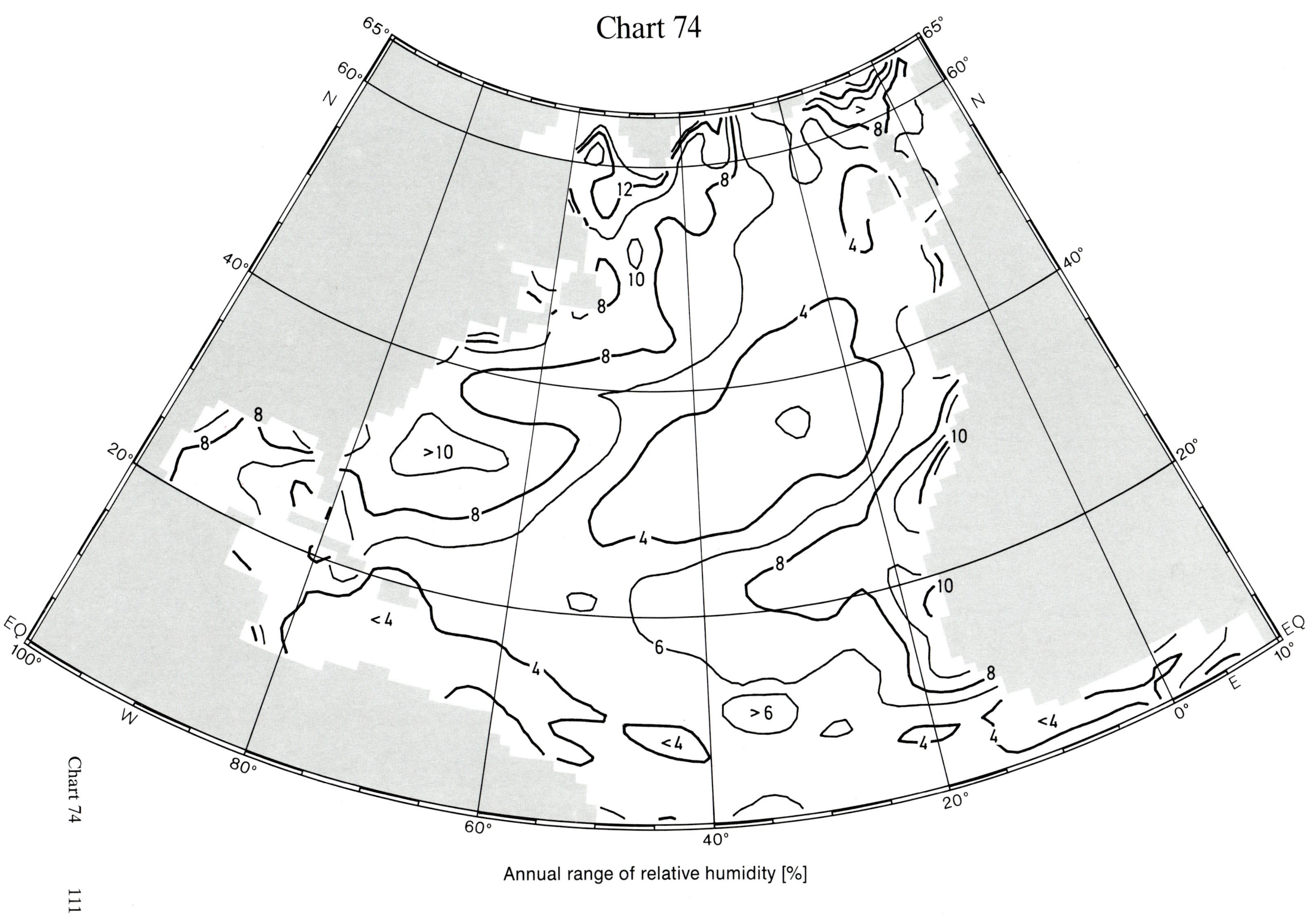

Annual range of relative humidity [%]

Chart 75

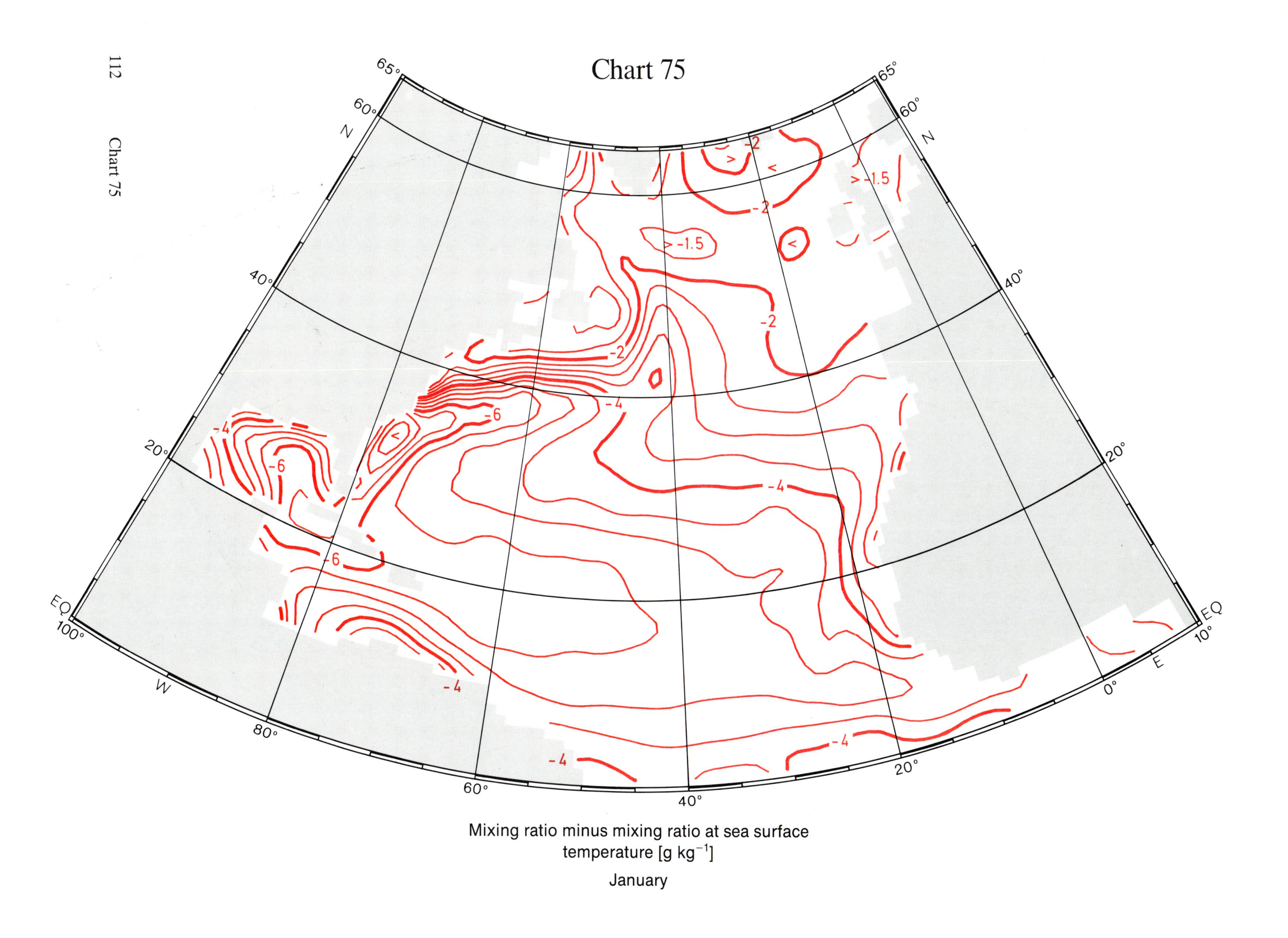

Mixing ratio minus mixing ratio at sea surface temperature [g kg^{-1}]
January

Chart 76

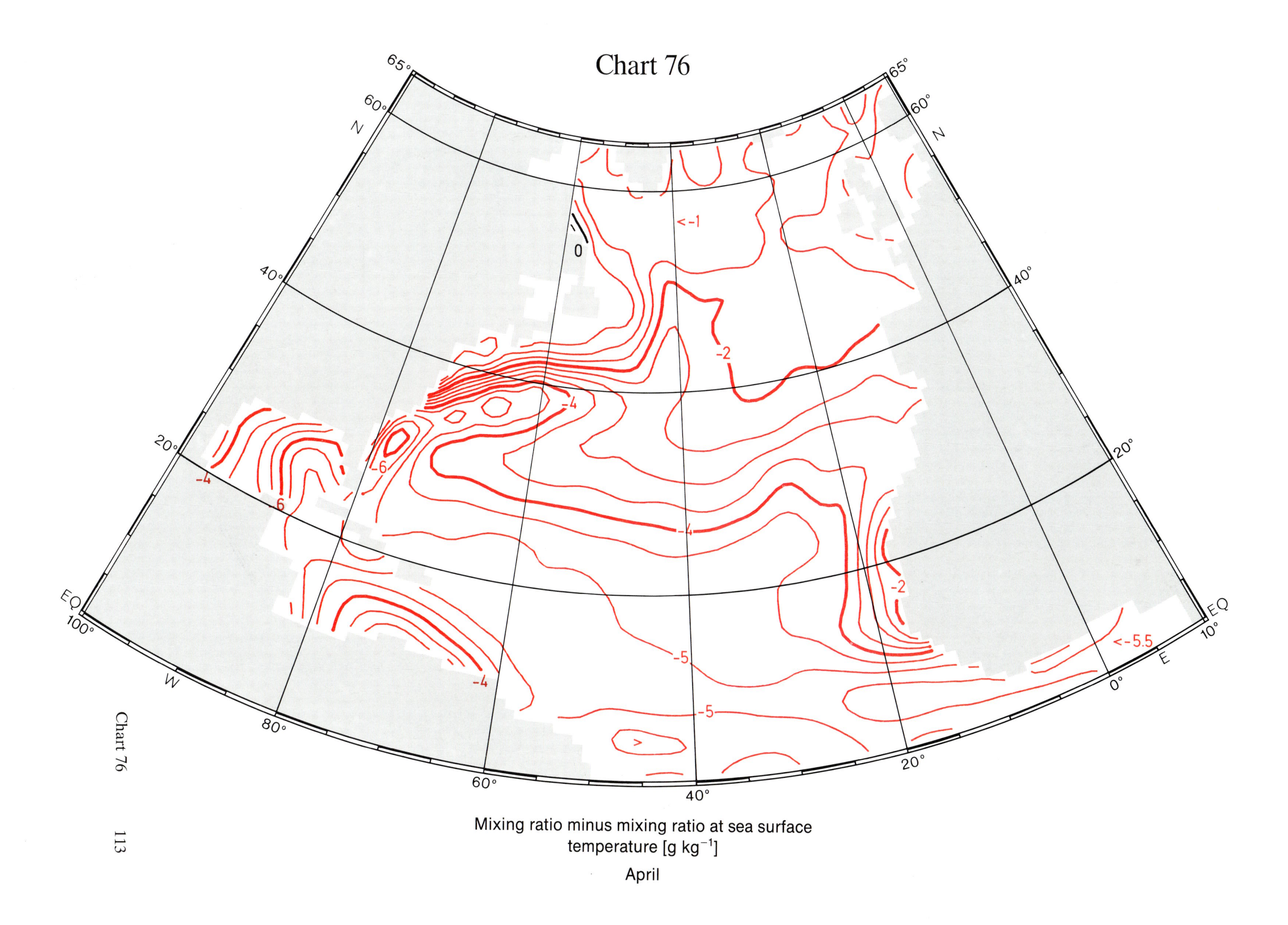

Mixing ratio minus mixing ratio at sea surface temperature [g kg^{-1}]
April

Chart 77

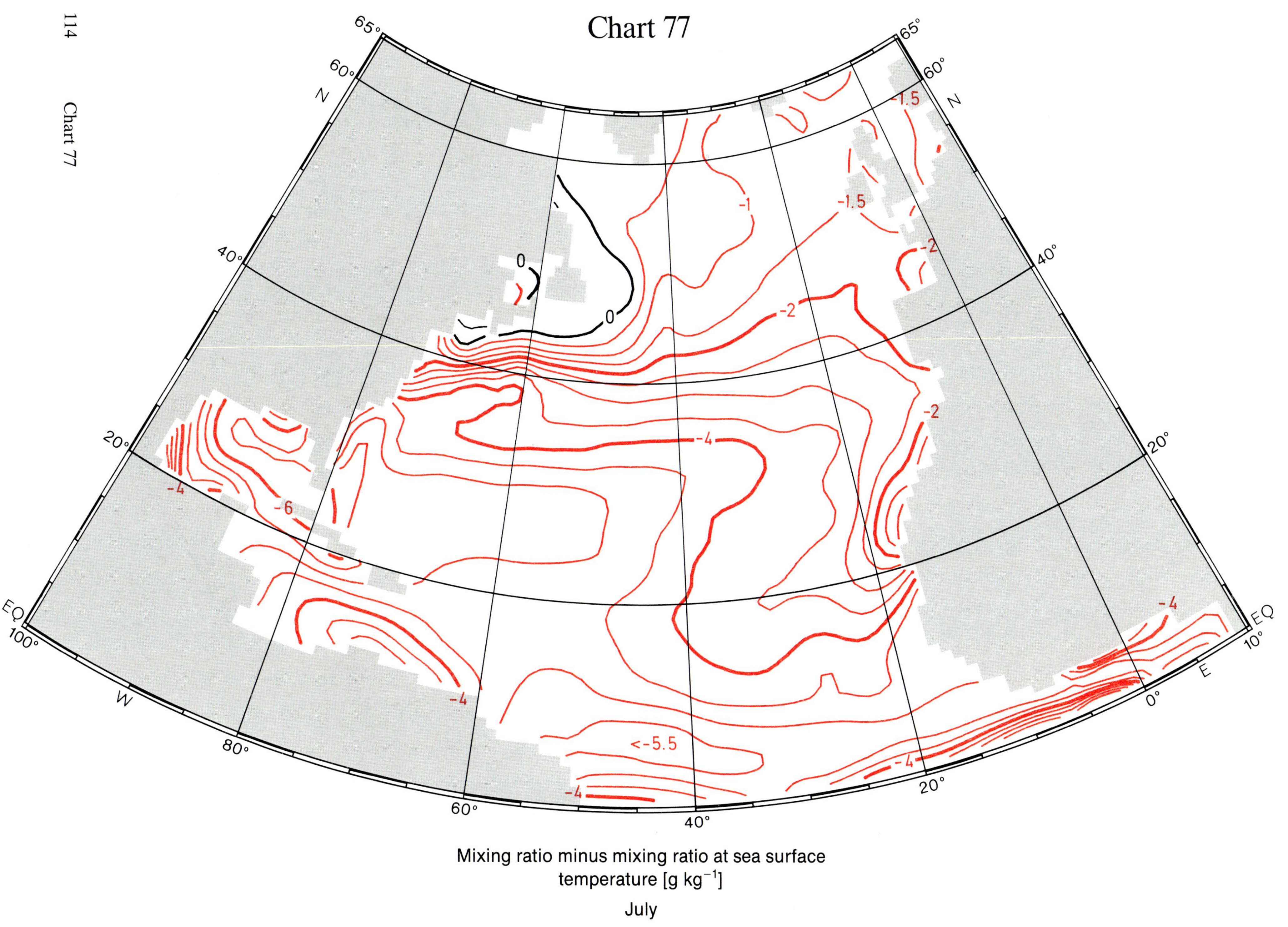

Mixing ratio minus mixing ratio at sea surface temperature [g kg^{-1}]
July

Chart 78

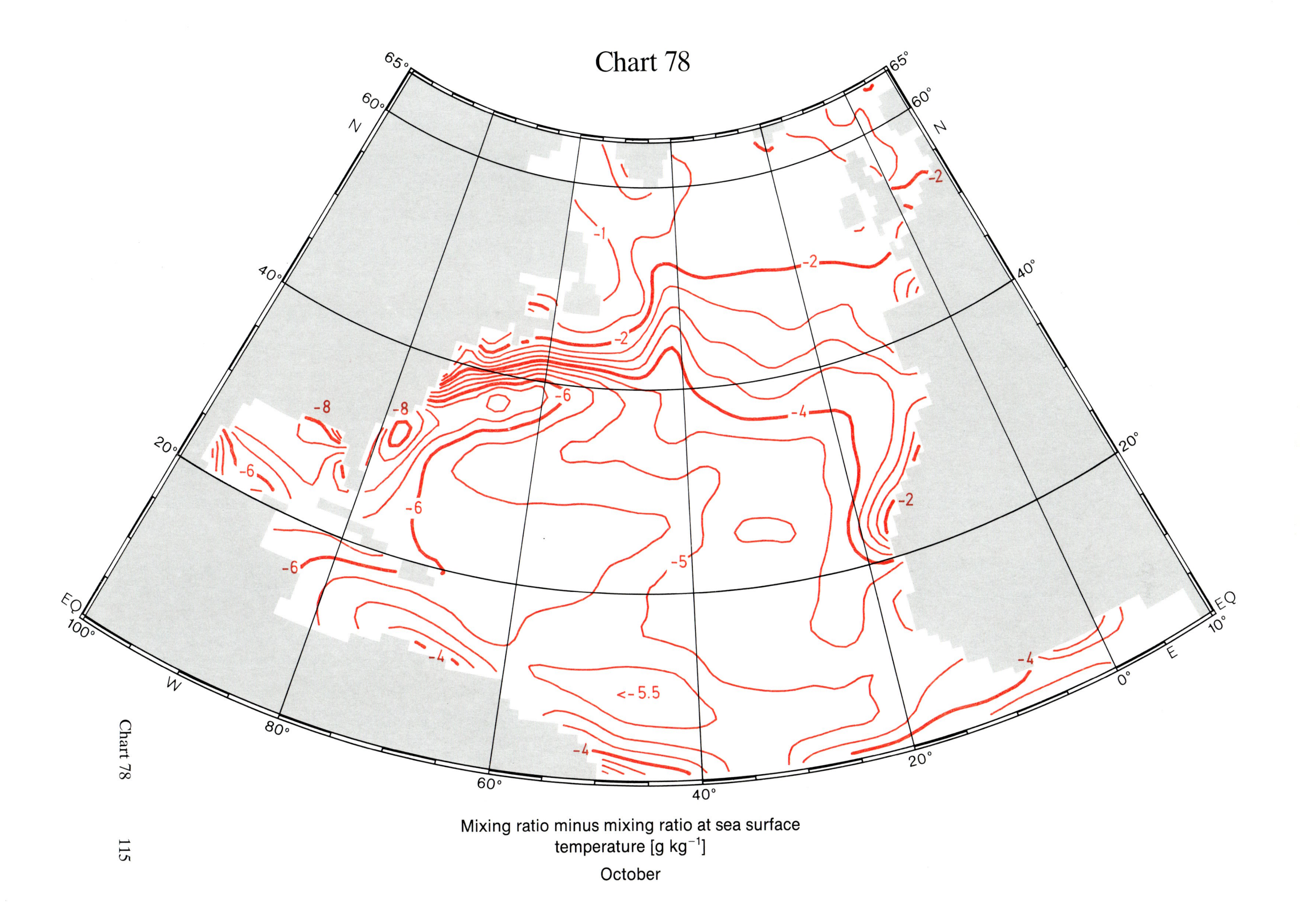

Mixing ratio minus mixing ratio at sea surface temperature [g kg^{-1}]
October

Chart 79

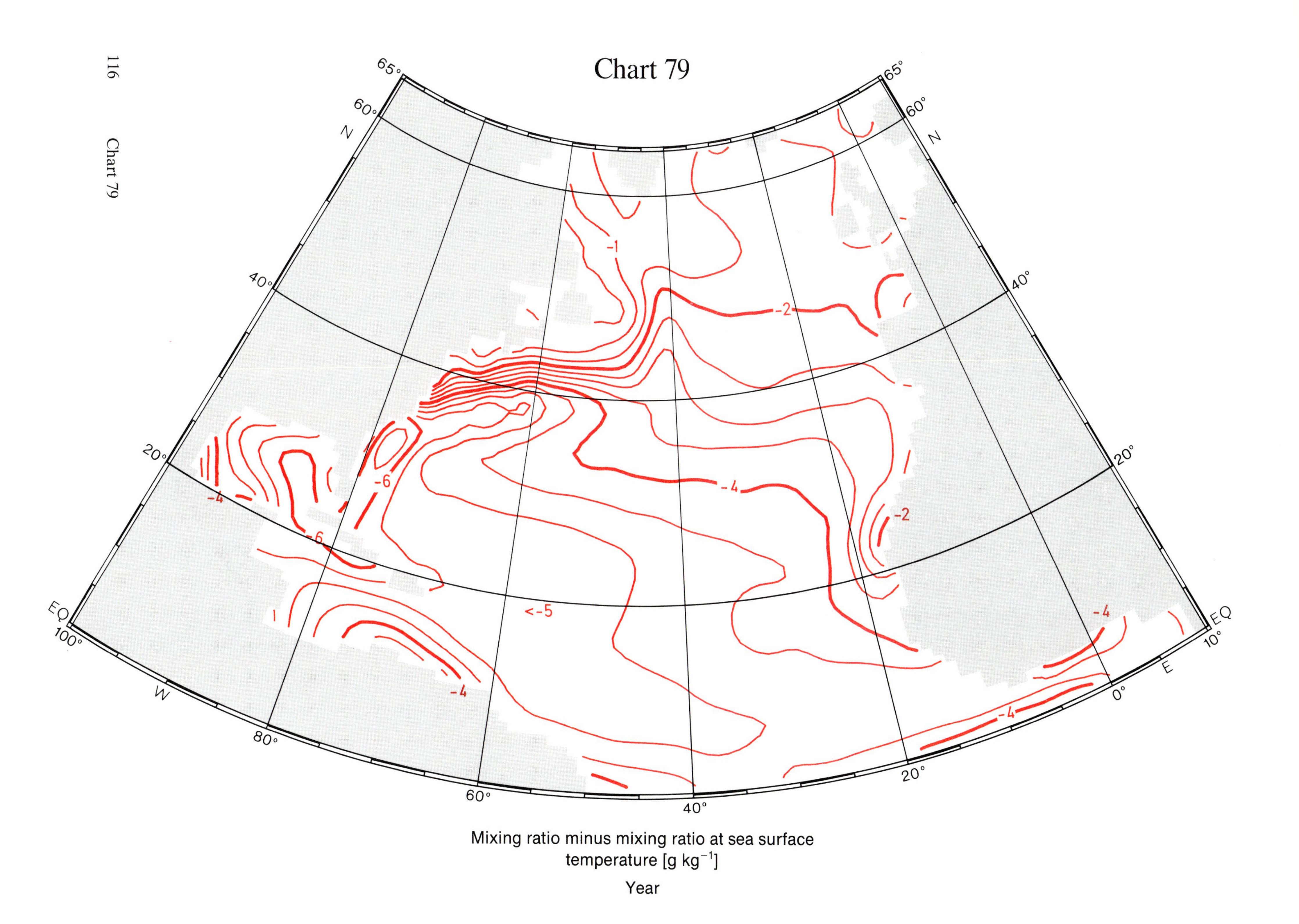

Mixing ratio minus mixing ratio at sea surface temperature [g kg^{-1}]

Year

Chart 80

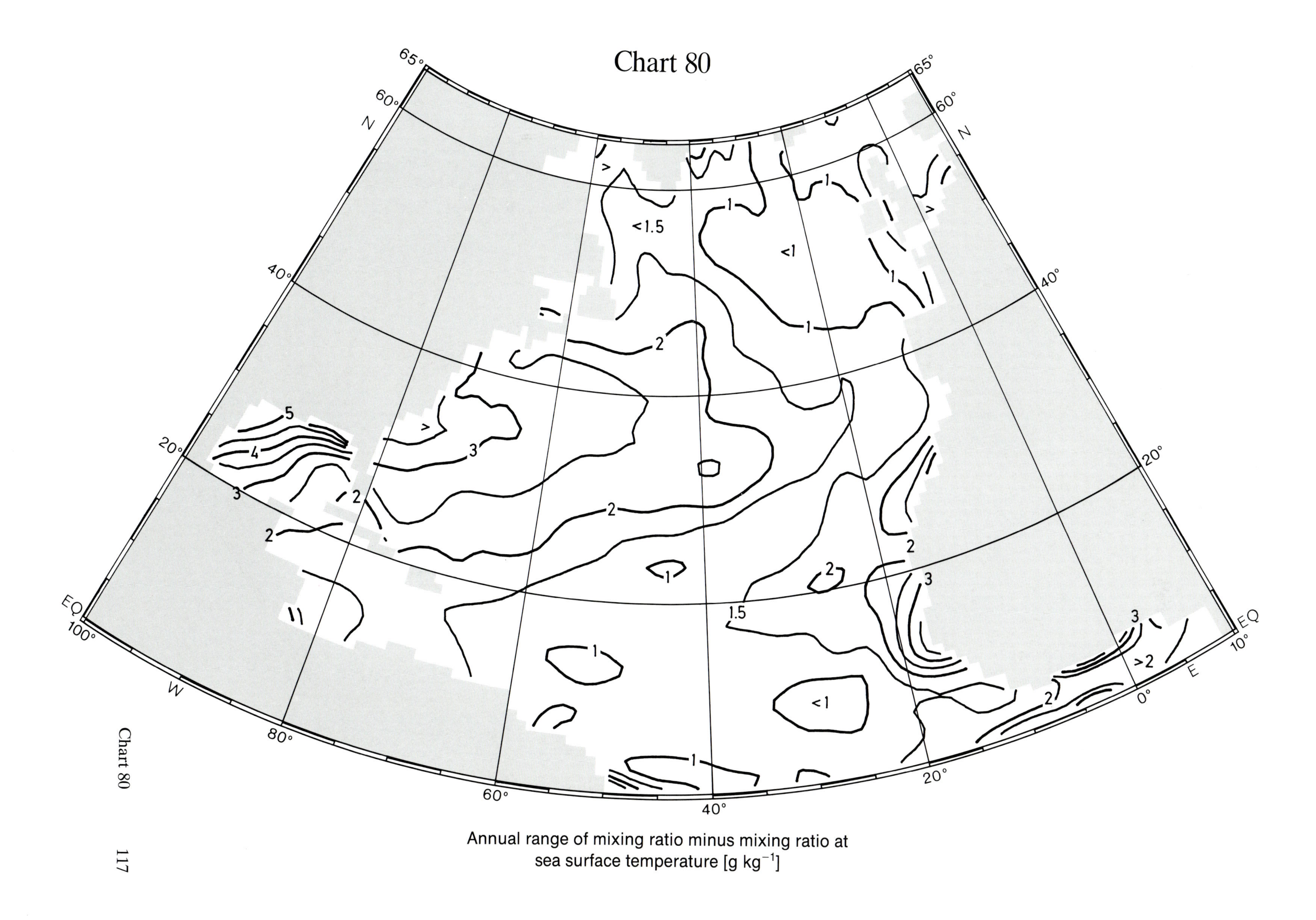

Annual range of mixing ratio minus mixing ratio at sea surface temperature [$g\ kg^{-1}$]

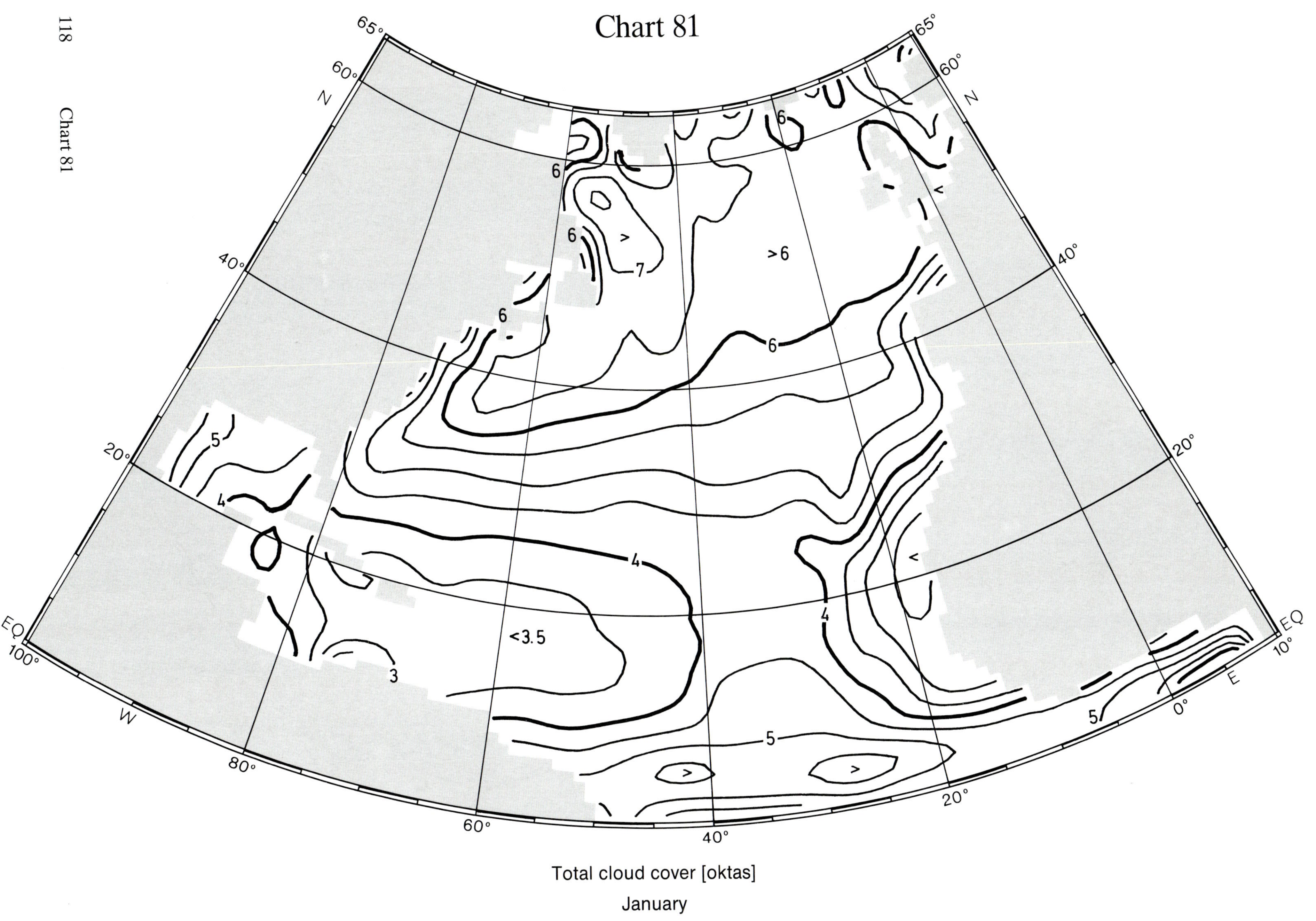
Chart 81
65°
60°
N
40°
20°
EQ
100°
W
80°
60°
40°
20°
0°
E
10°
6
>6
7
5
4
<3.5
3
Total cloud cover [oktas]
January

Chart 82

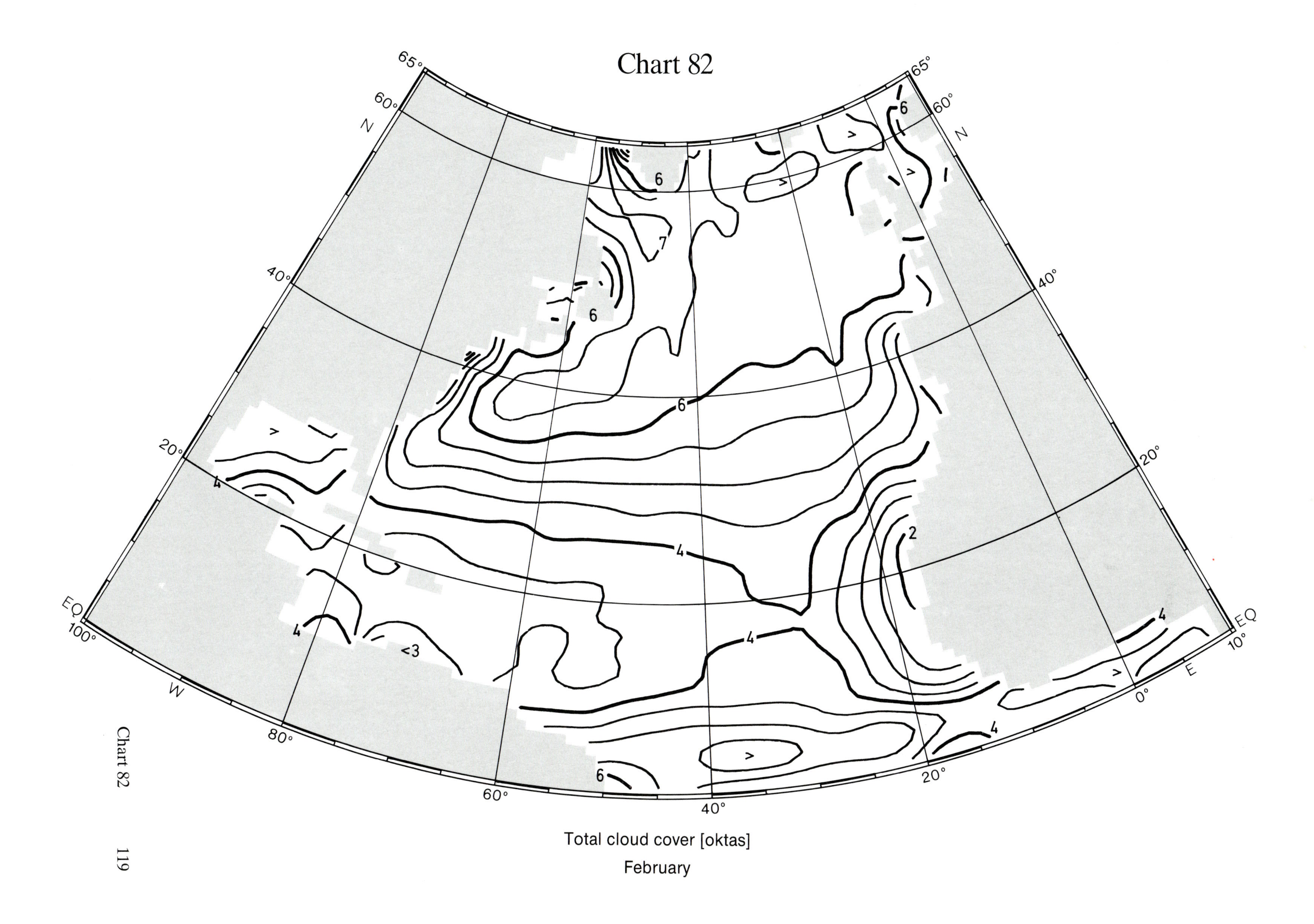

Total cloud cover [oktas]
February

Chart 83

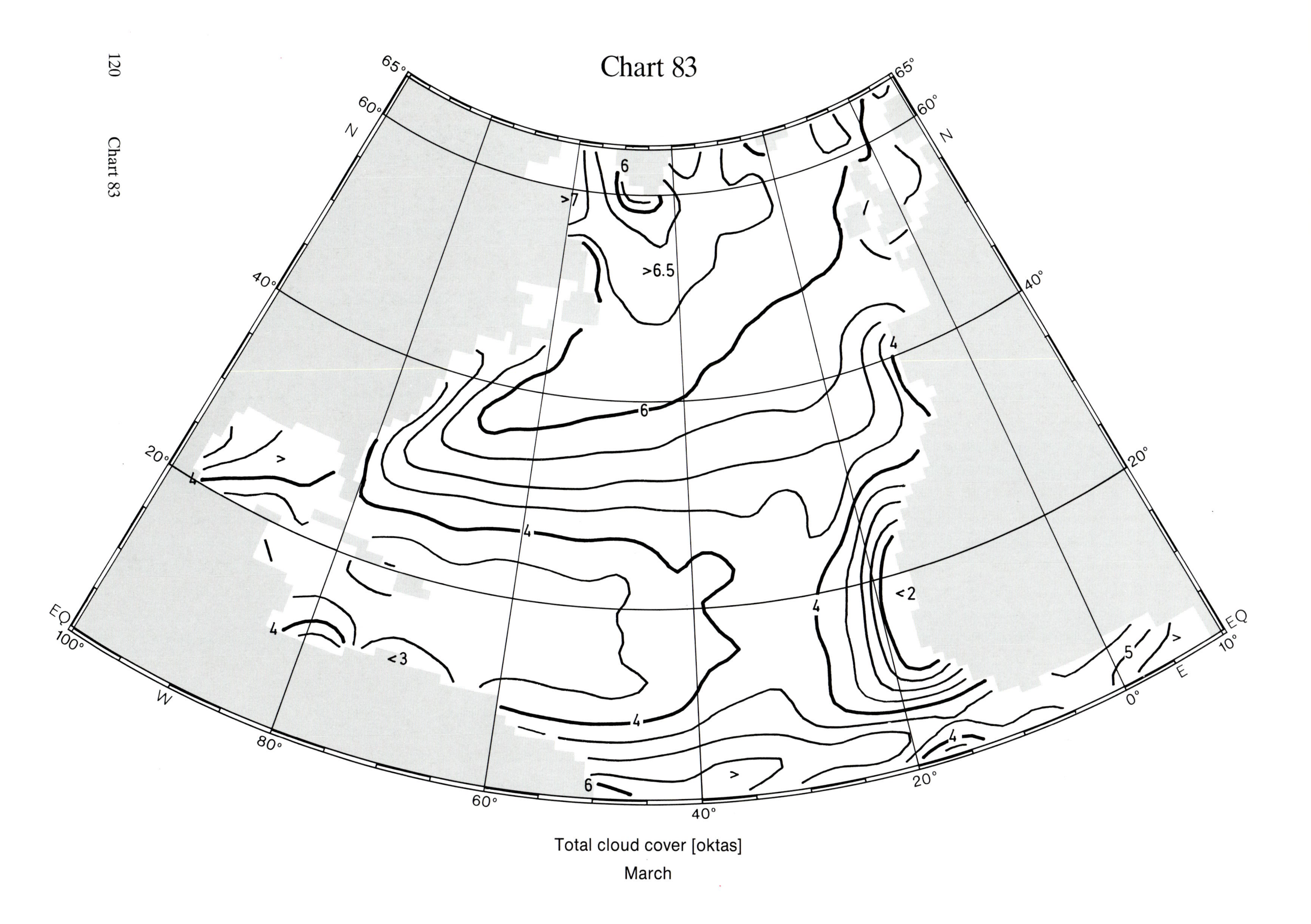

Total cloud cover [oktas]
March

Chart 84

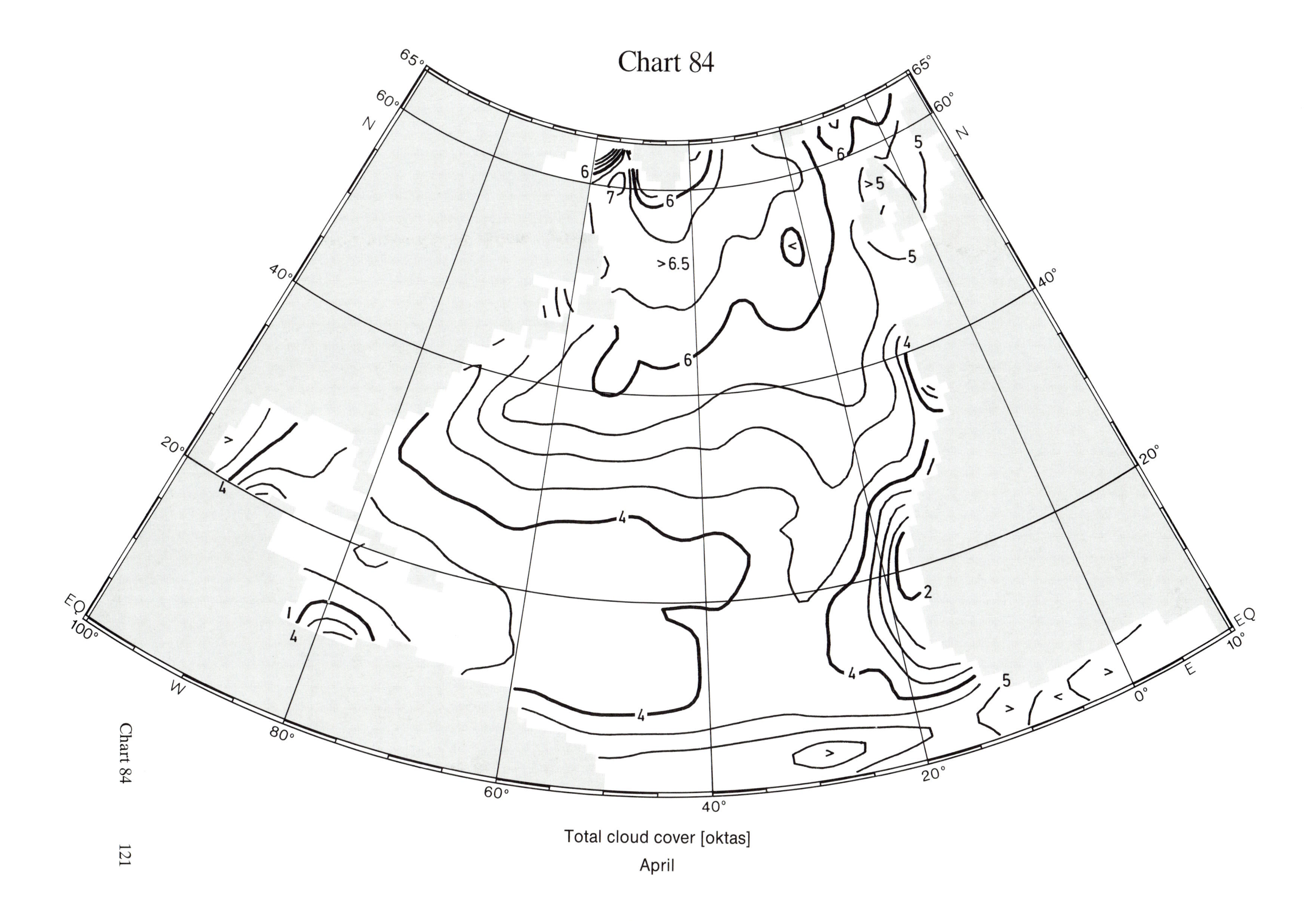

Total cloud cover [oktas]
April

Chart 85

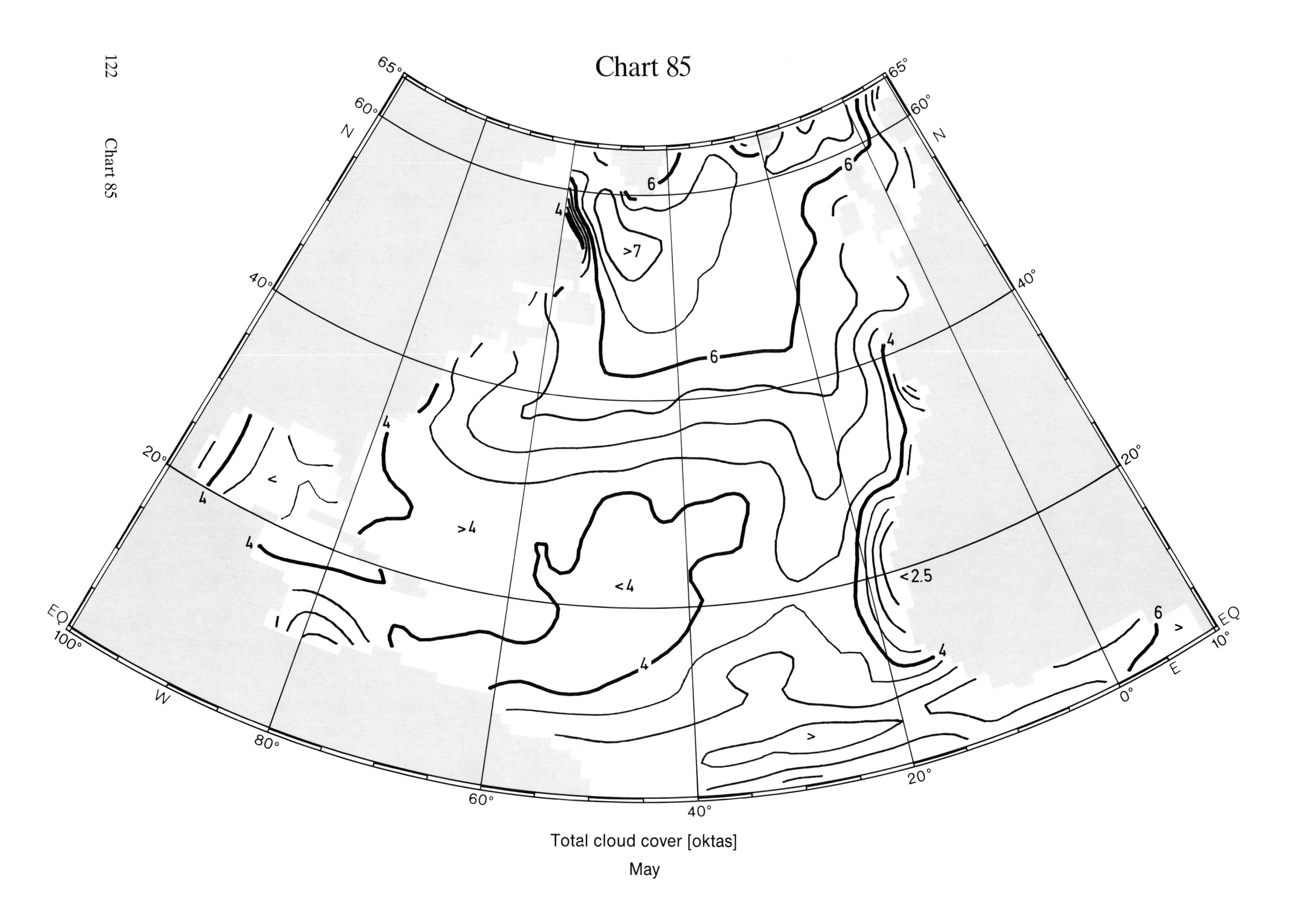

Total cloud cover [oktas]
May

Chart 86

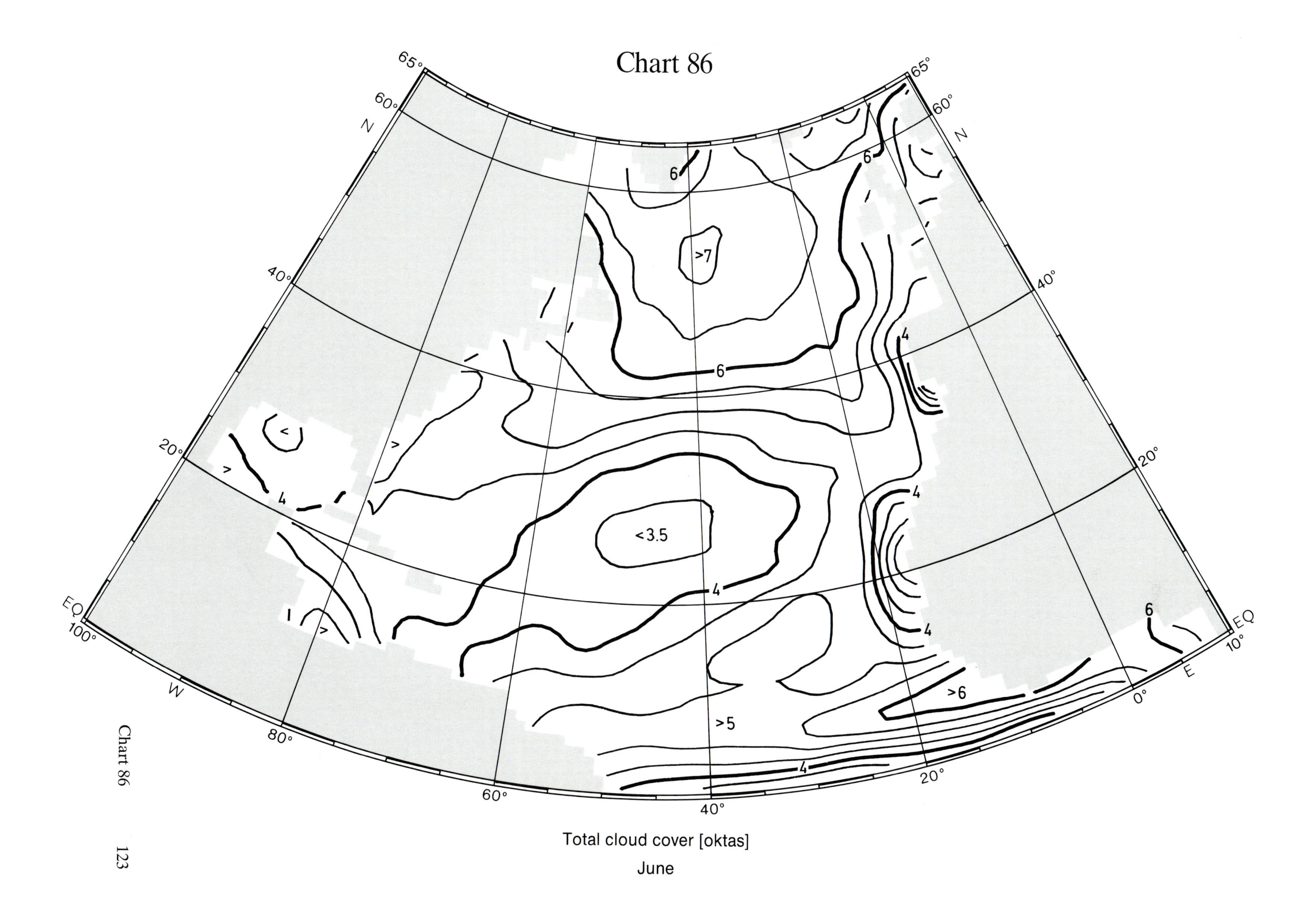

Total cloud cover [oktas]

June

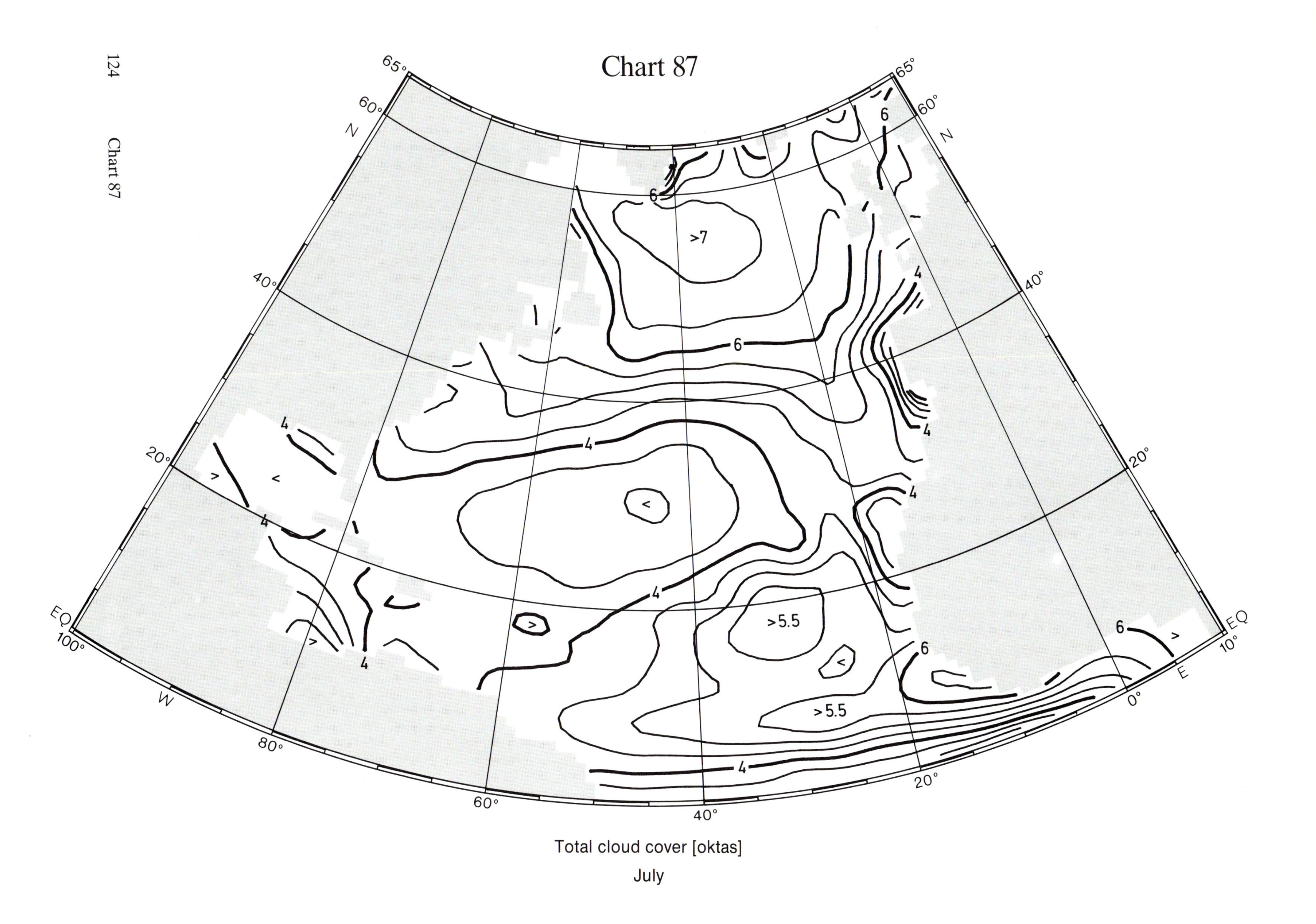
Chart 87
65°
60°
N
40°
20°
EQ
100°
W
80°
60°
40°
20°
0°
E
10°
>7
>5.5
>5.5
6
4
Total cloud cover [oktas]
July

Chart 88

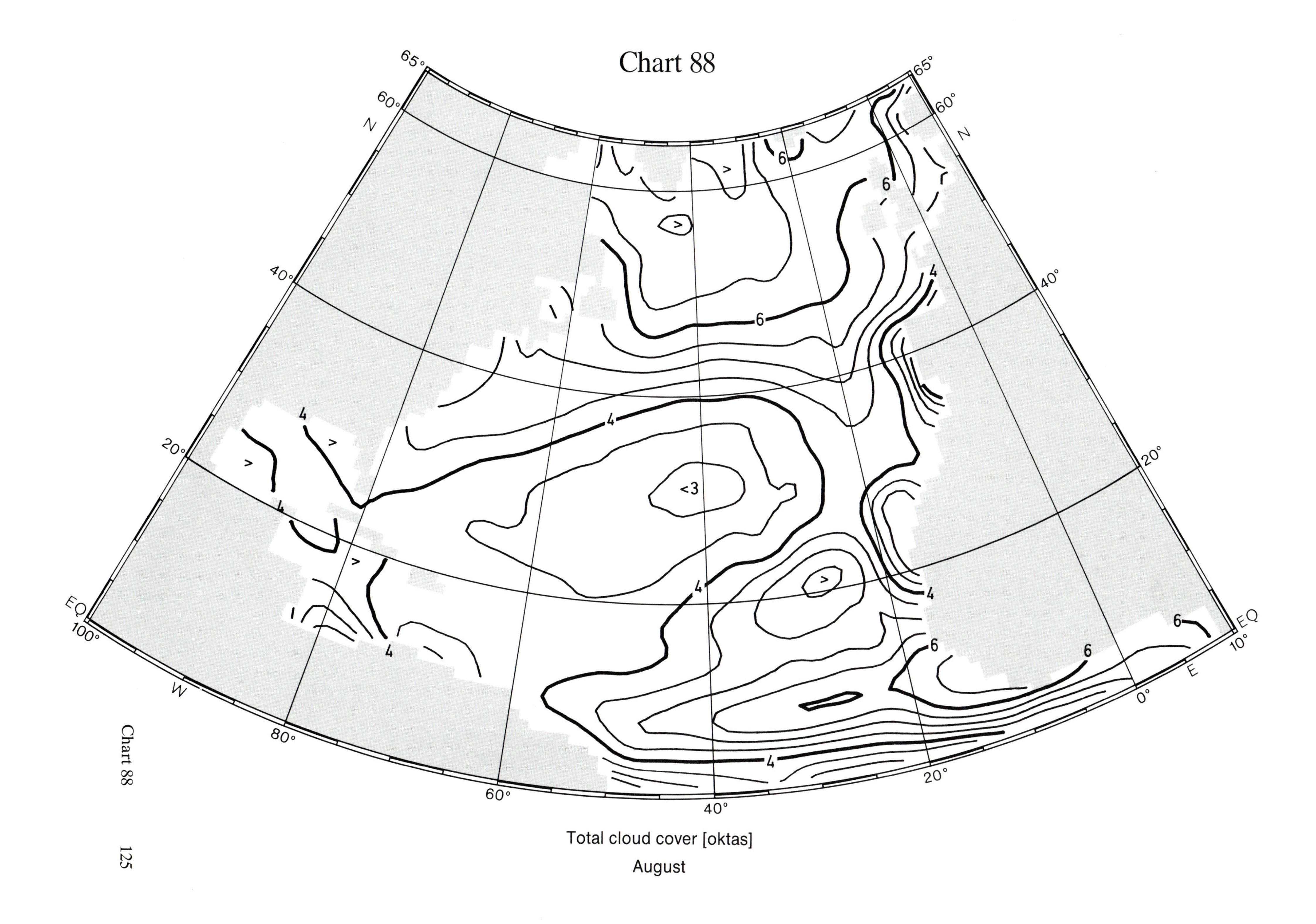

Total cloud cover [oktas]
August

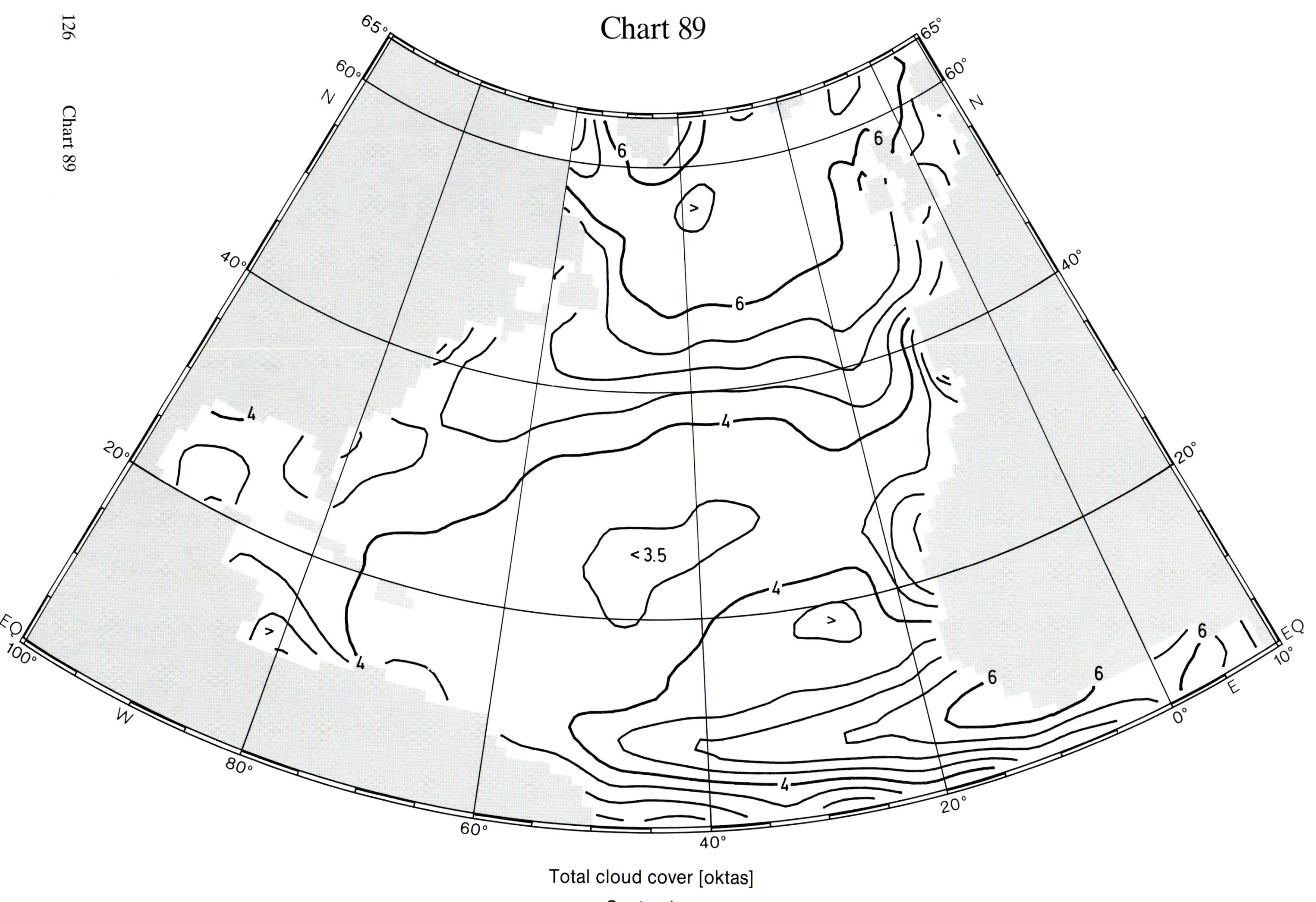

Chart 89
65°
60°
N
40°
20°
EQ
100°
W
80°
60°
40°
20°
0°
E
10°
6
4
<3.5
Total cloud cover [oktas]
September

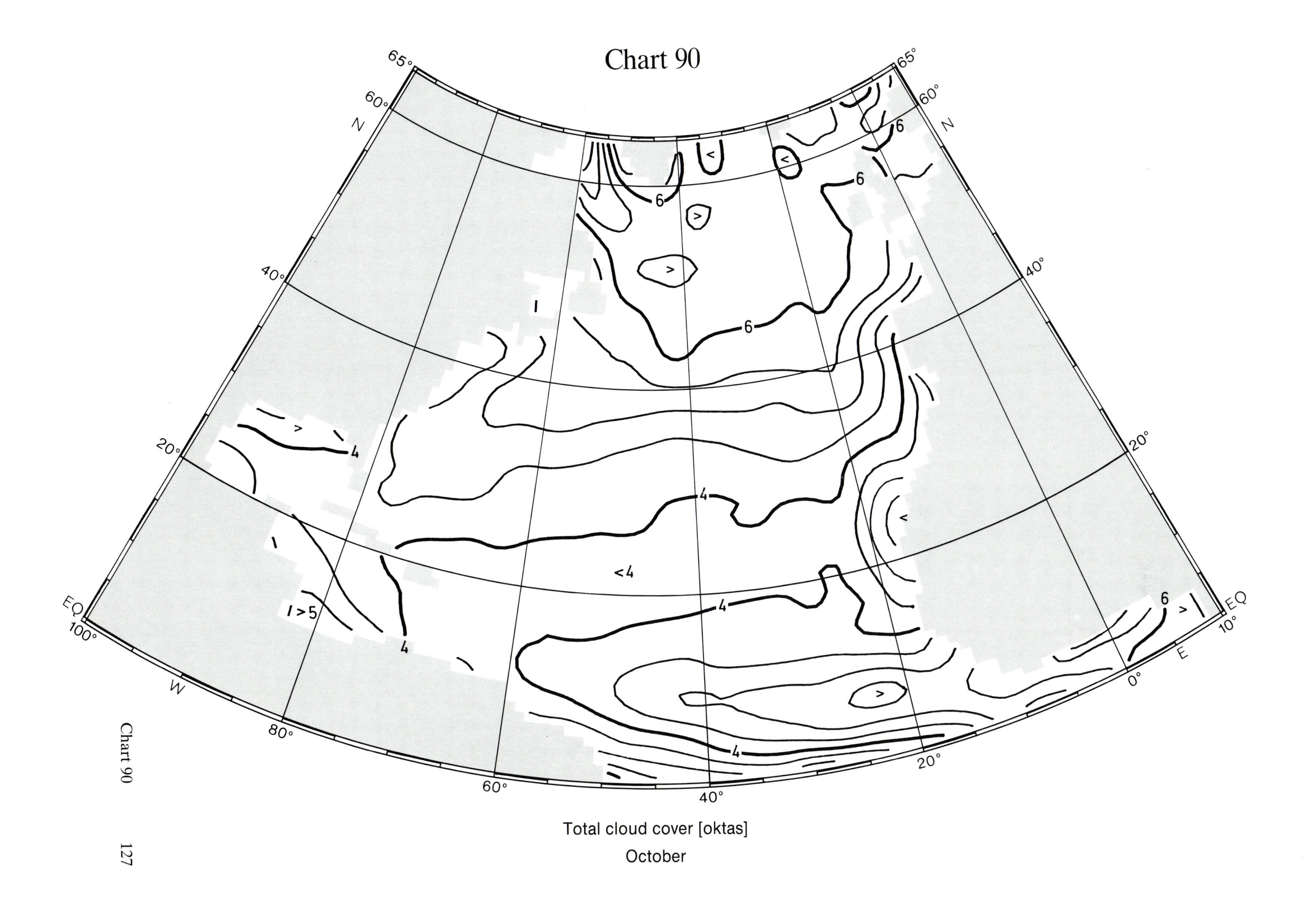

Chart 90
65°
60°
N
40°
20°
EQ
100°
W
80°
60°
40°
20°
0°
E
10°
6
4
<4
1>5
Total cloud cover [oktas]
October

Chart 91

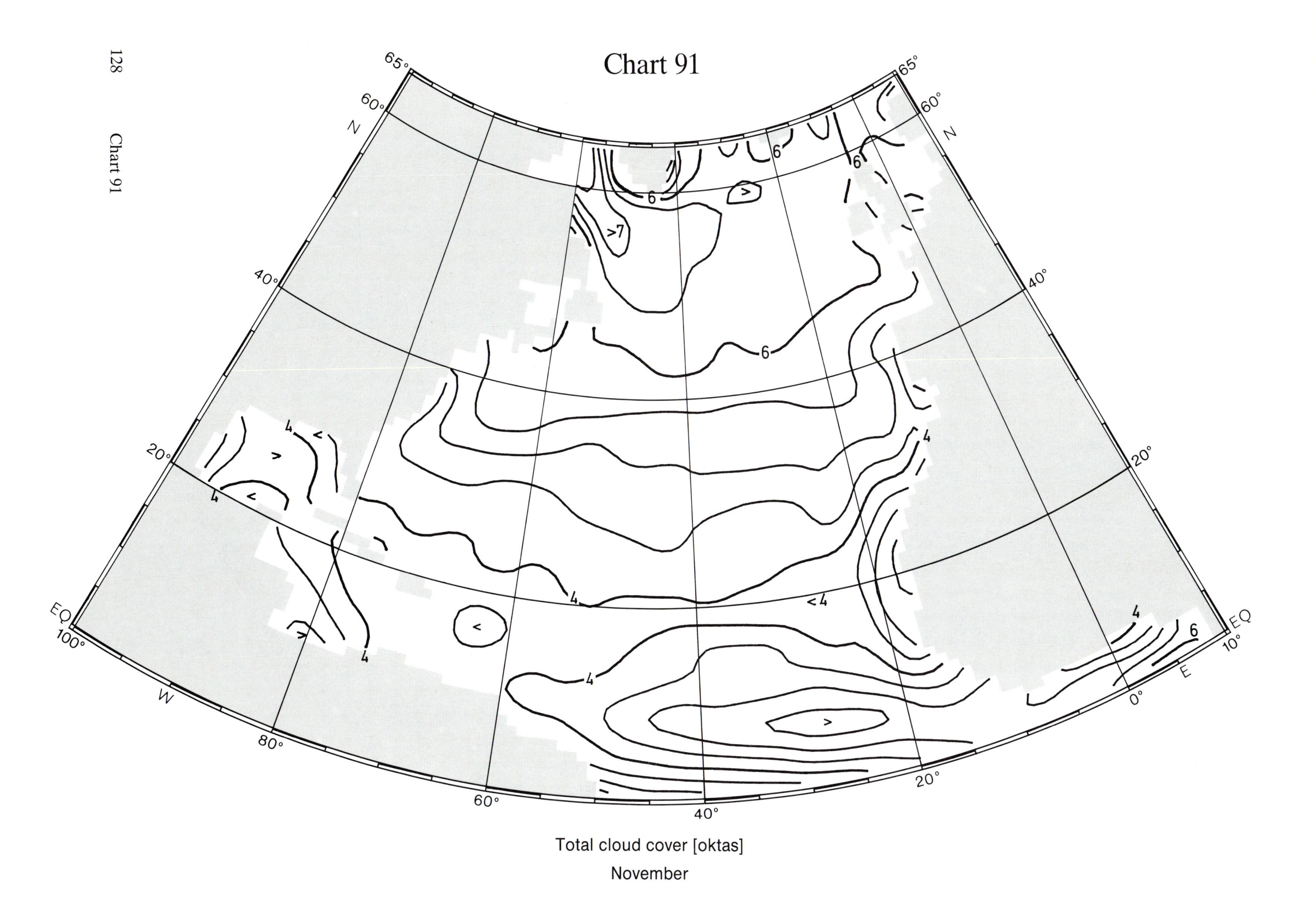

Total cloud cover [oktas]
November

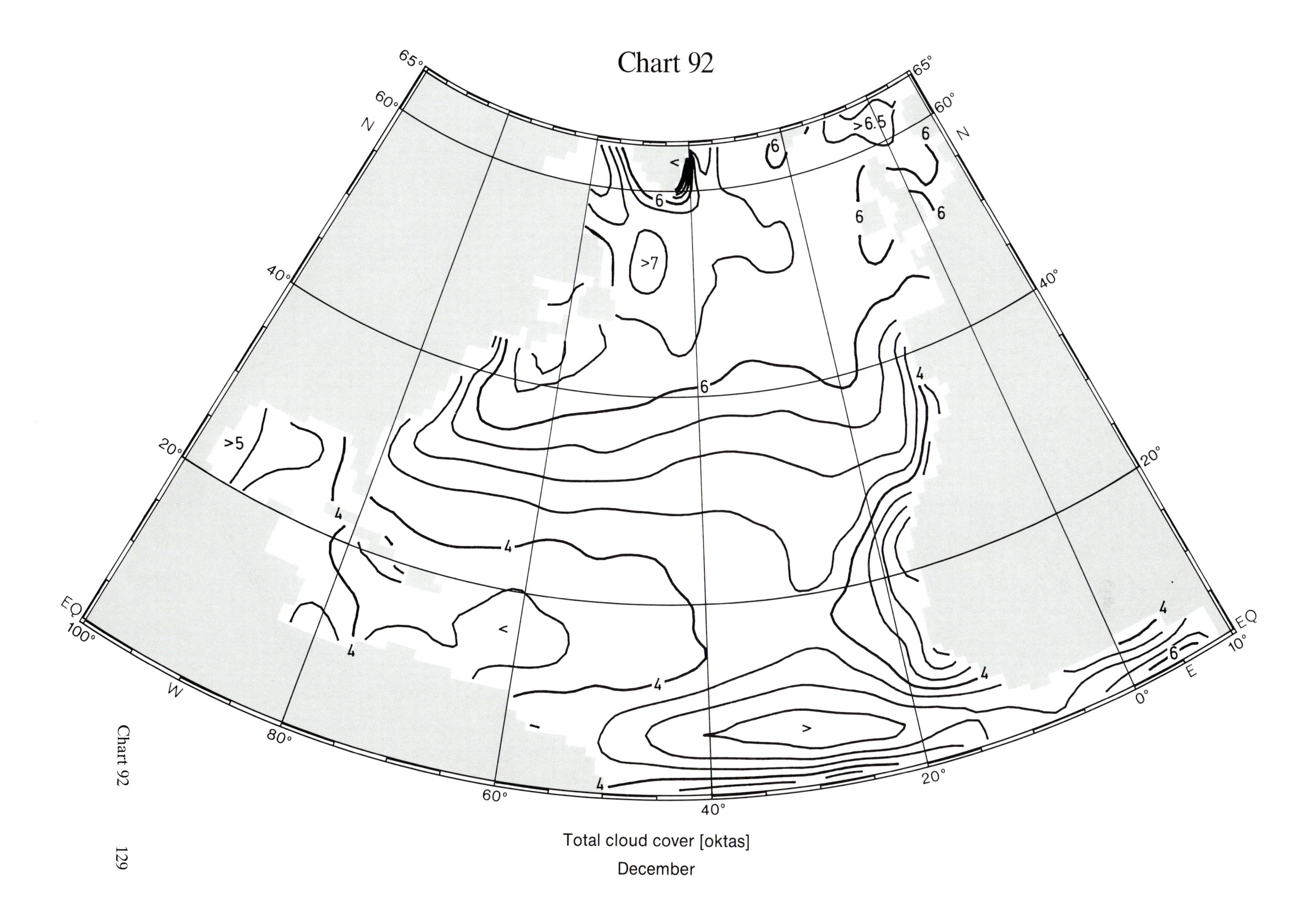

Total cloud cover [oktas]
December

Chart 93

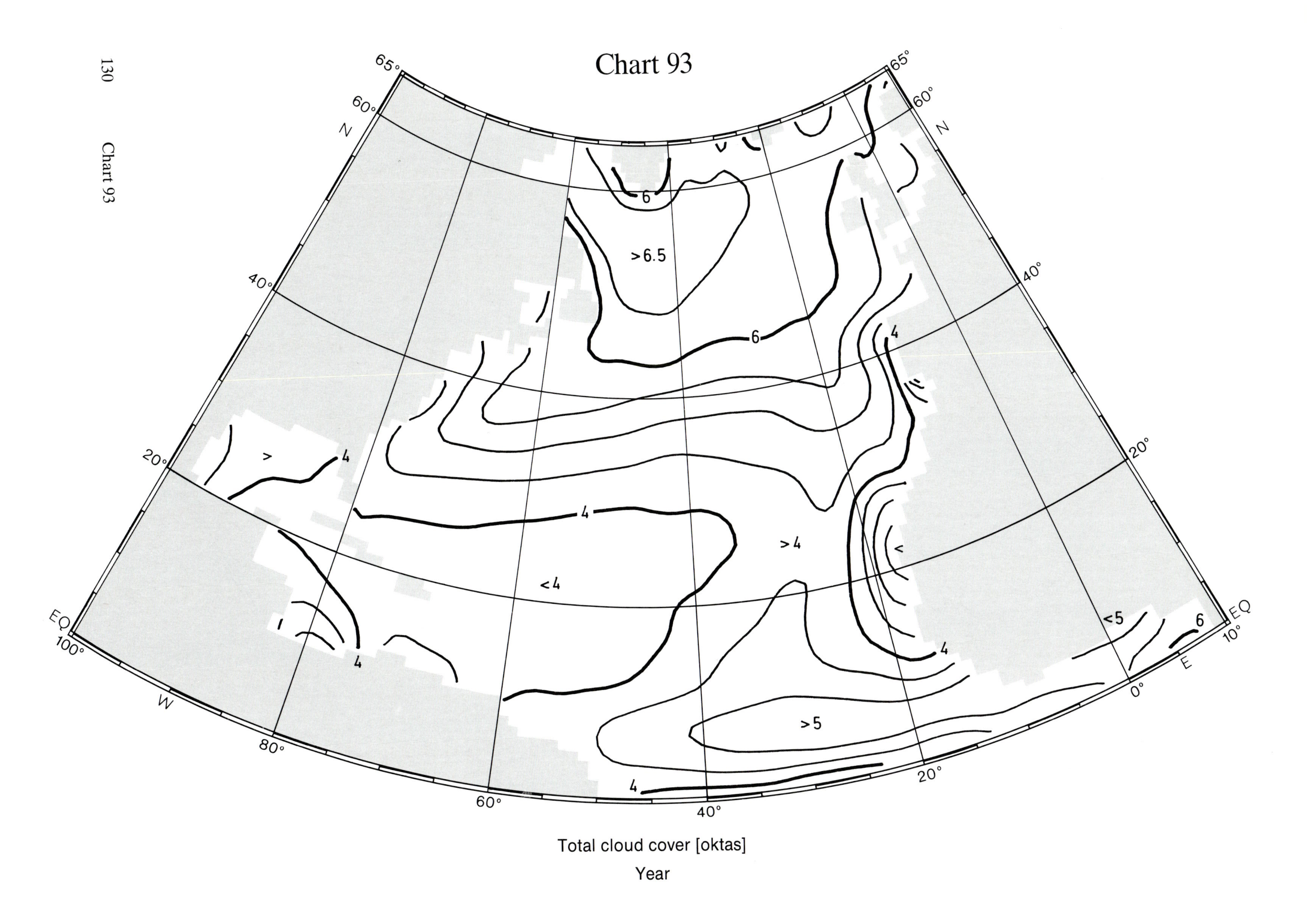

Total cloud cover [oktas]

Year

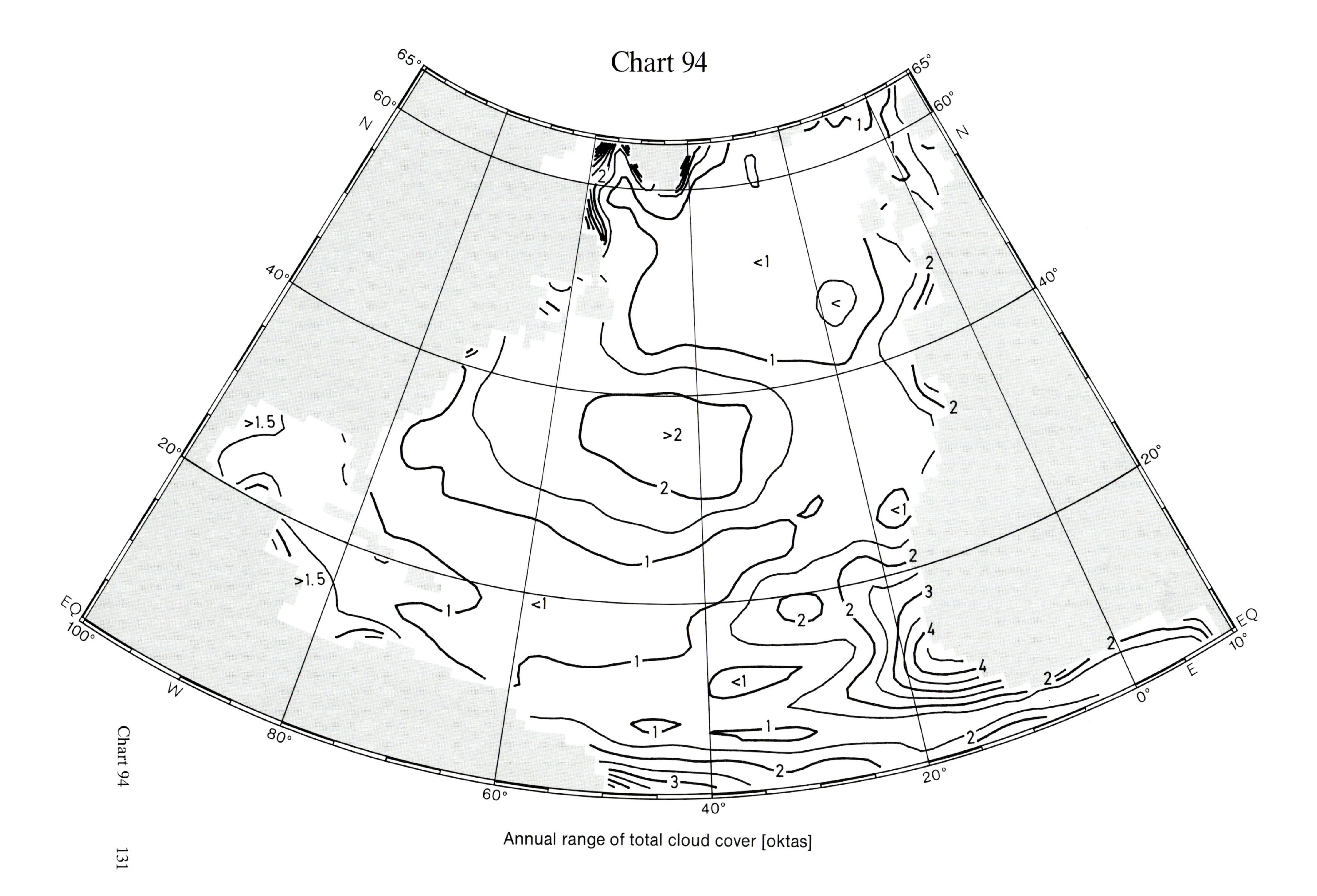

Chart 94
65°
60°
N
40°
20°
EQ
100°
W
80°
60°
40°
20°
0°
E
10°
>2
>1.5
<1
Annual range of total cloud cover [oktas]

Chart 95

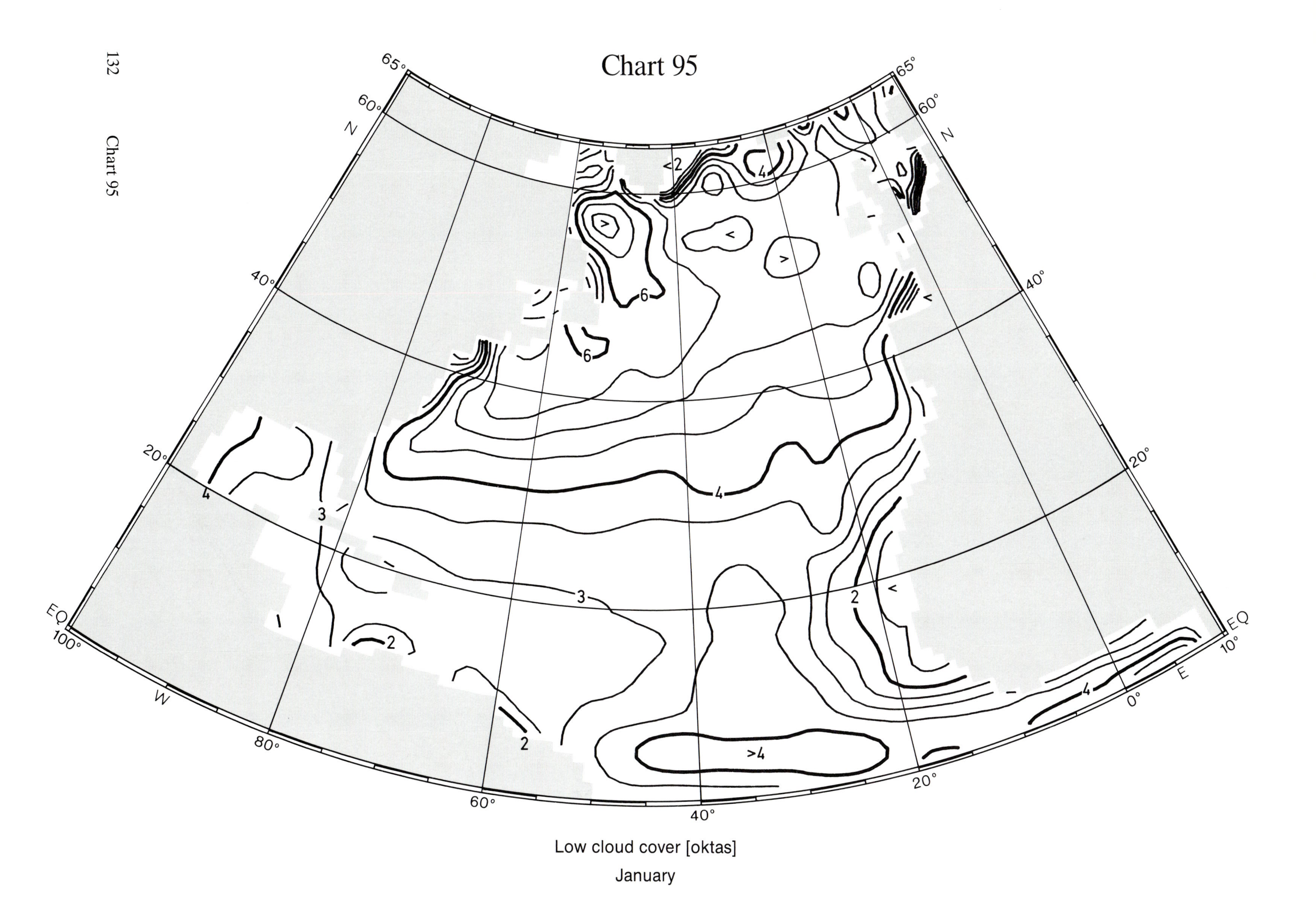

Low cloud cover [oktas]
January

Chart 96

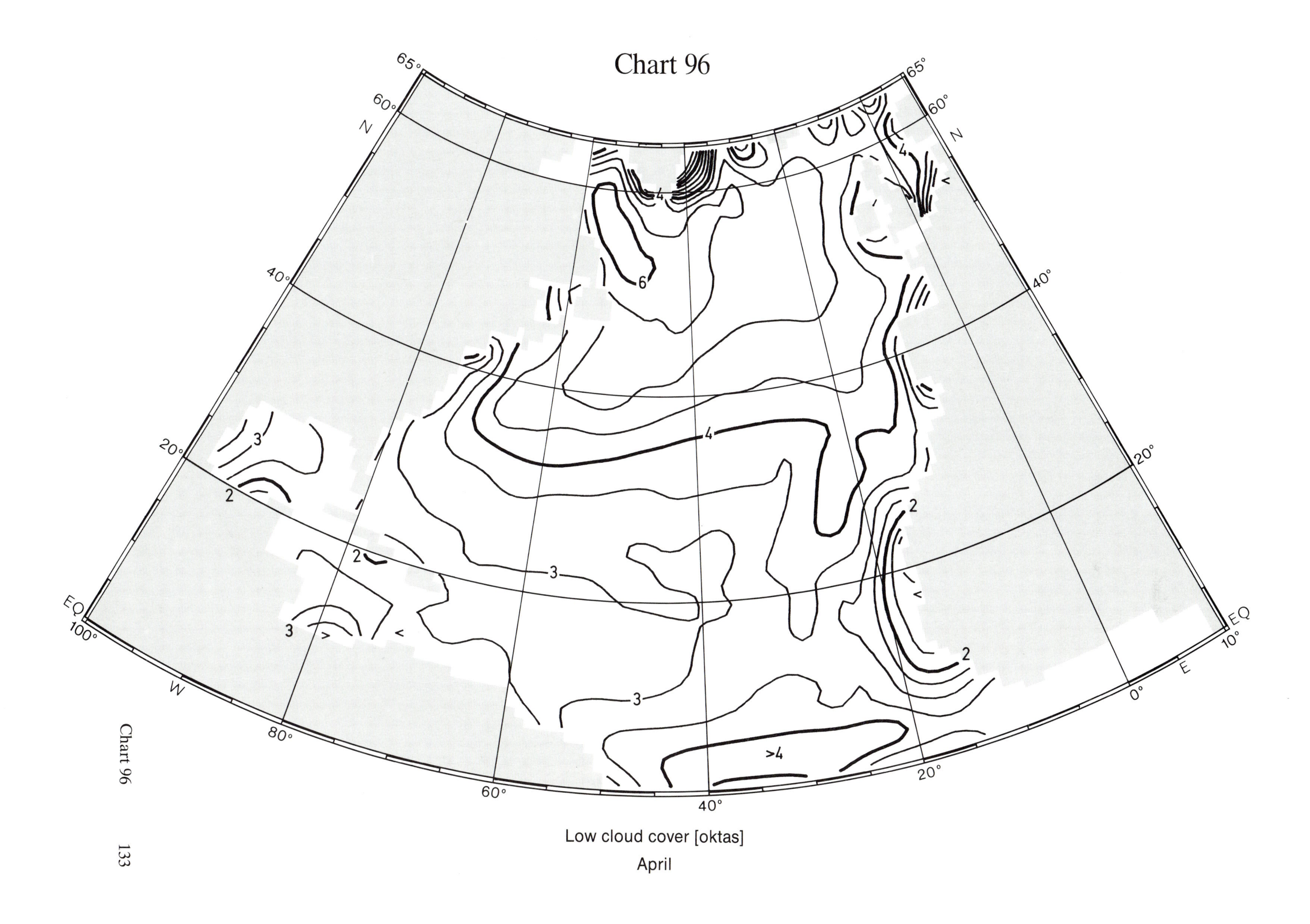

Low cloud cover [oktas]
April

Chart 97

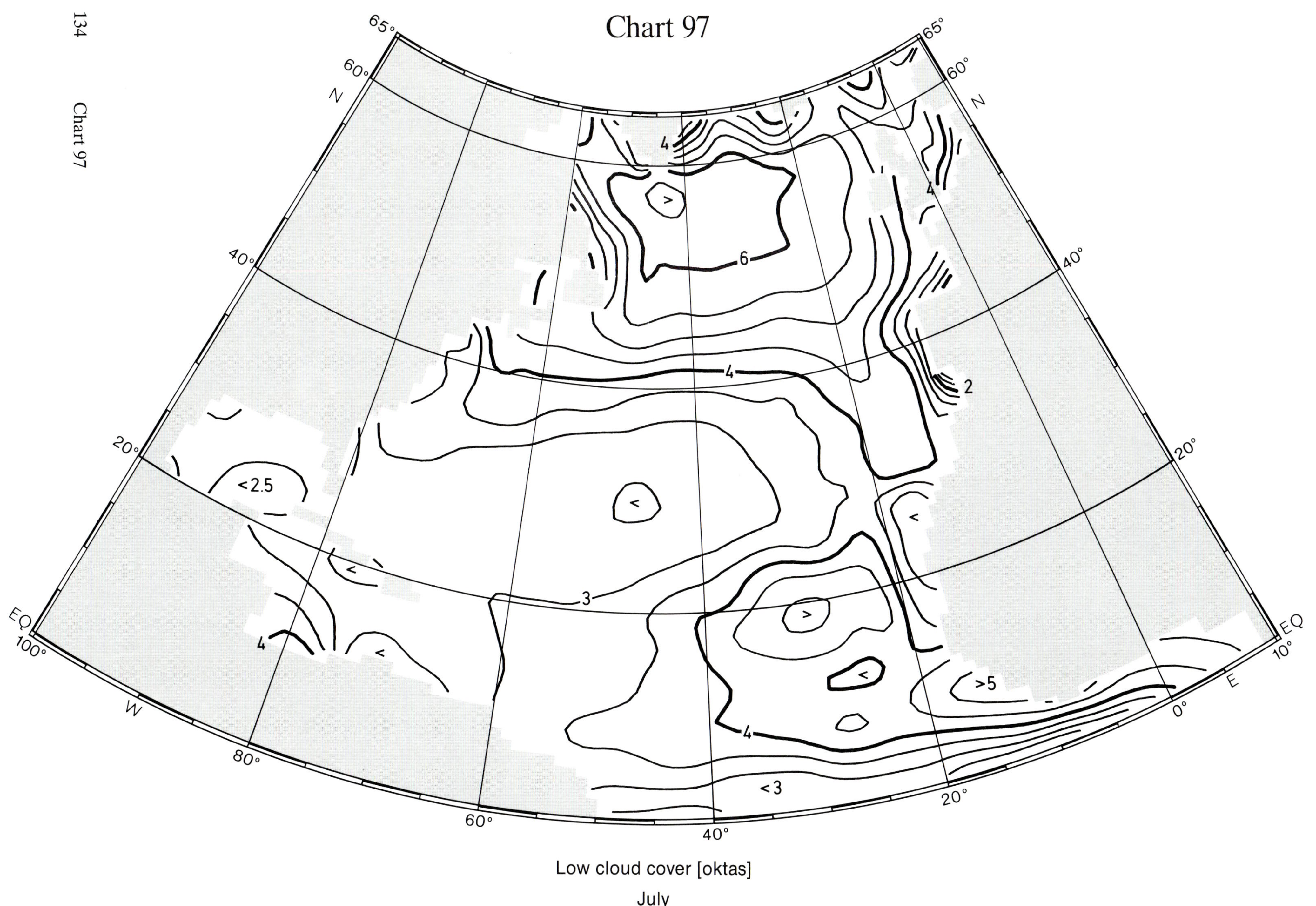

Low cloud cover [oktas]
July

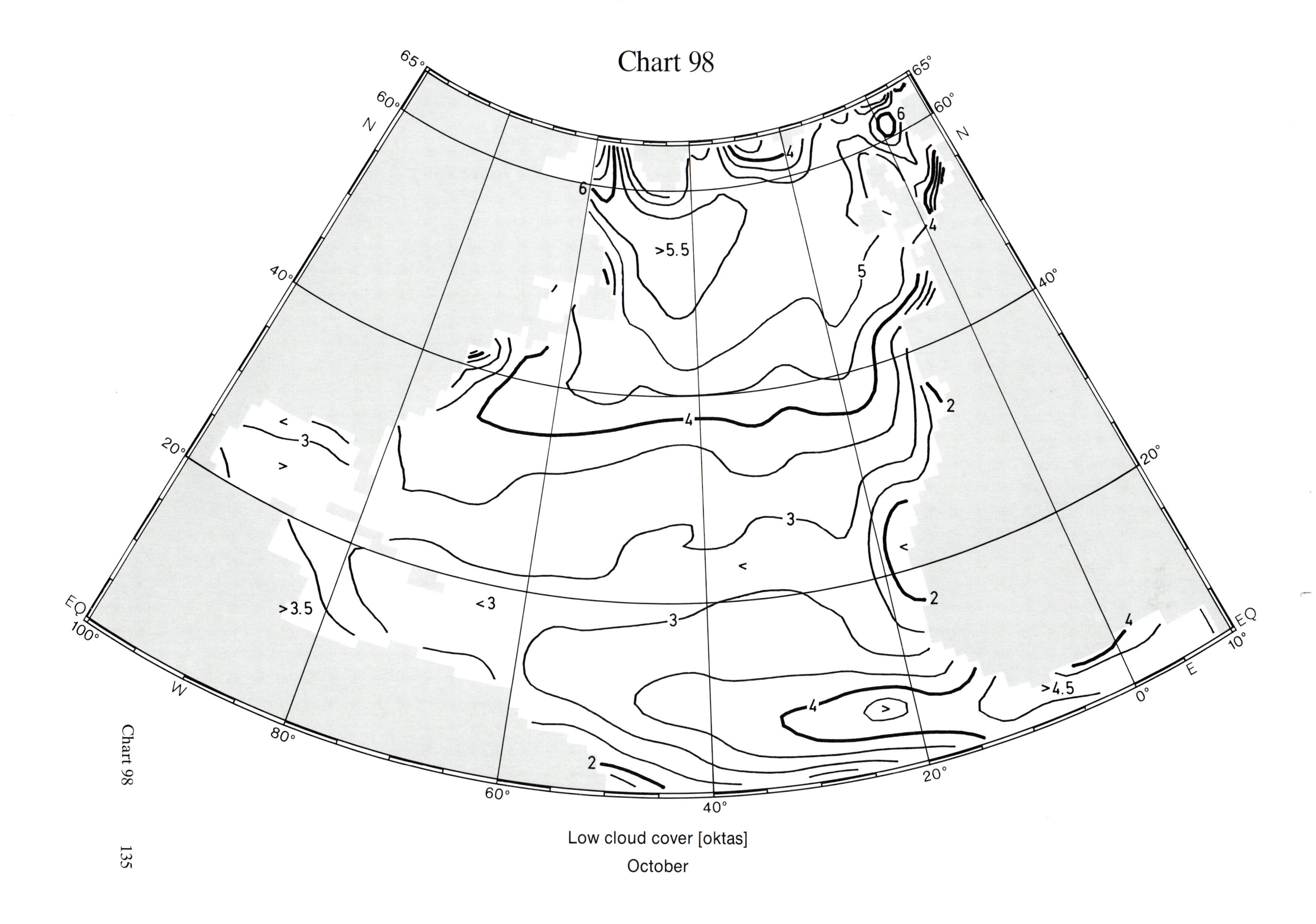
Chart 98
Low cloud cover [oktas]
October
65°
60°
40°
20°
EQ
10°
100°
0°
20°
40°
60°
80°
N
E
W
>5.5
>4.5
>3.5
<3
6
5
4
3
2

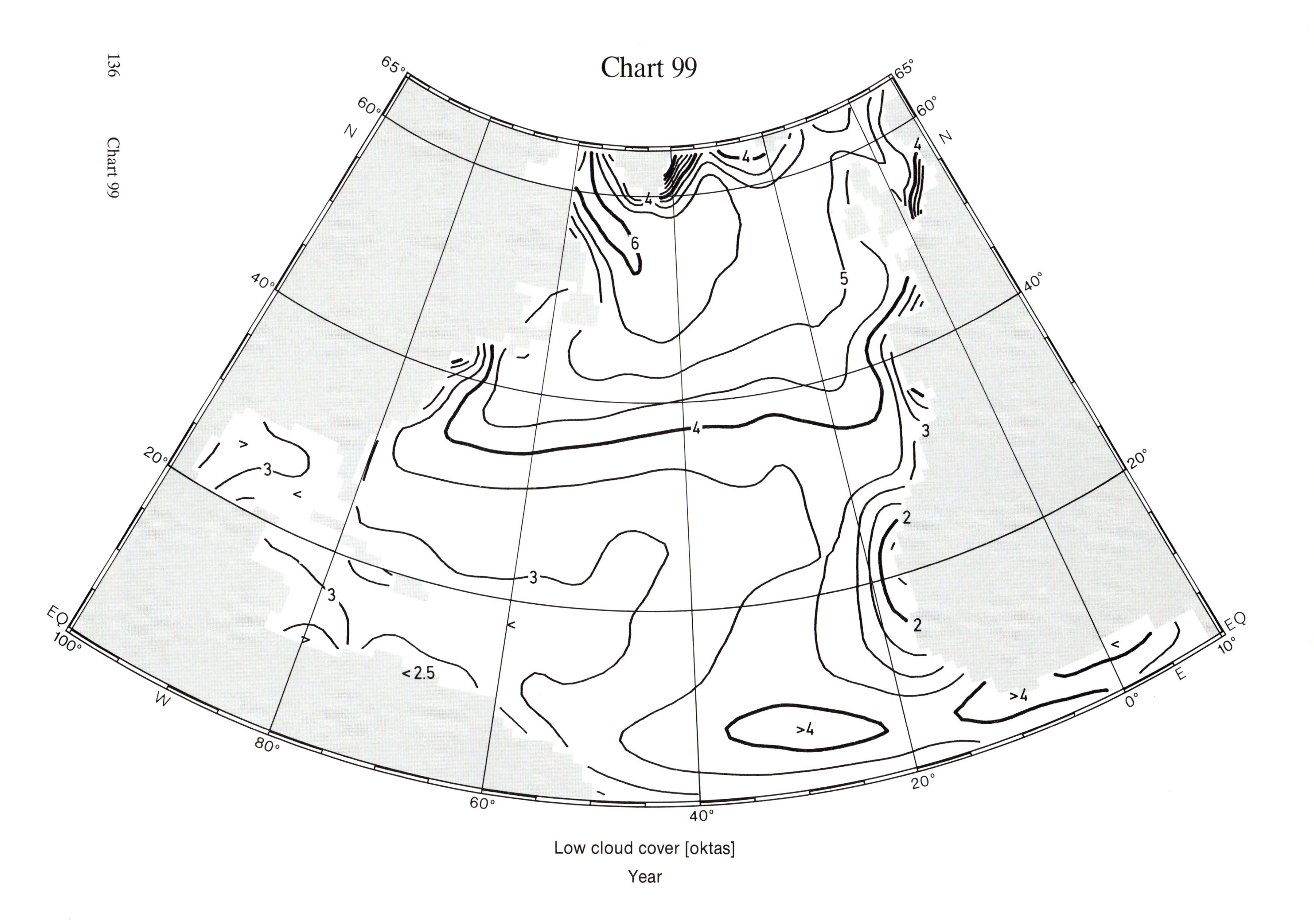
Chart 99
65°
60°
N
40°
20°
EQ
100°
W
80°
60°
40°
20°
0°
E
10°
6
5
4
3
2
>4
<2.5
Low cloud cover [oktas]
Year

Chart 100

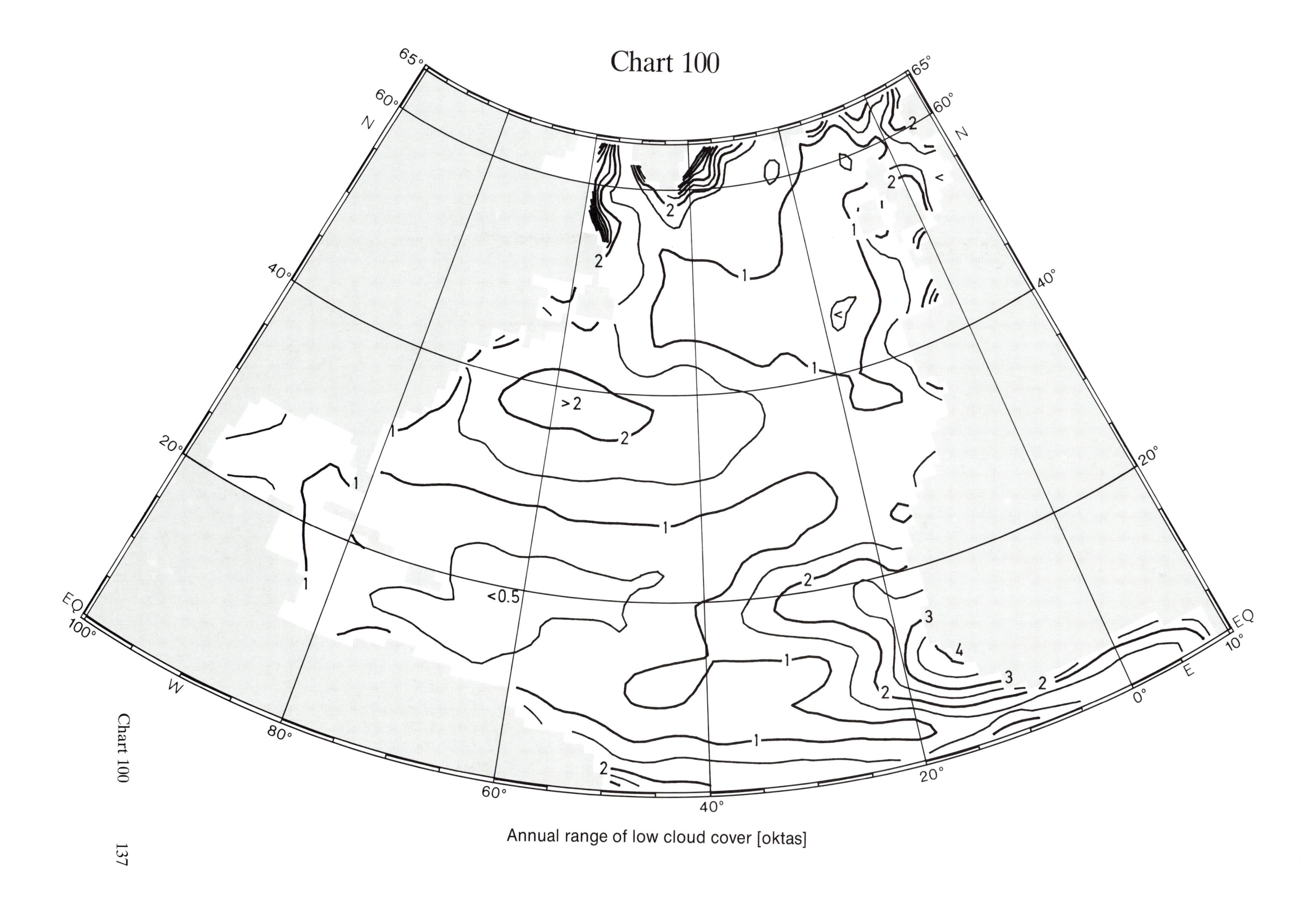

Annual range of low cloud cover [oktas]

Chart 101

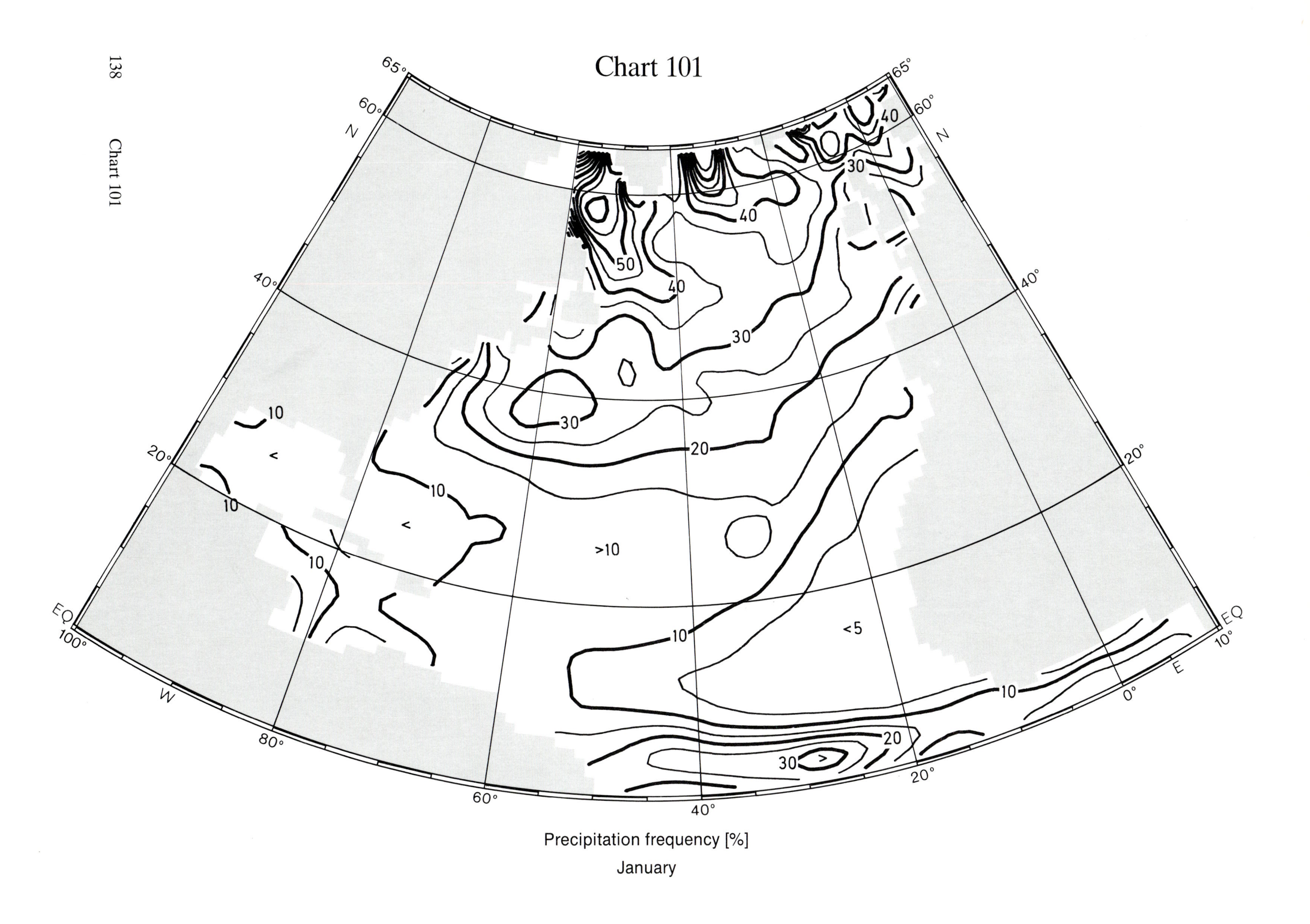

Precipitation frequency [%]
January

Chart 102

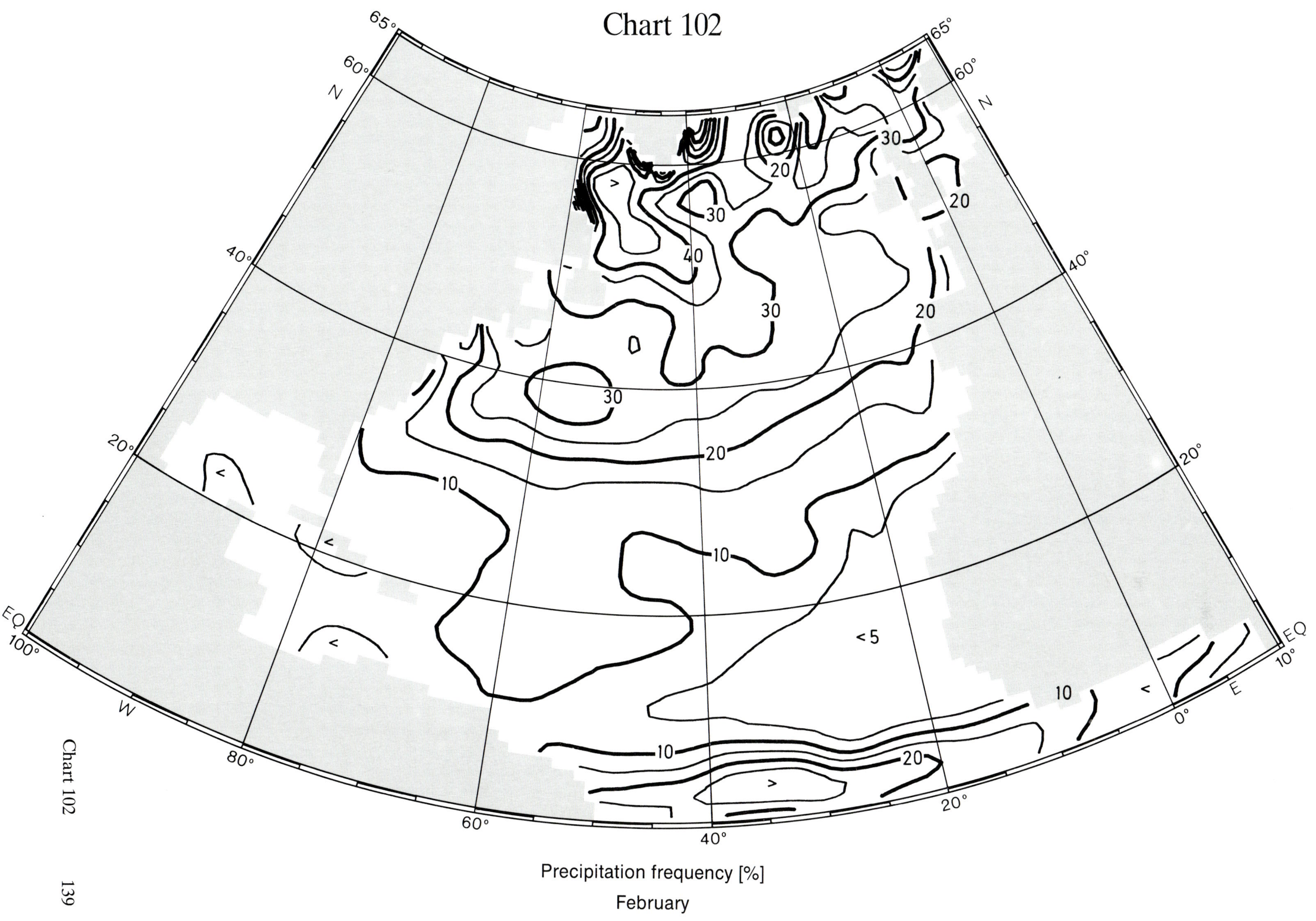

65°
60°
N
40°
20°
EQ
100°
W
80°
60°
40°
20°
0°
E
10°
30
20
40
10
<5
Precipitation frequency [%]
February

Chart 103

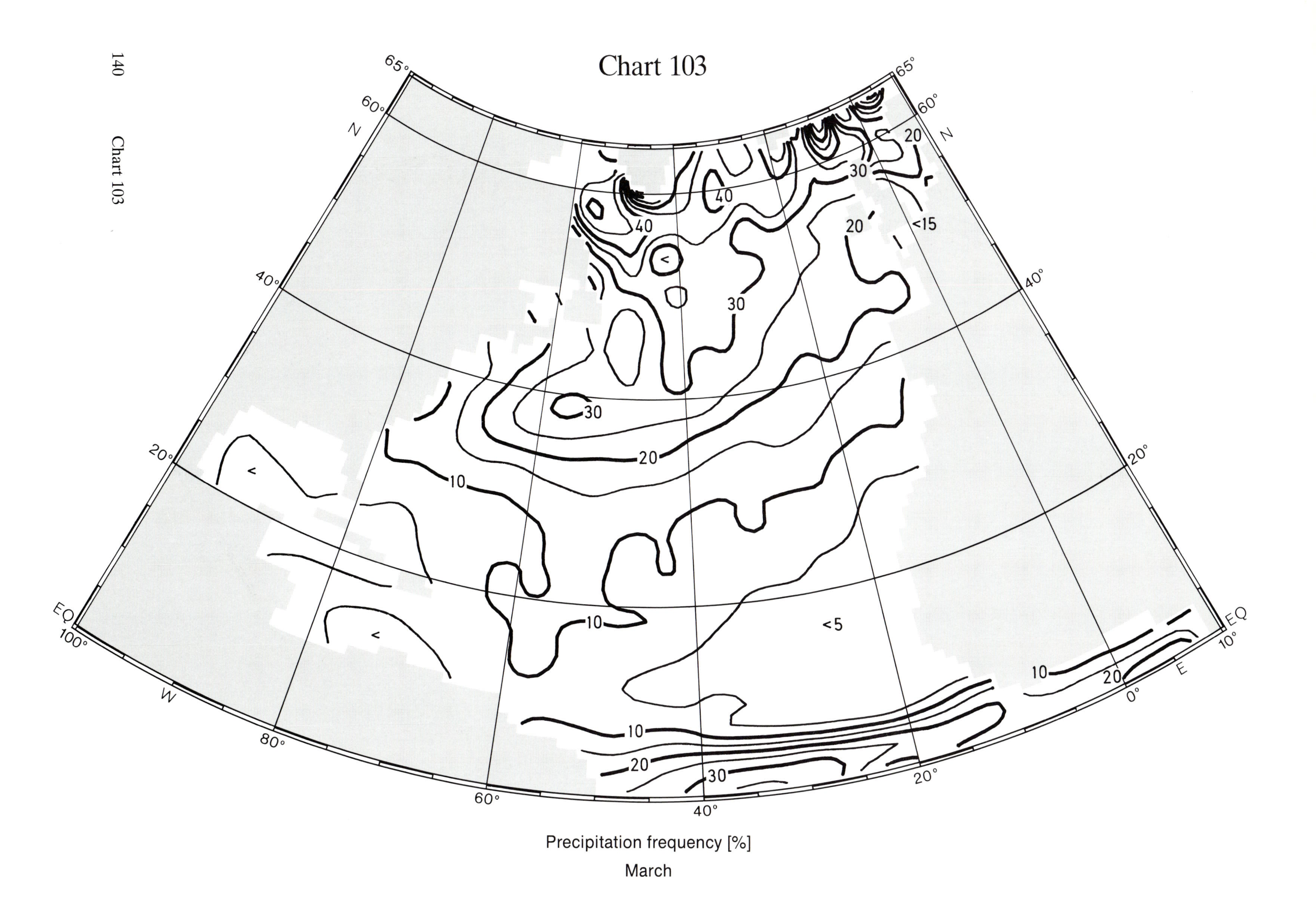

Precipitation frequency [%]

March

Chart 104

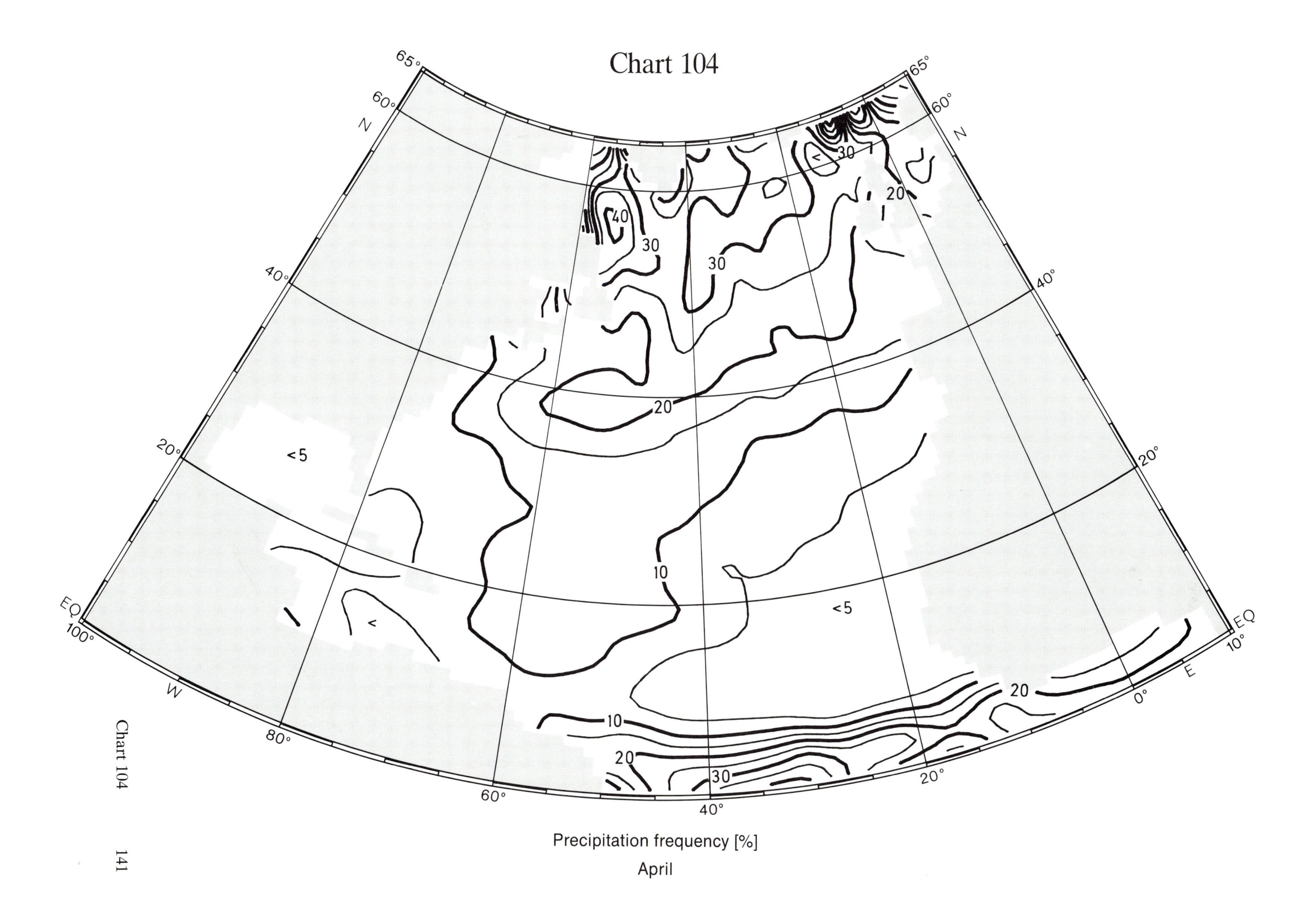

Precipitation frequency [%]
April

Chart 105

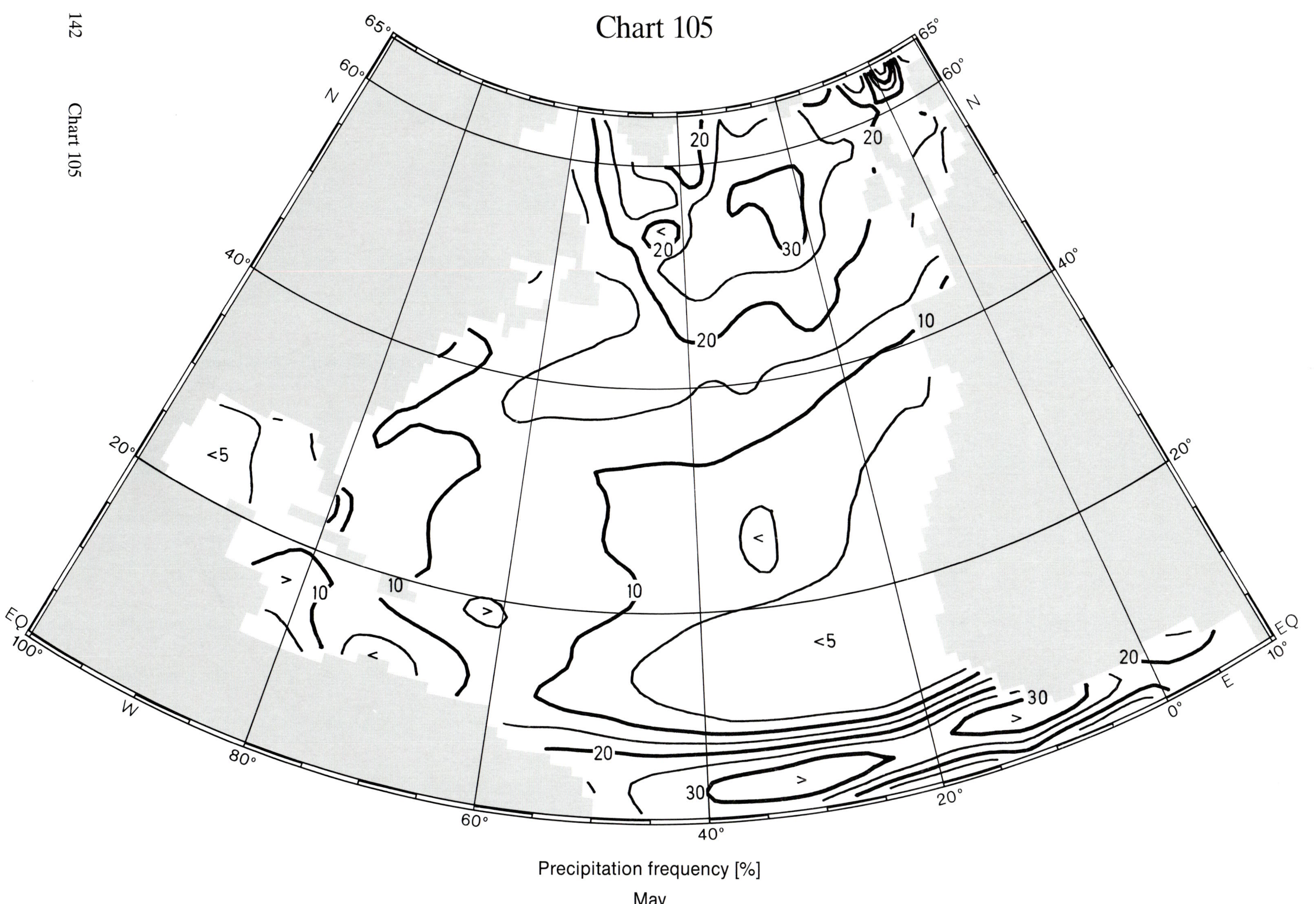

Precipitation frequency [%]
May

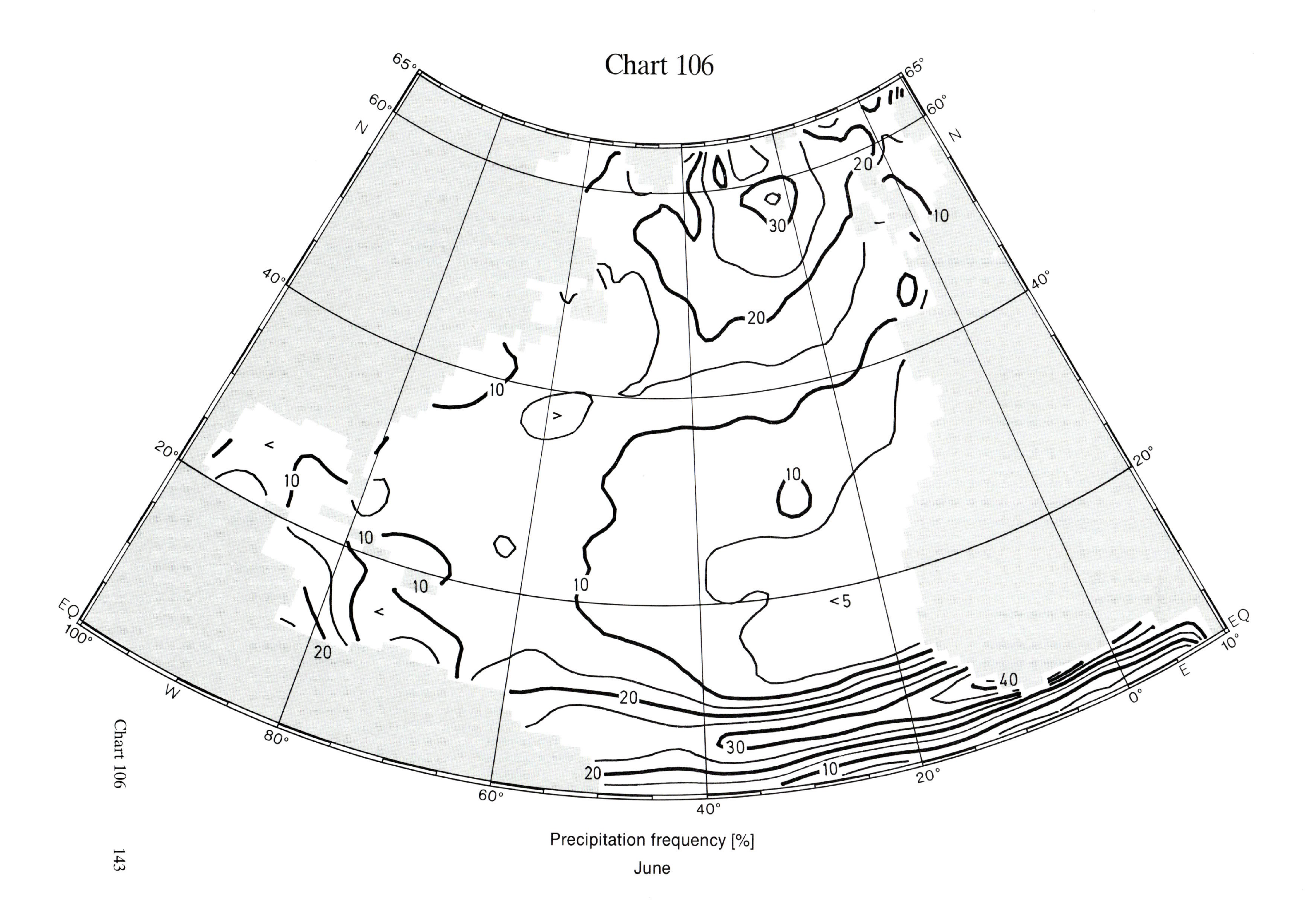
Chart 106
65°
60°
N
40°
20°
EQ
100°
W
80°
60°
40°
20°
0°
E
10°
10
20
30
40
<5
Precipitation frequency [%]
June

Chart 107

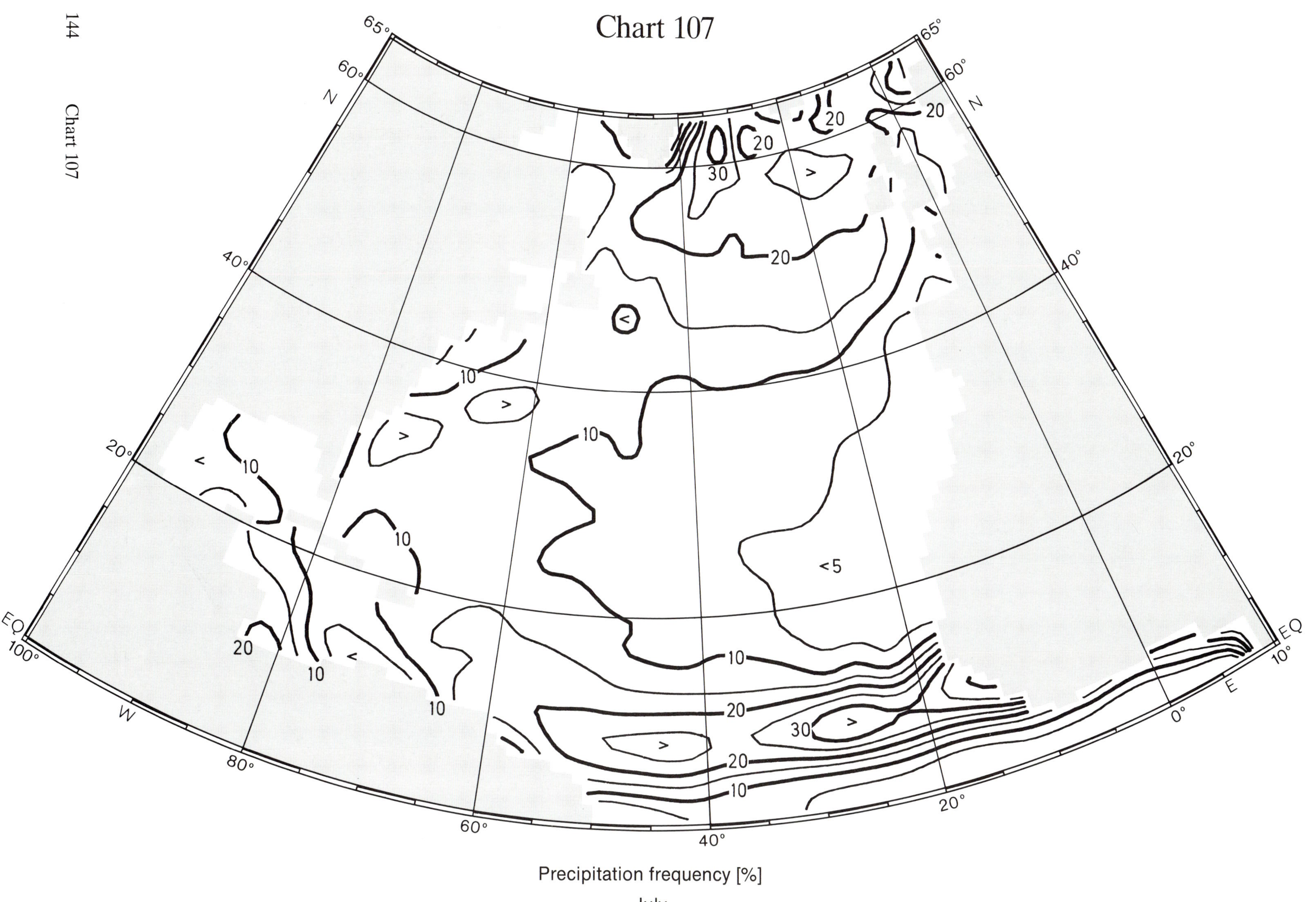

Precipitation frequency [%]

July

Chart 108

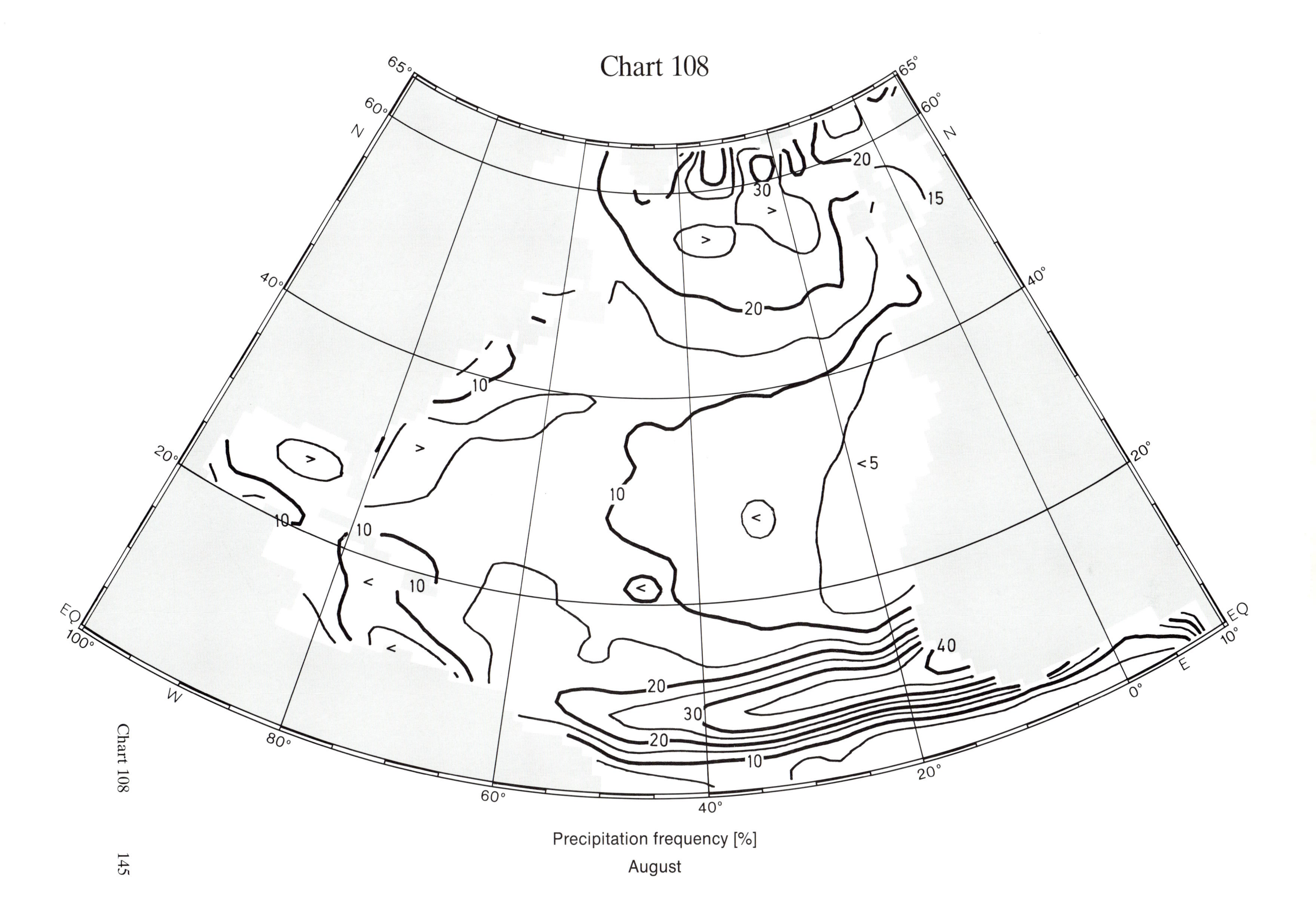

Precipitation frequency [%]
August

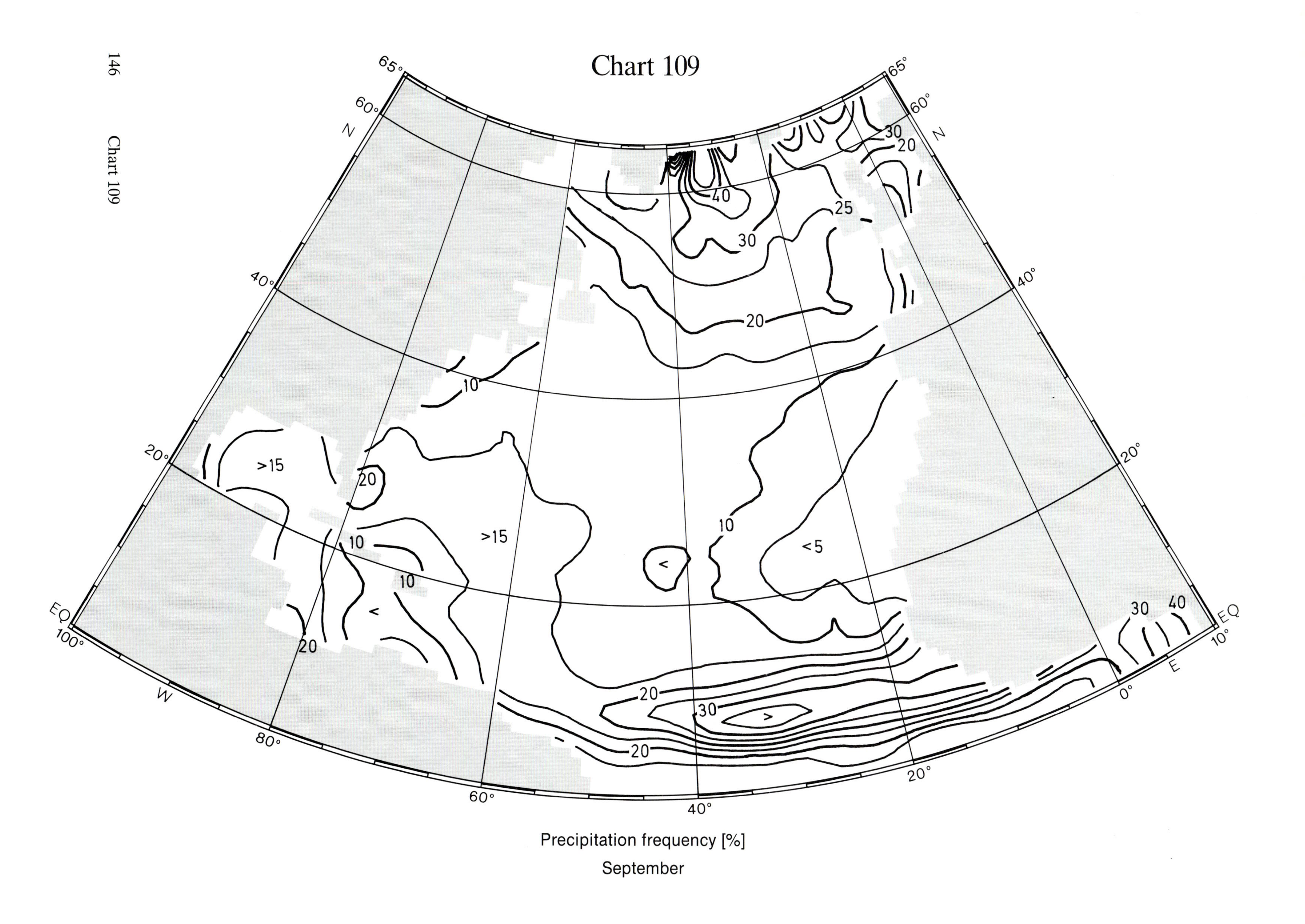
Chart 109
65°
60°
N
40°
20°
EQ
100°
W
80°
60°
40°
20°
0°
E
10°
>15
<5
<
>
10
20
25
30
40
Precipitation frequency [%]
September

Chart 110

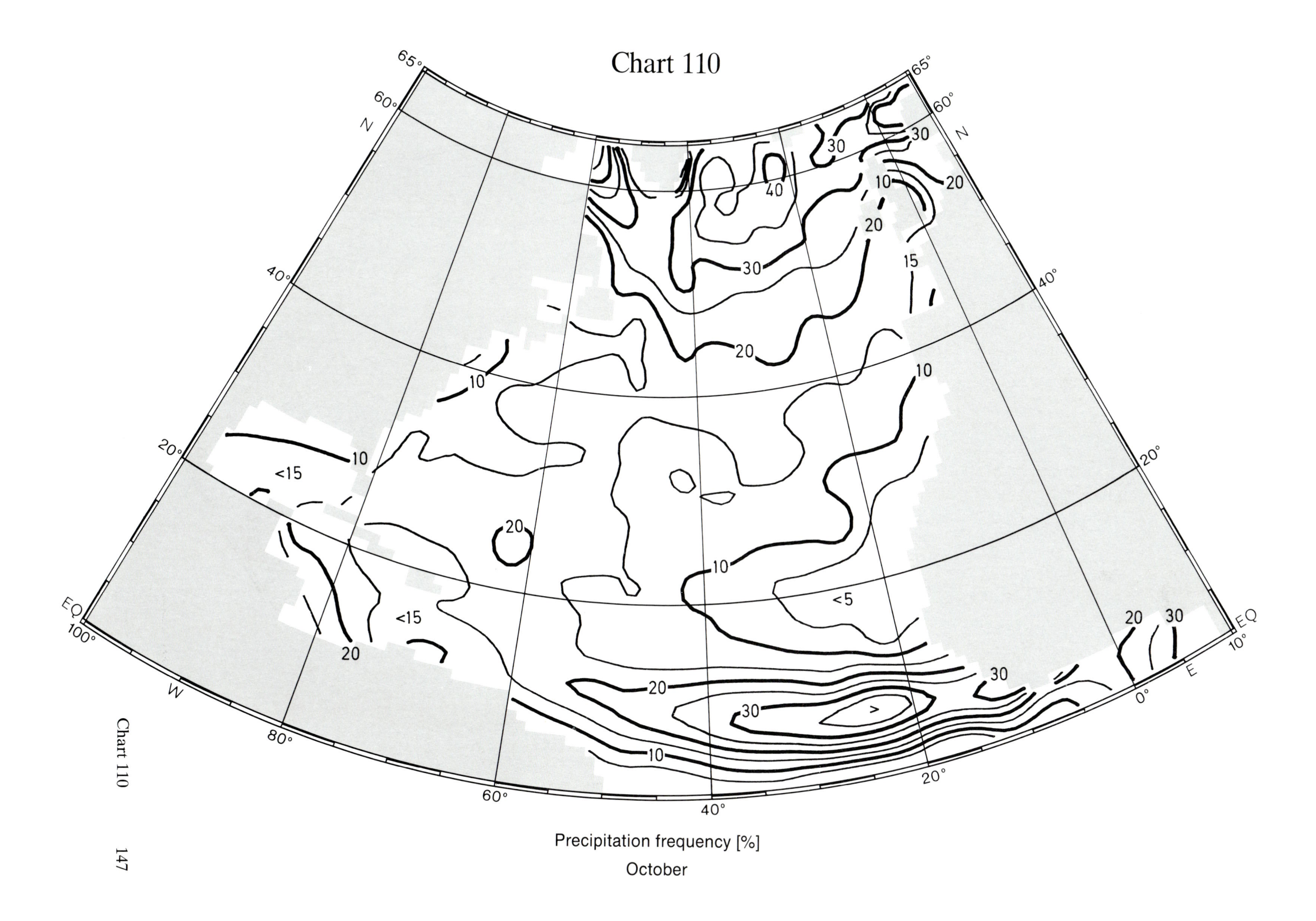

Precipitation frequency [%]
October

Chart 111

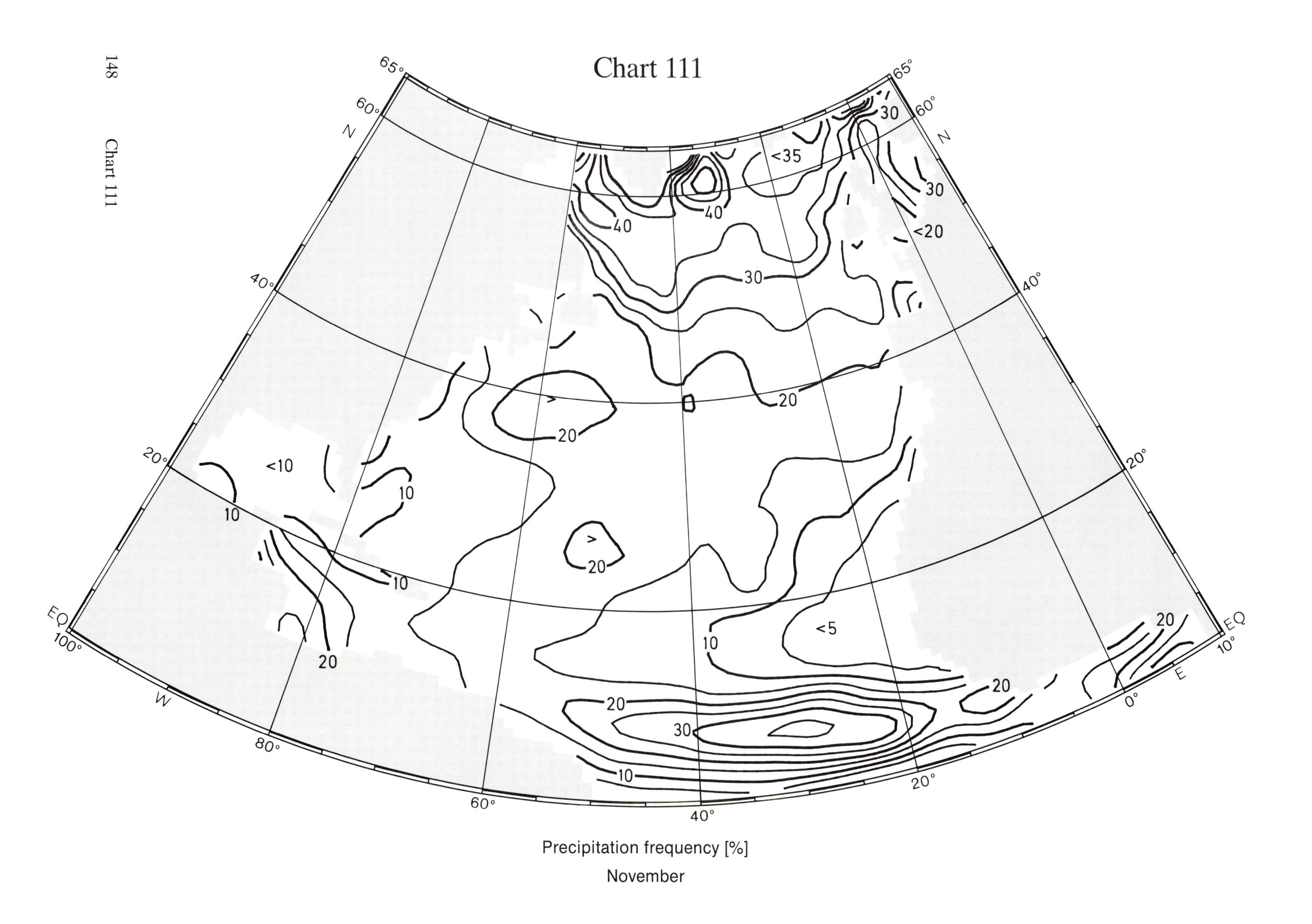

Precipitation frequency [%]
November

Chart 112

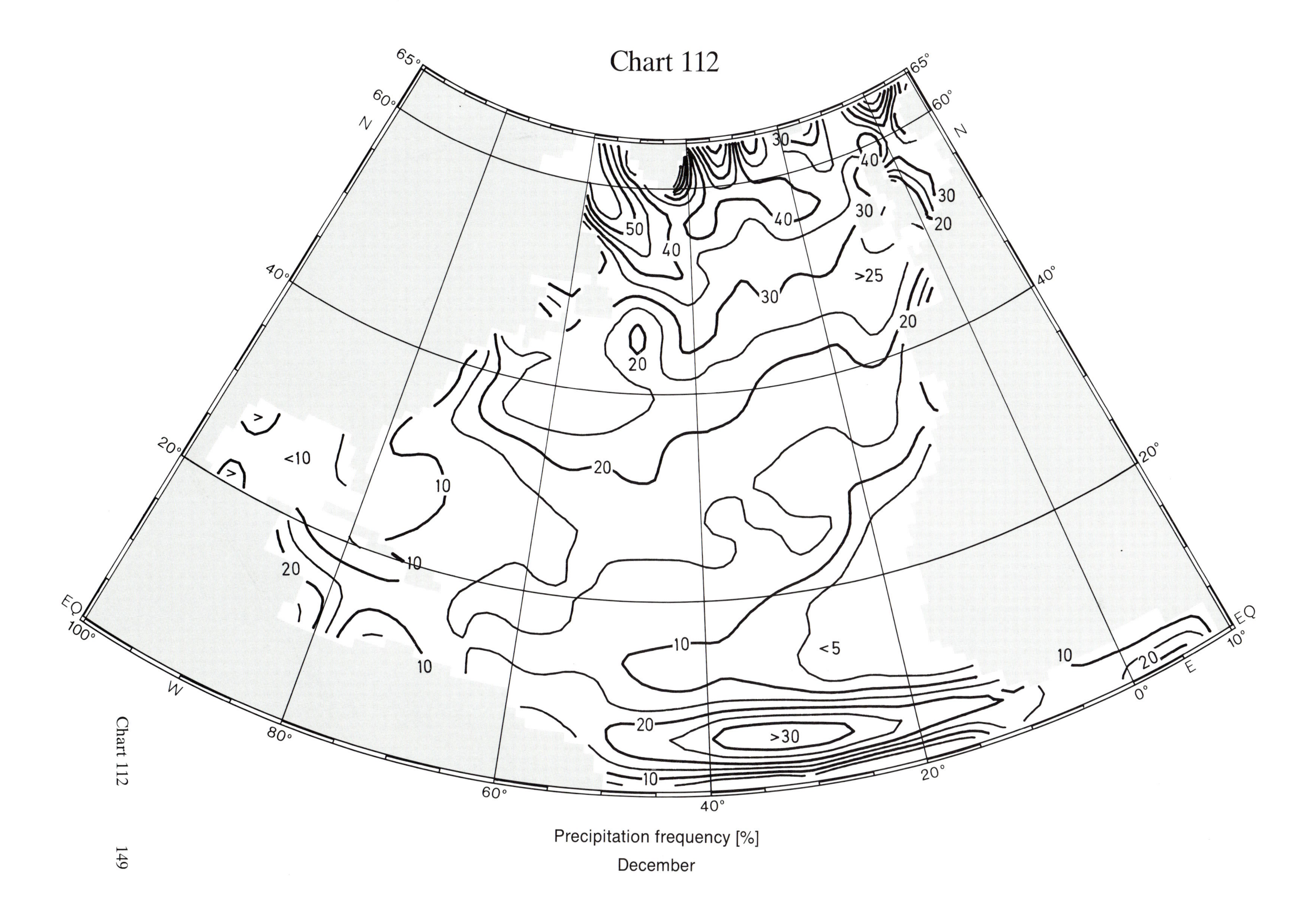

Precipitation frequency [%]
December

Chart 113

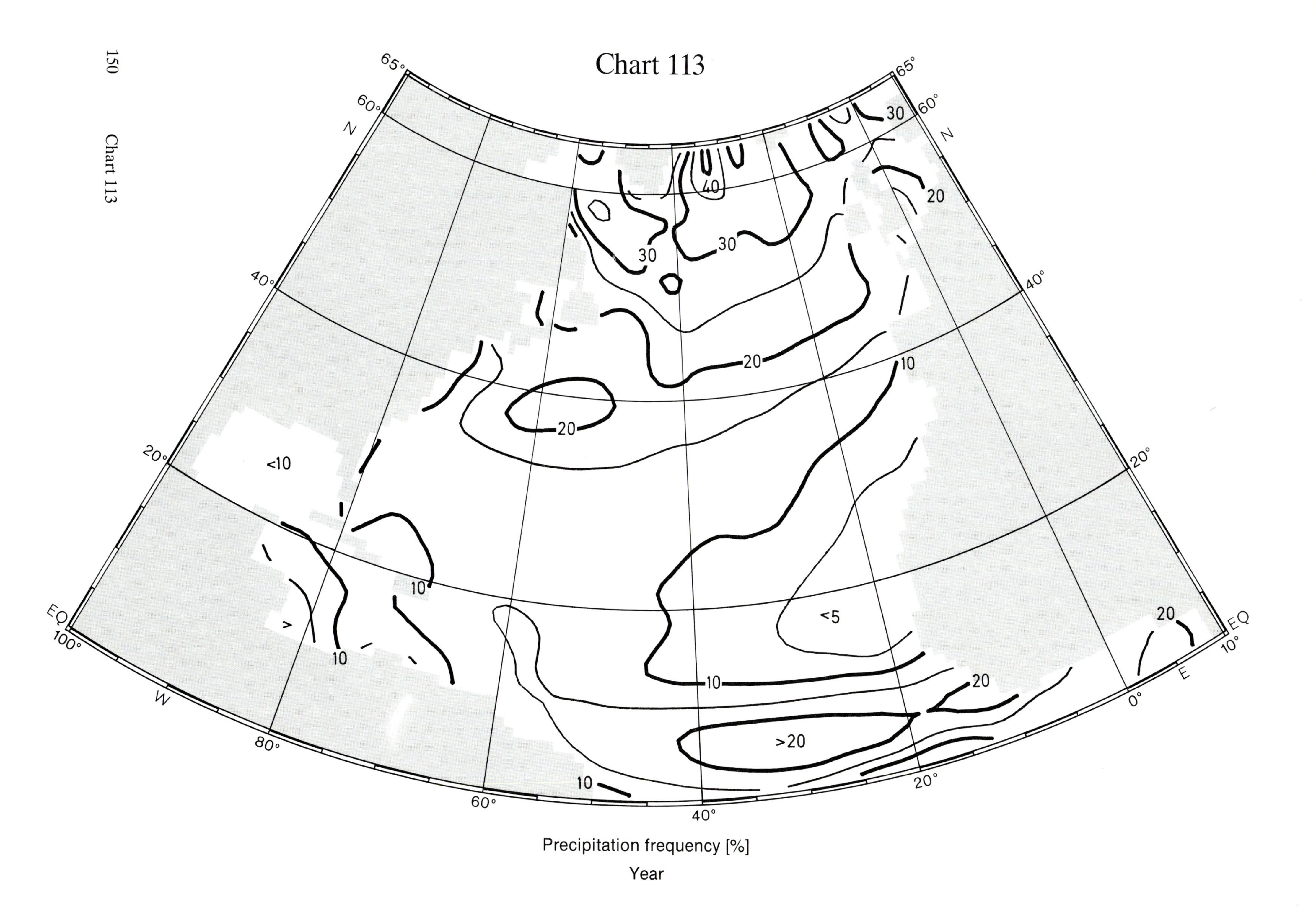

Precipitation frequency [%]

Year

Chart 114

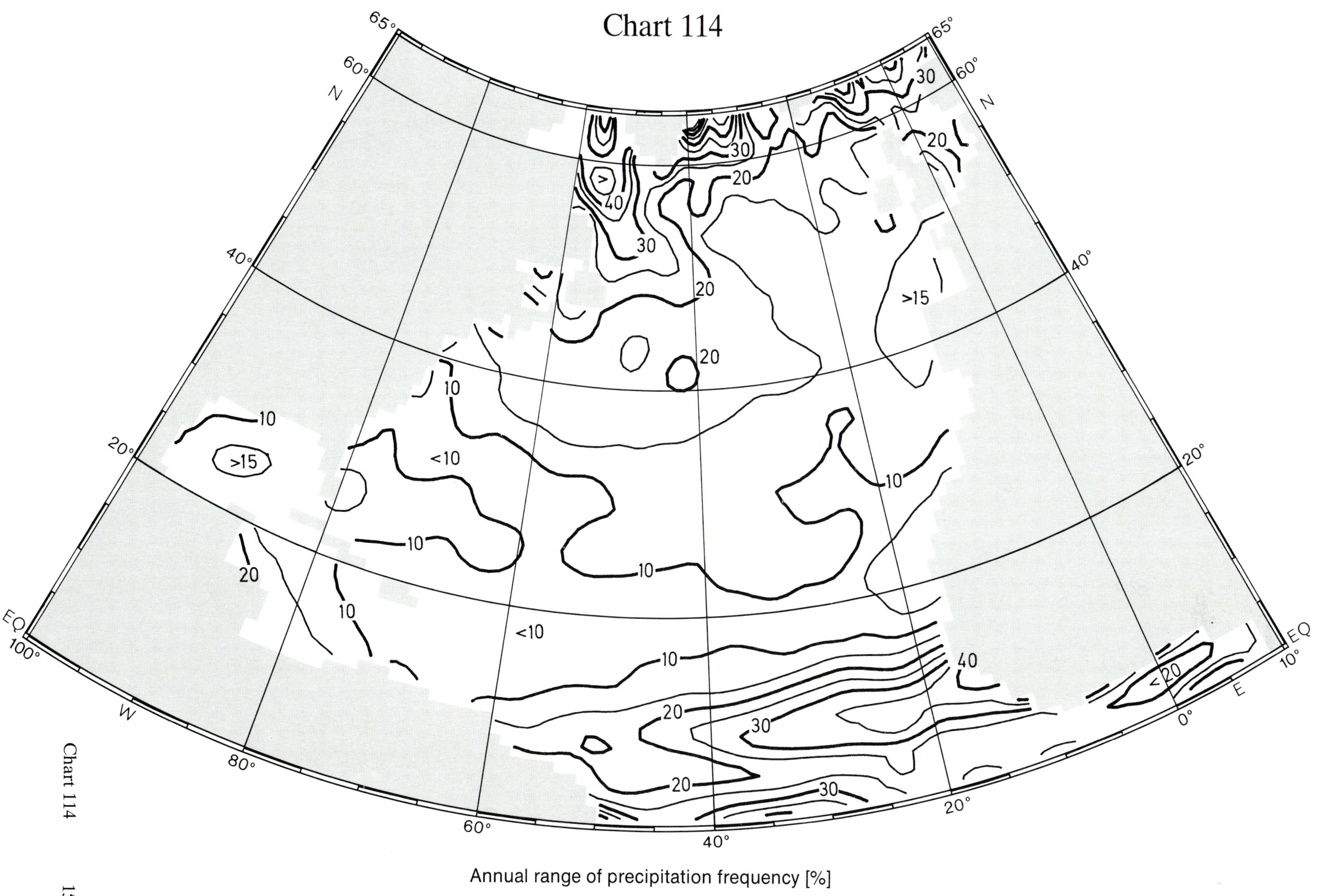

Annual range of precipitation frequency [%]

Chart 115

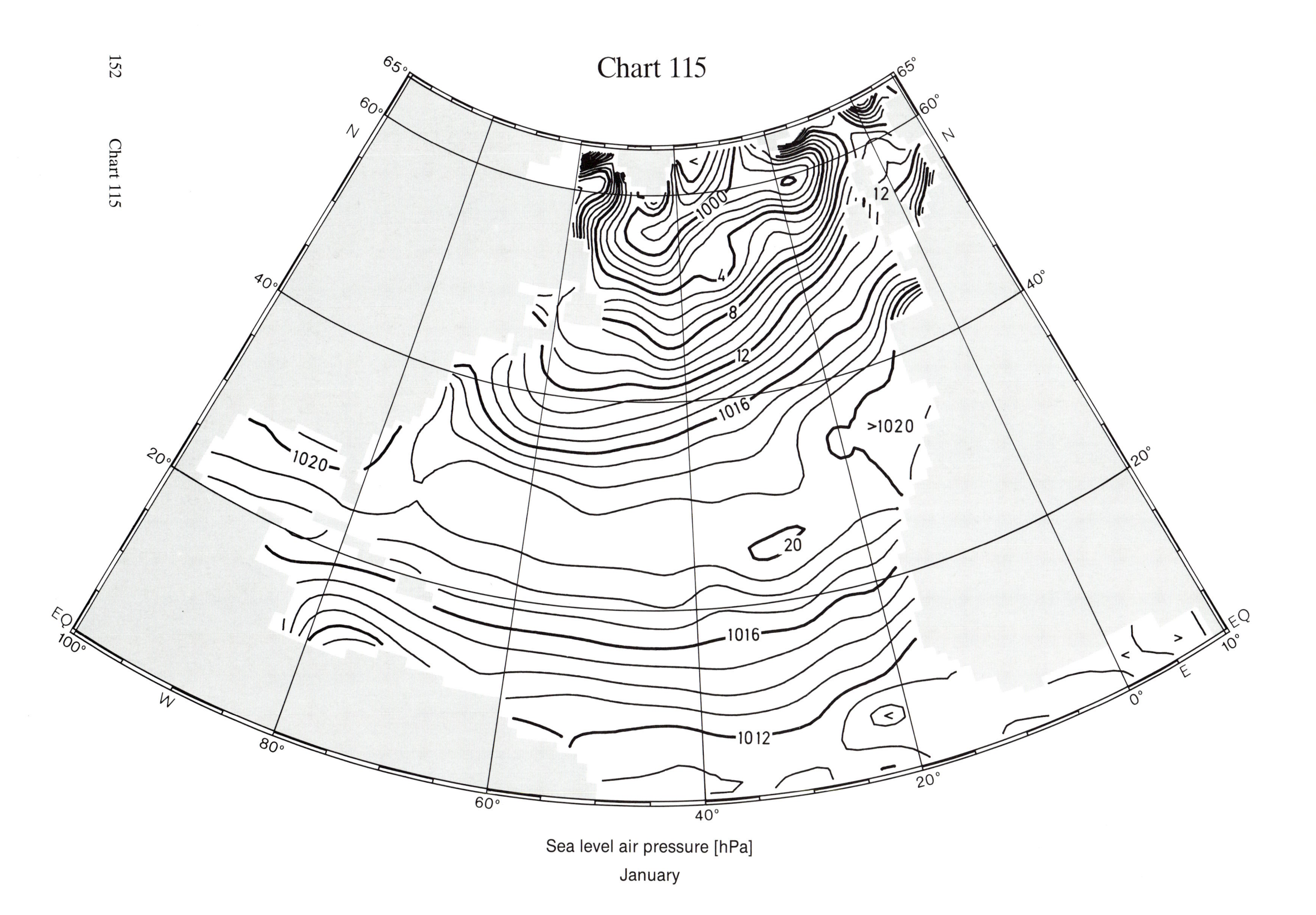

Sea level air pressure [hPa]

January

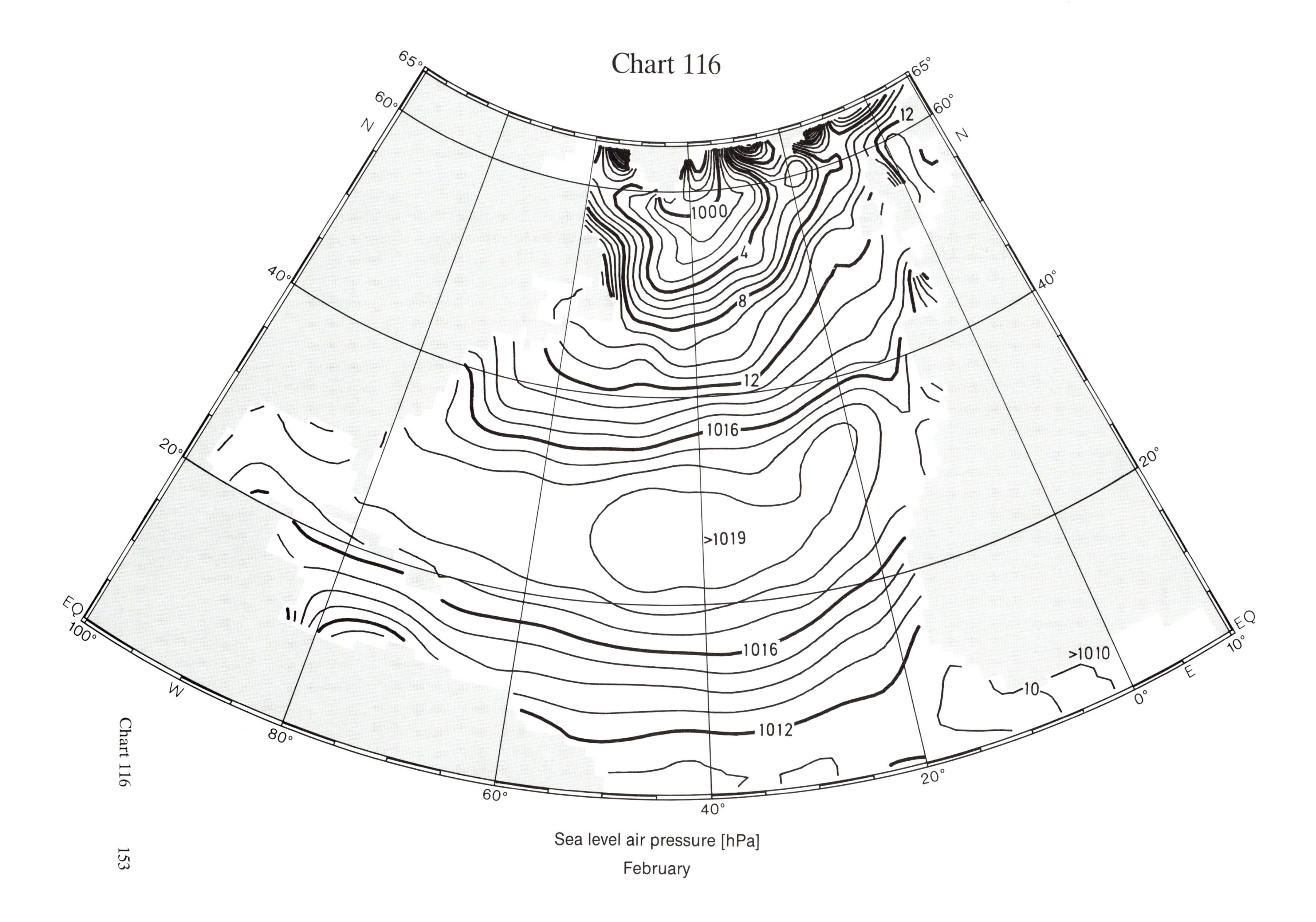
Chart 116
1000
4
8
12
1016
>1019
1016
1012
>1010
10
12
65°
60°
N
40°
20°
EQ
100°
W
80°
60°
40°
20°
0°
E
10°
Sea level air pressure [hPa]
February

Chart 117

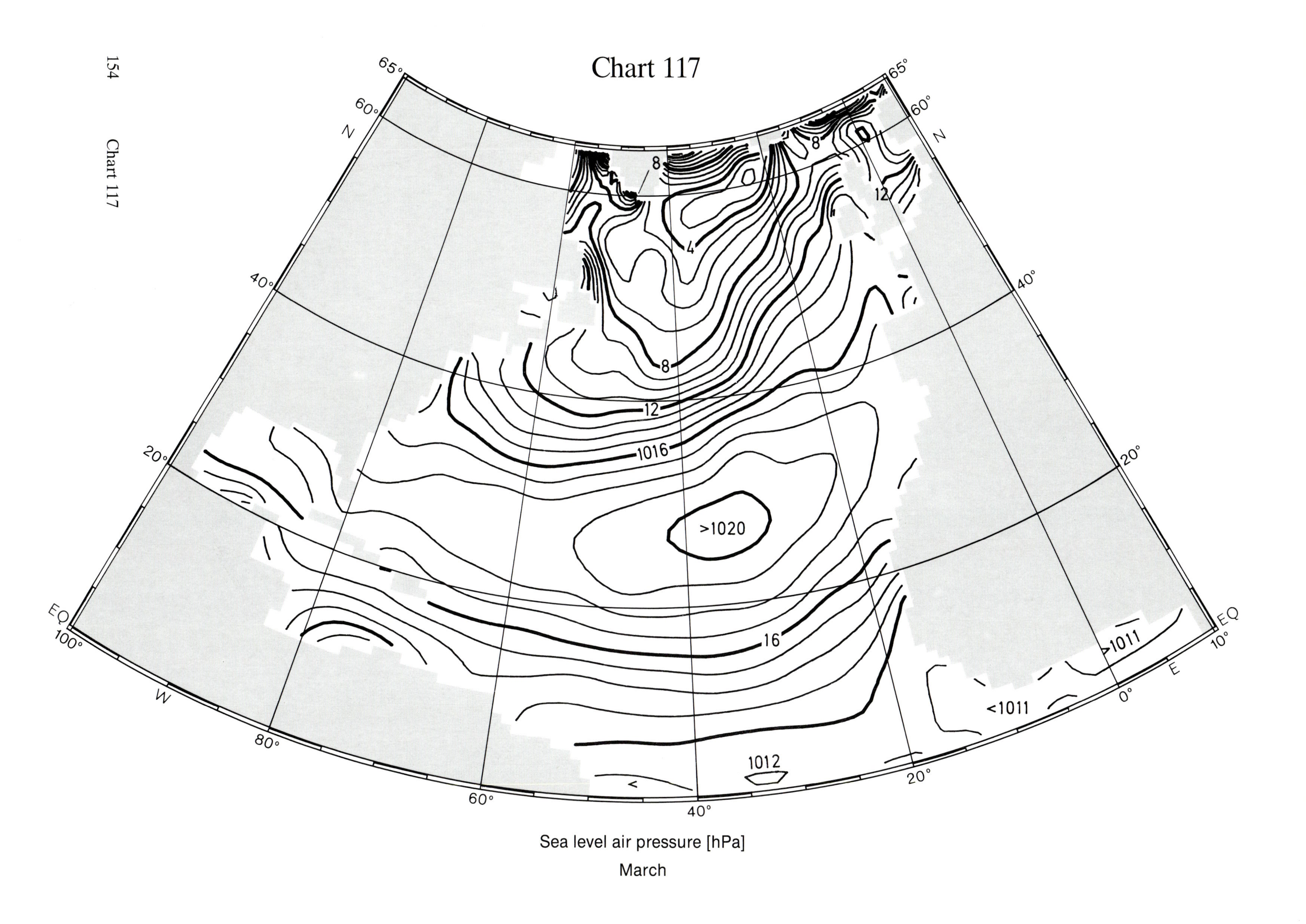

Sea level air pressure [hPa]

March

Chart 118

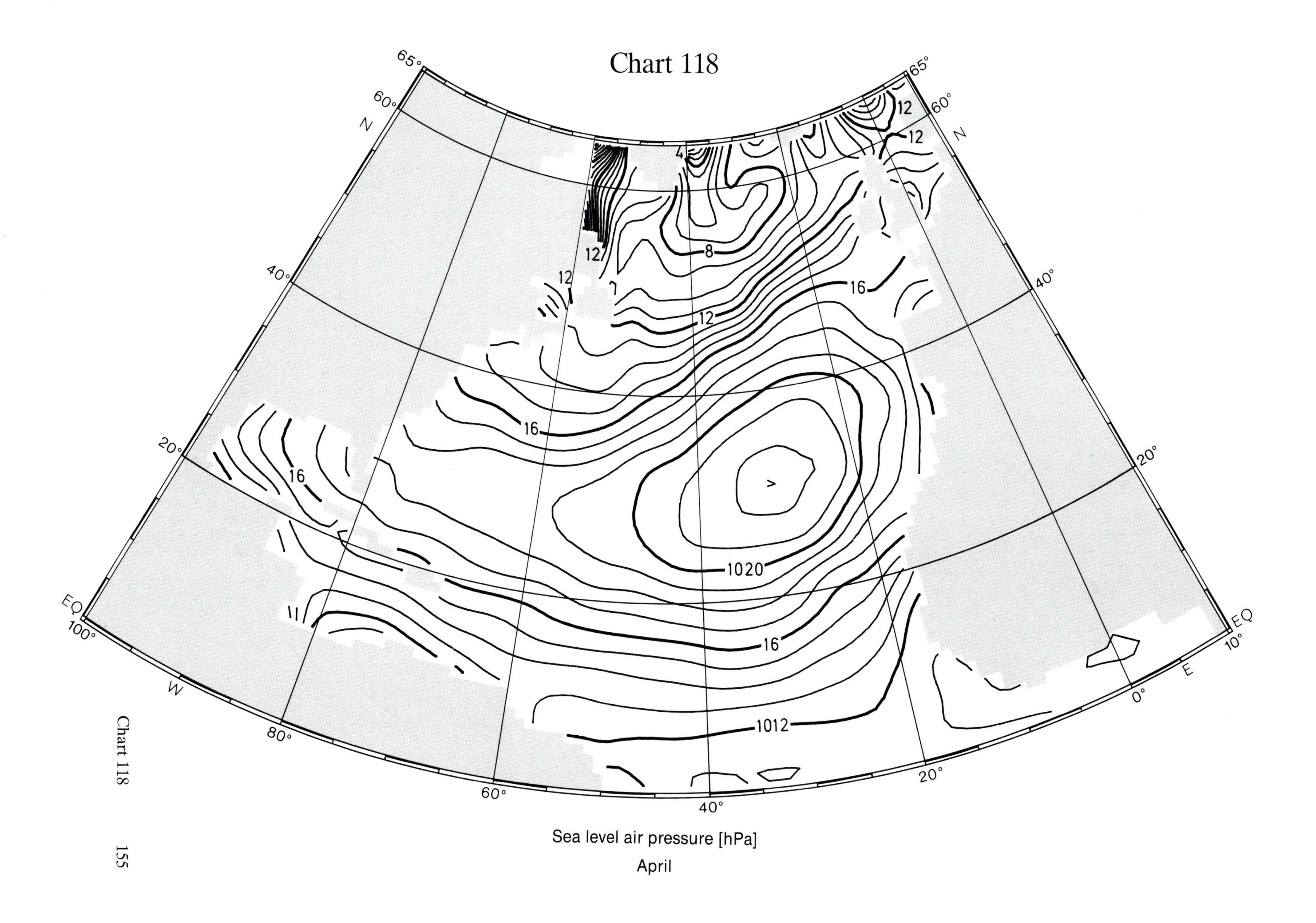

Sea level air pressure [hPa]
April

Chart 119

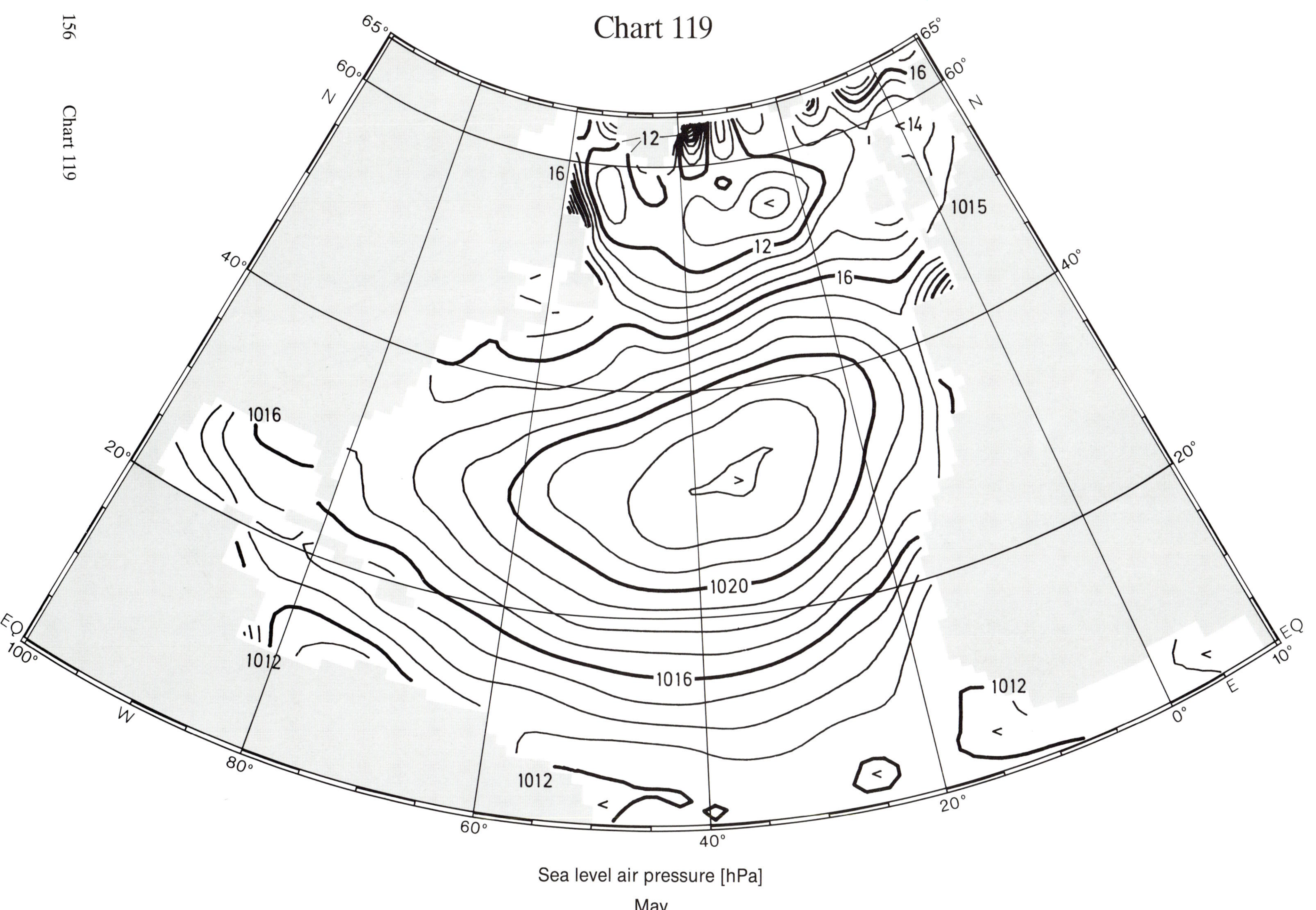

Sea level air pressure [hPa]

May

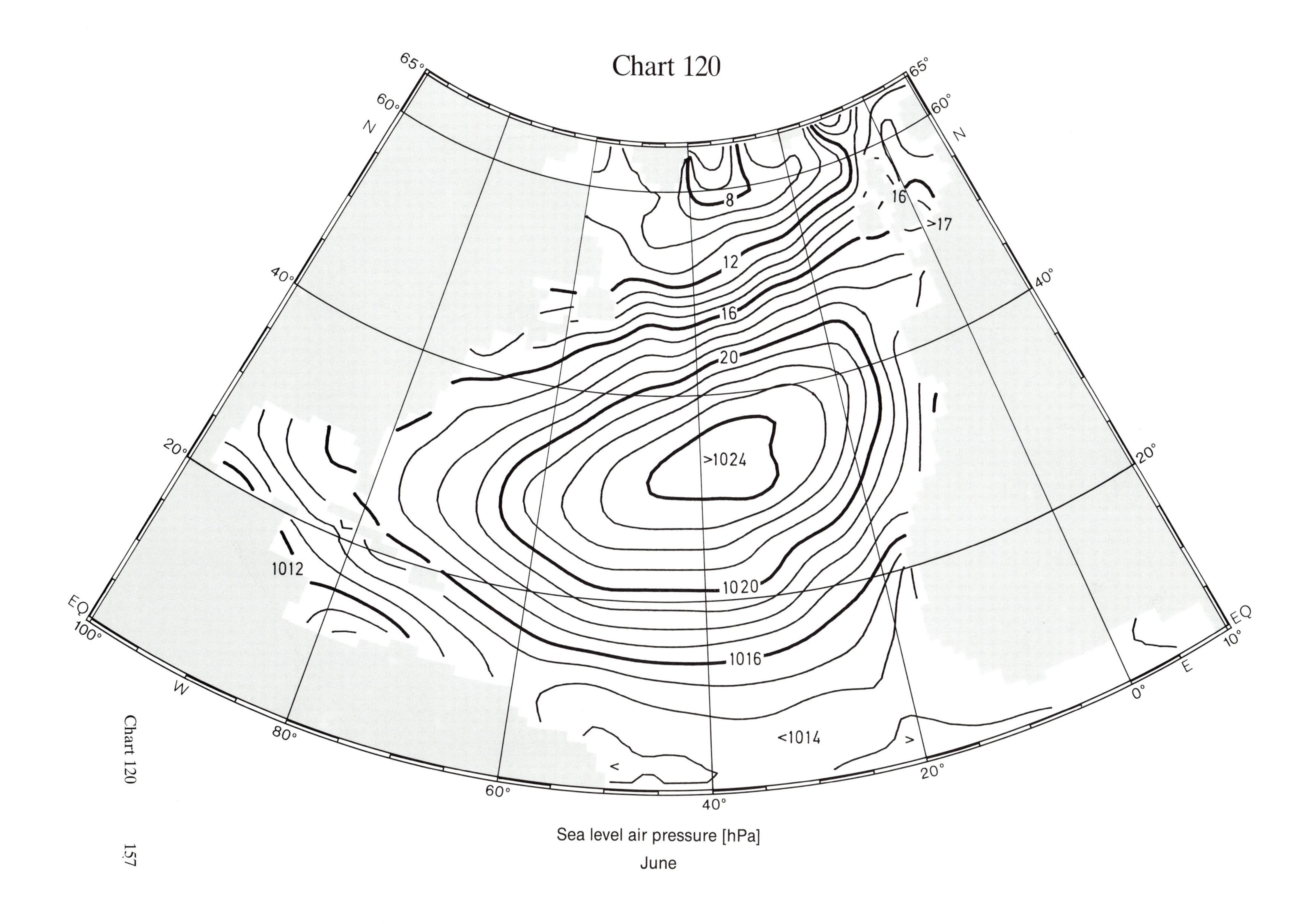
Chart 120
65°
60°
N
40°
20°
EQ
100°
W
80°
60°
40°
20°
0°
E
10°
8
16
>17
12
16
20
>1024
1012
1020
1016
<1014
Sea level air pressure [hPa]
June

Chart 121

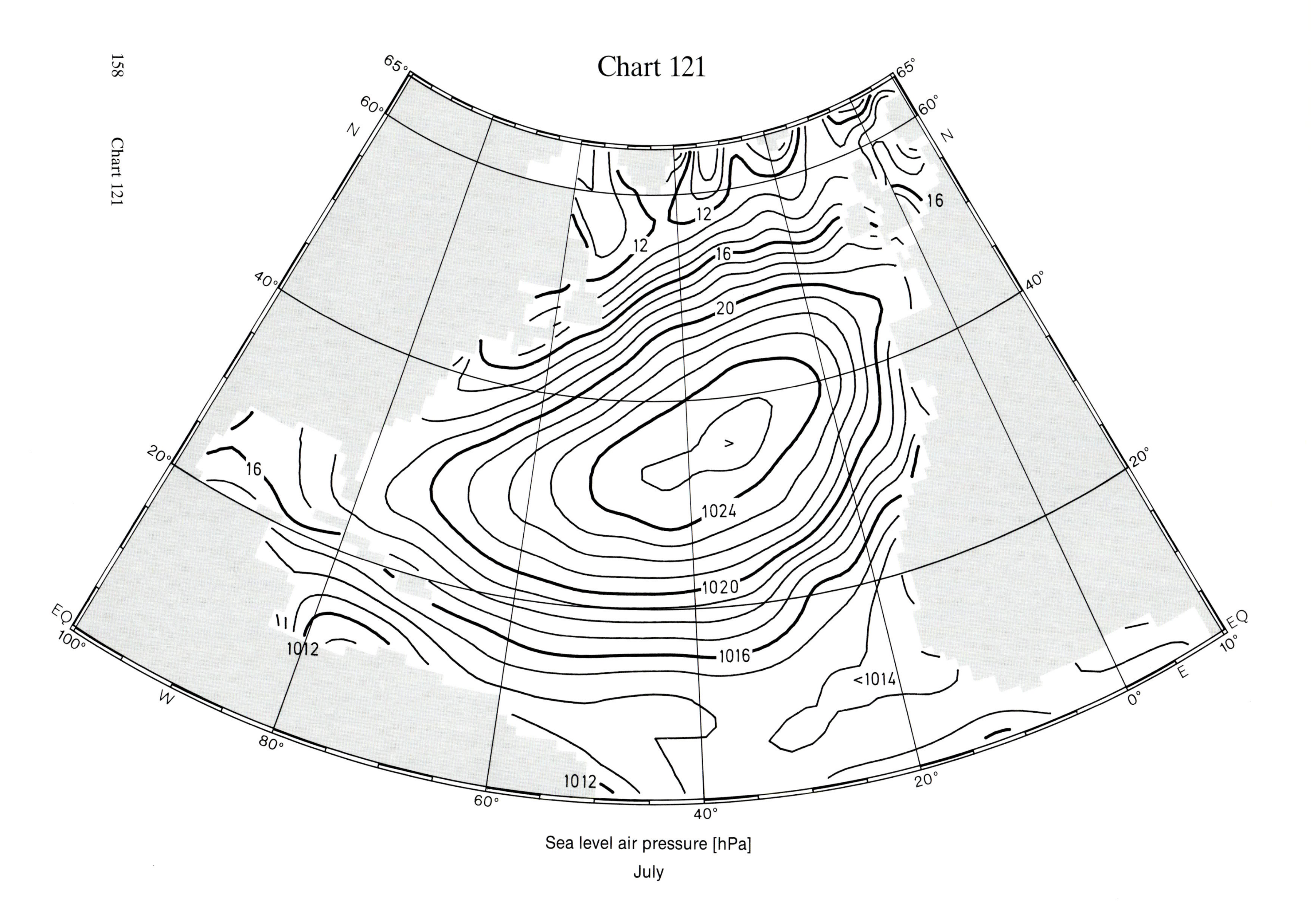

Sea level air pressure [hPa]

July

Chart 122

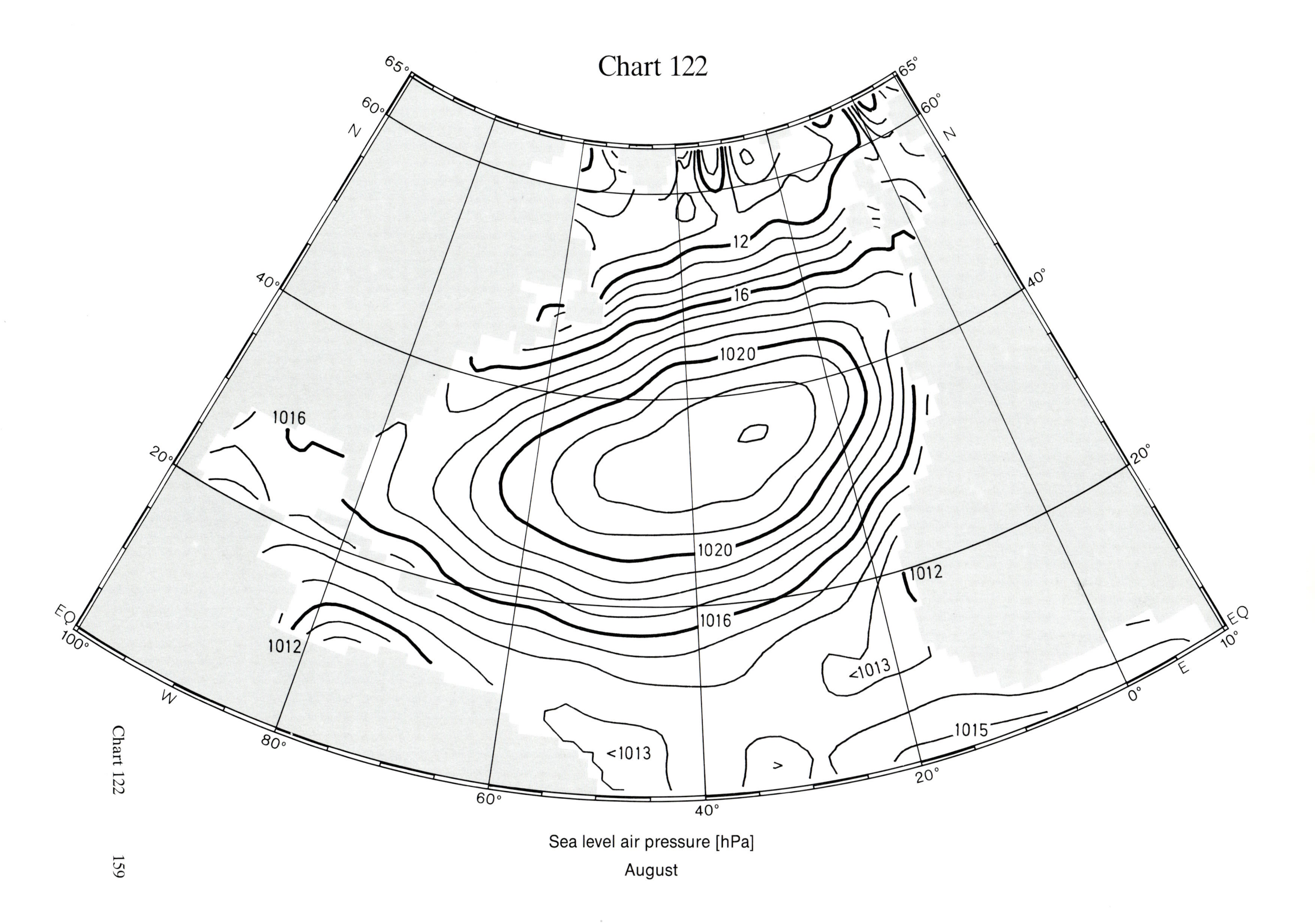

Sea level air pressure [hPa]
August

Chart 123

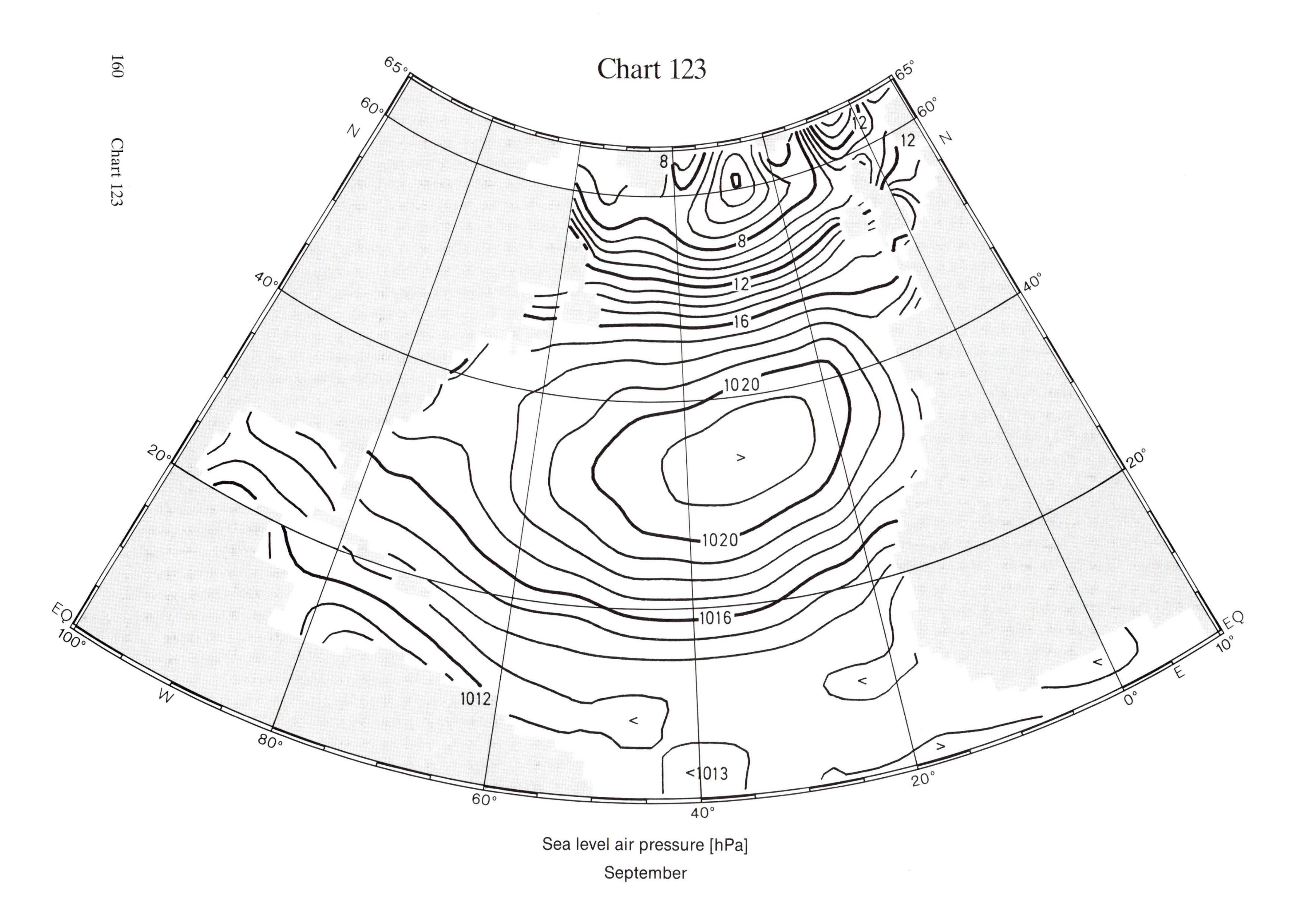

Sea level air pressure [hPa]
September

Chart 124

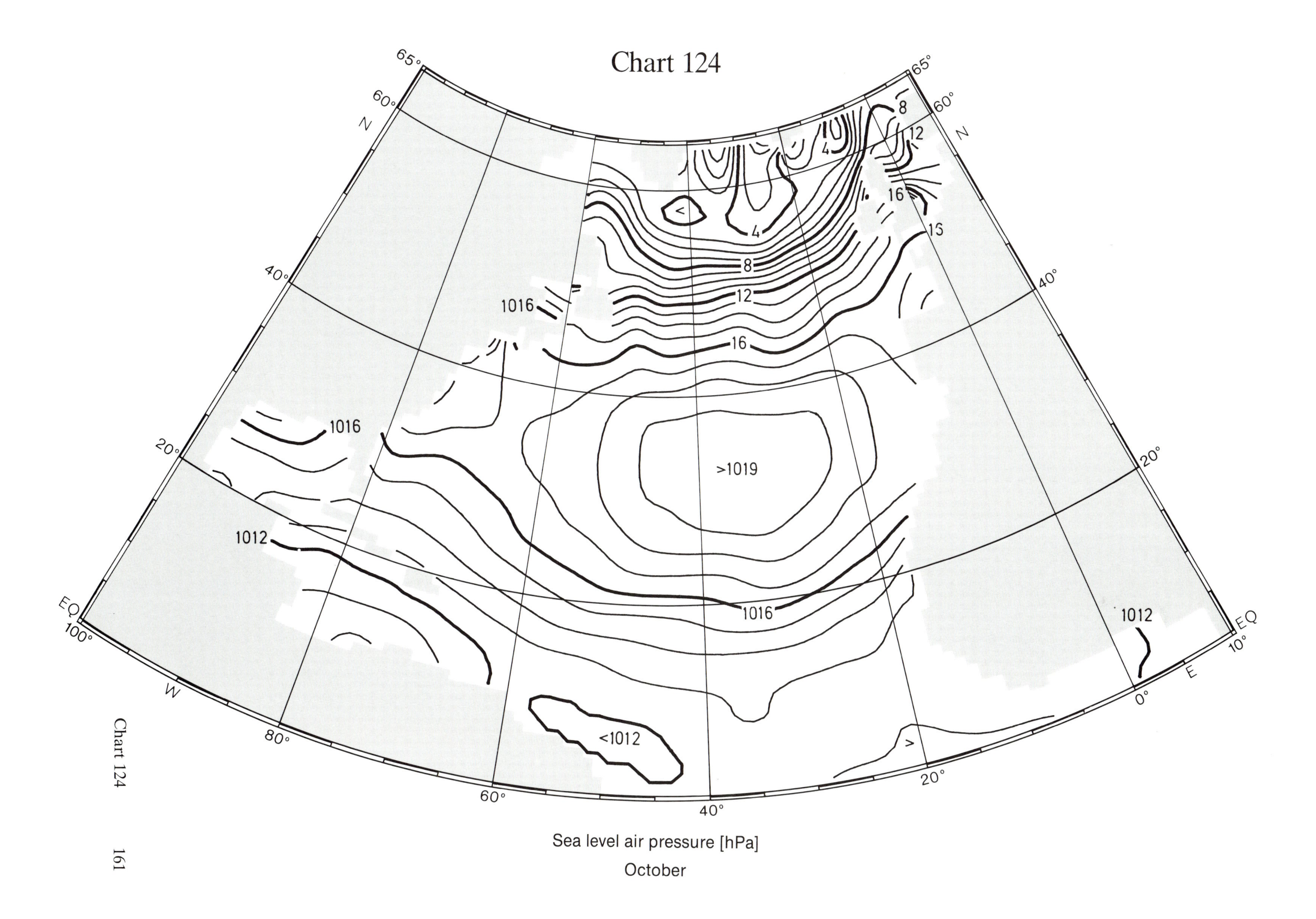

Sea level air pressure [hPa]
October

Chart 125

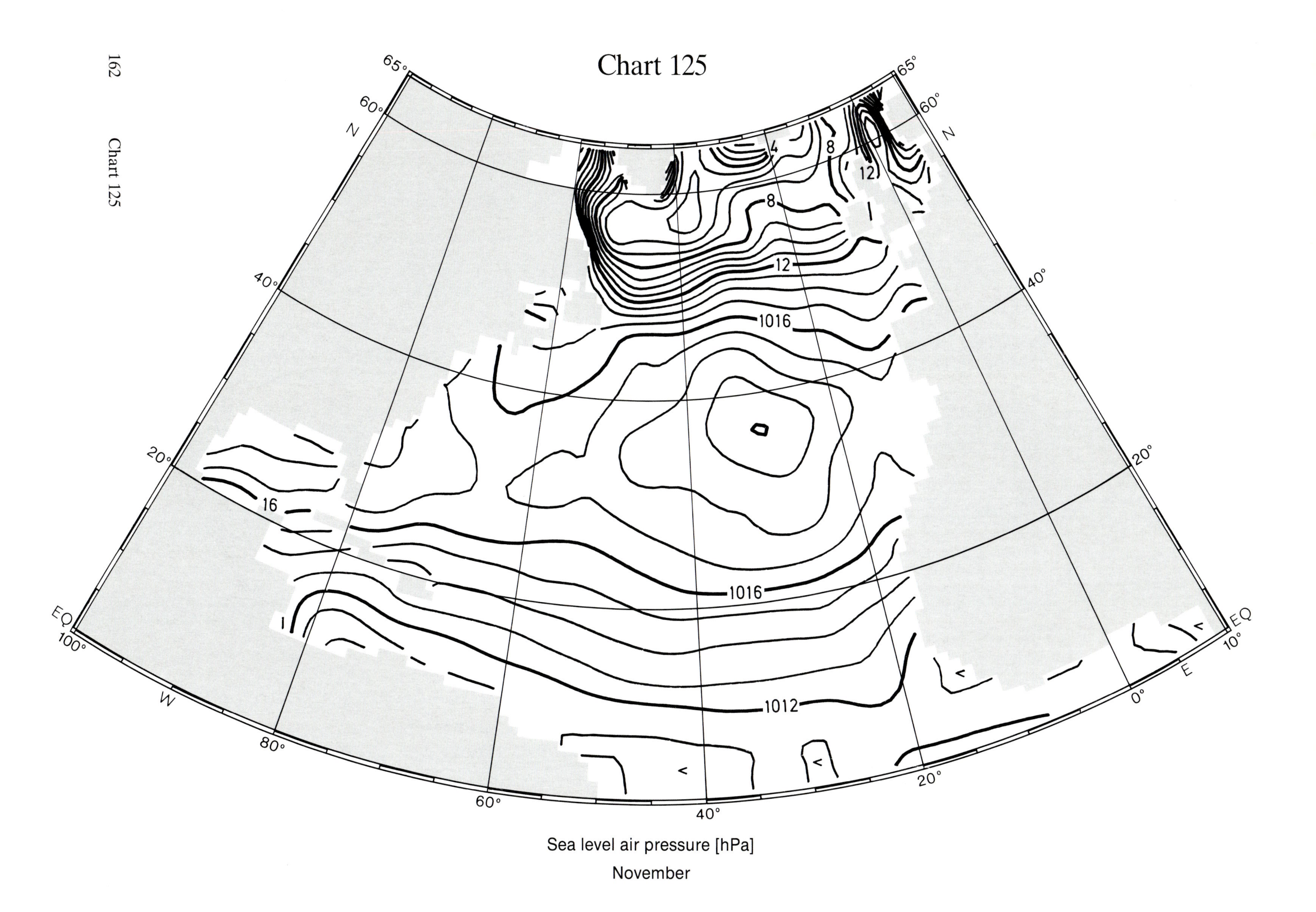

Sea level air pressure [hPa]

November

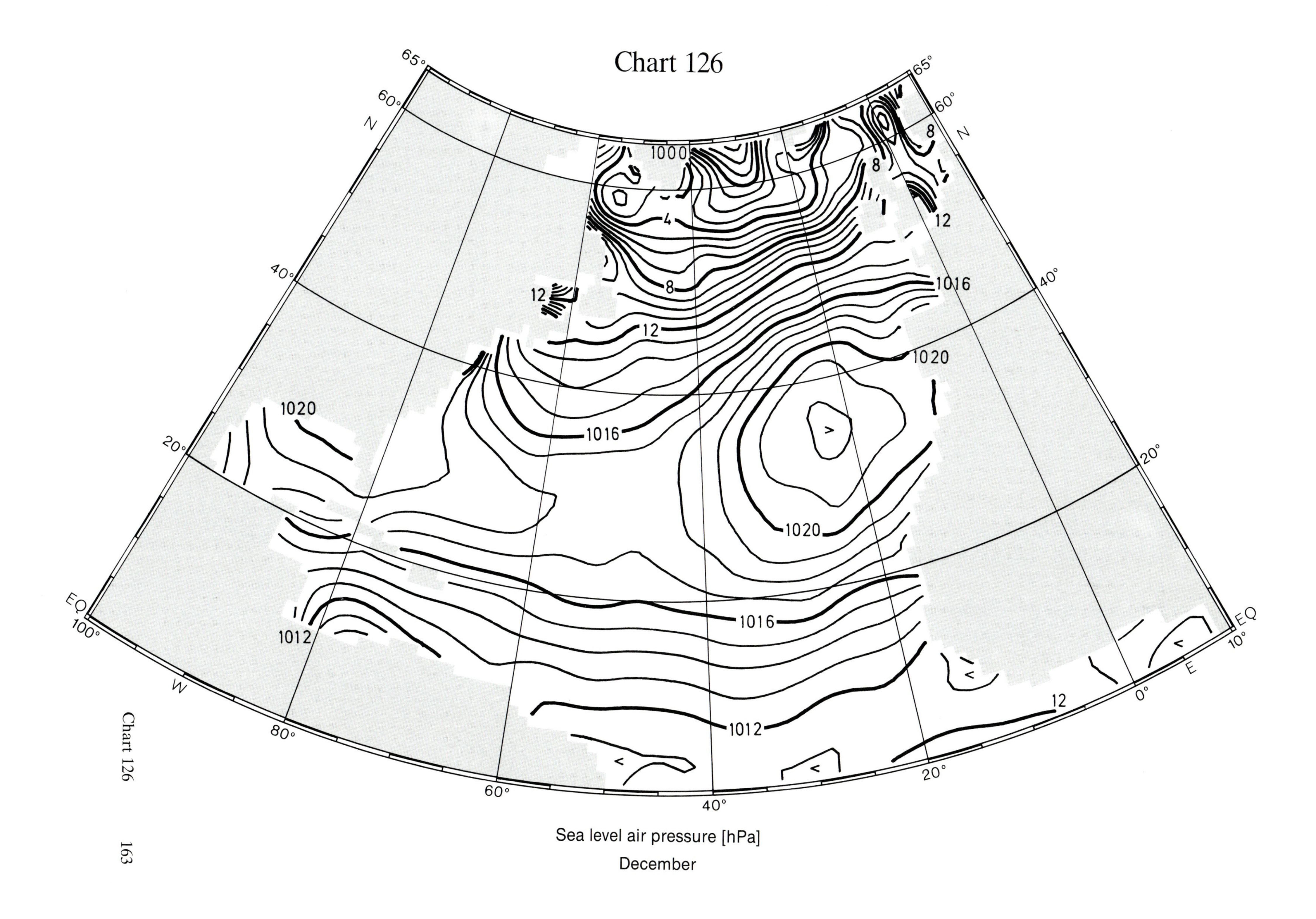
Chart 126
Sea level air pressure [hPa]
December
65°
60°
N
40°
20°
EQ
100°
W
80°
60°
40°
20°
0°
E
10°
1000
4
8
12
1012
1016
1020

Chart 127

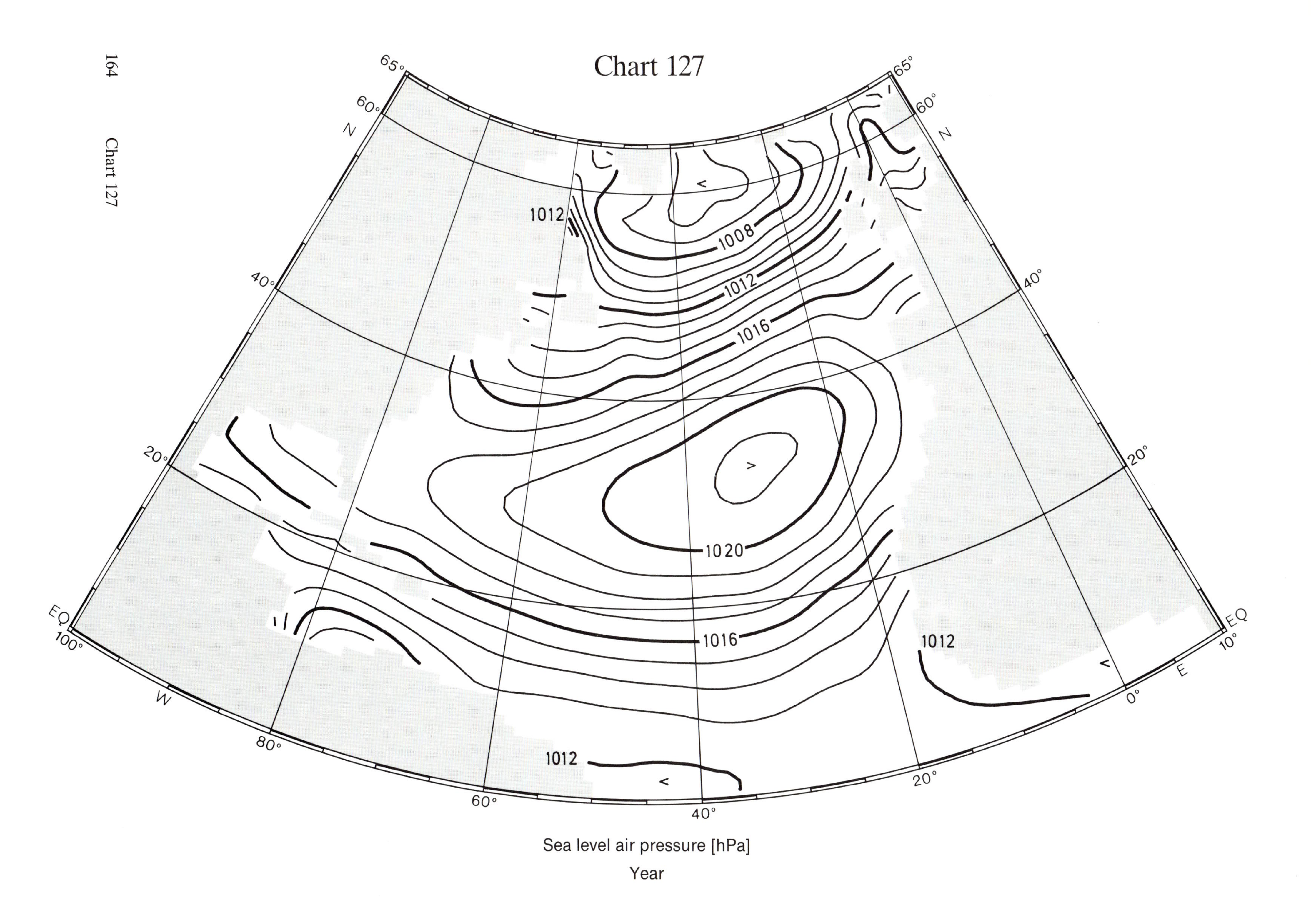

Sea level air pressure [hPa]
Year

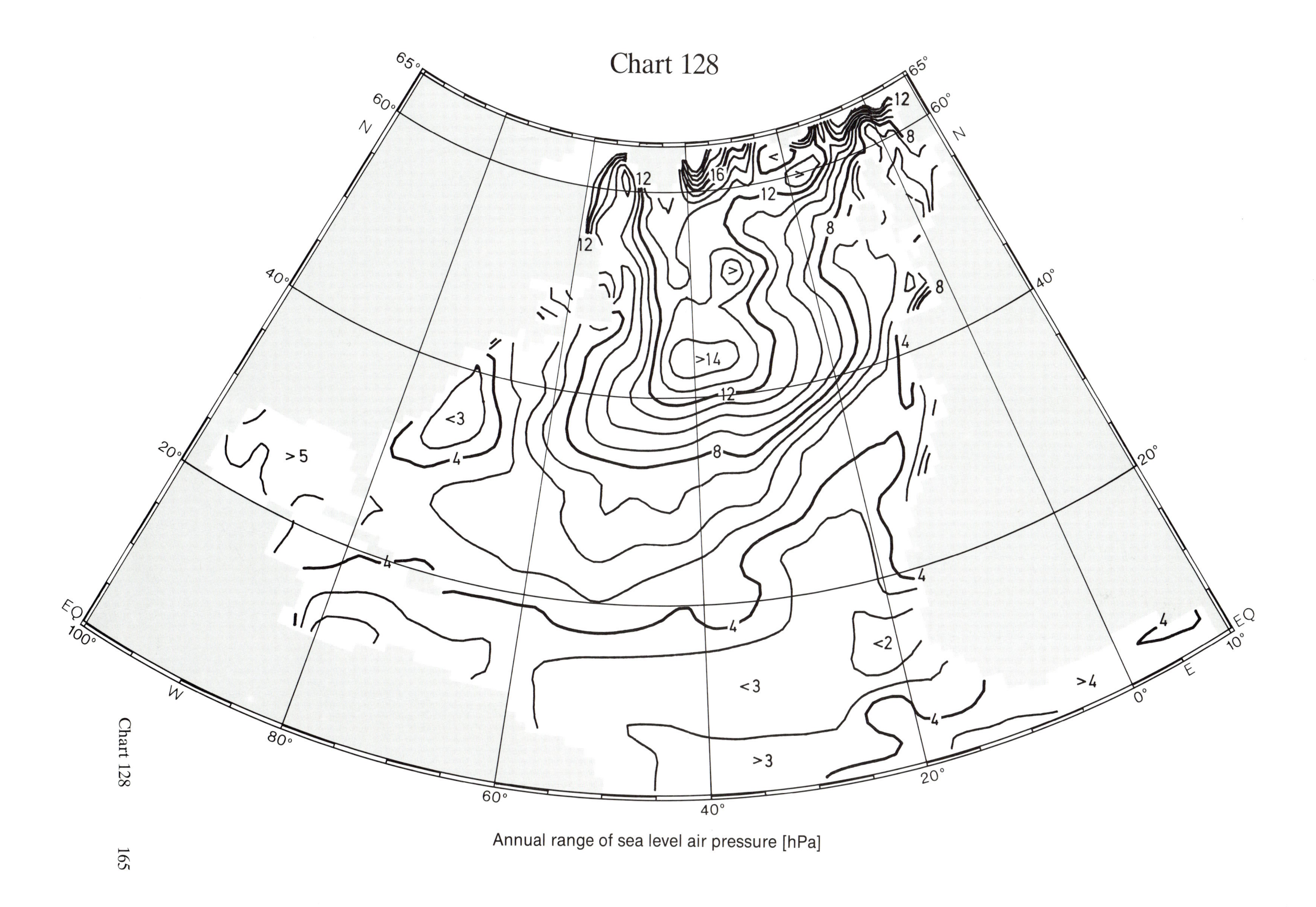
Chart 128
65°
60°
N
40°
20°
EQ
100°
W
80°
60°
40°
20°
0°
E
10°
12
16
8
>14
>5
>4
>3
<3
<2
4
Annual range of sea level air pressure [hPa]

Chart 129

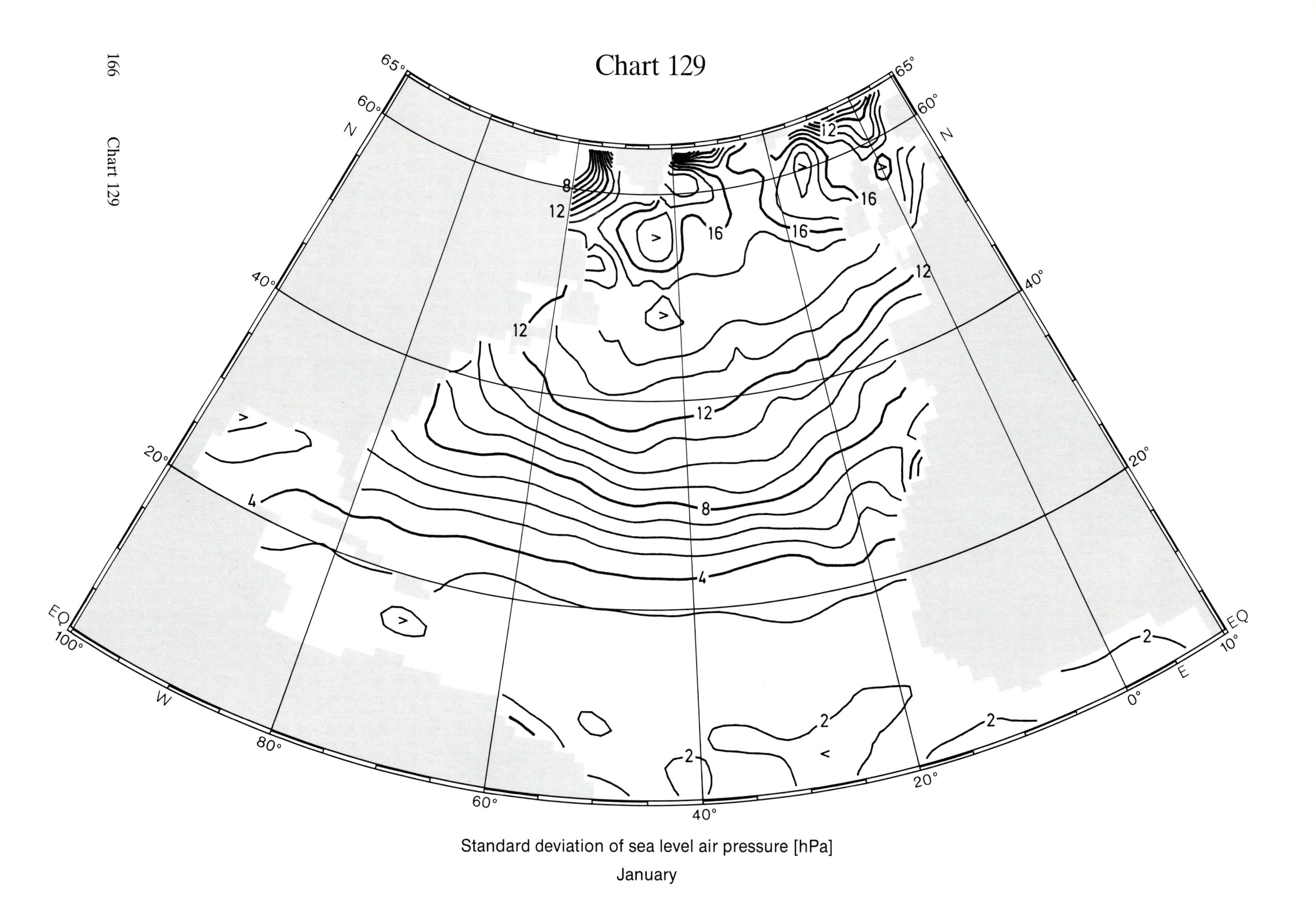

Standard deviation of sea level air pressure [hPa]
January

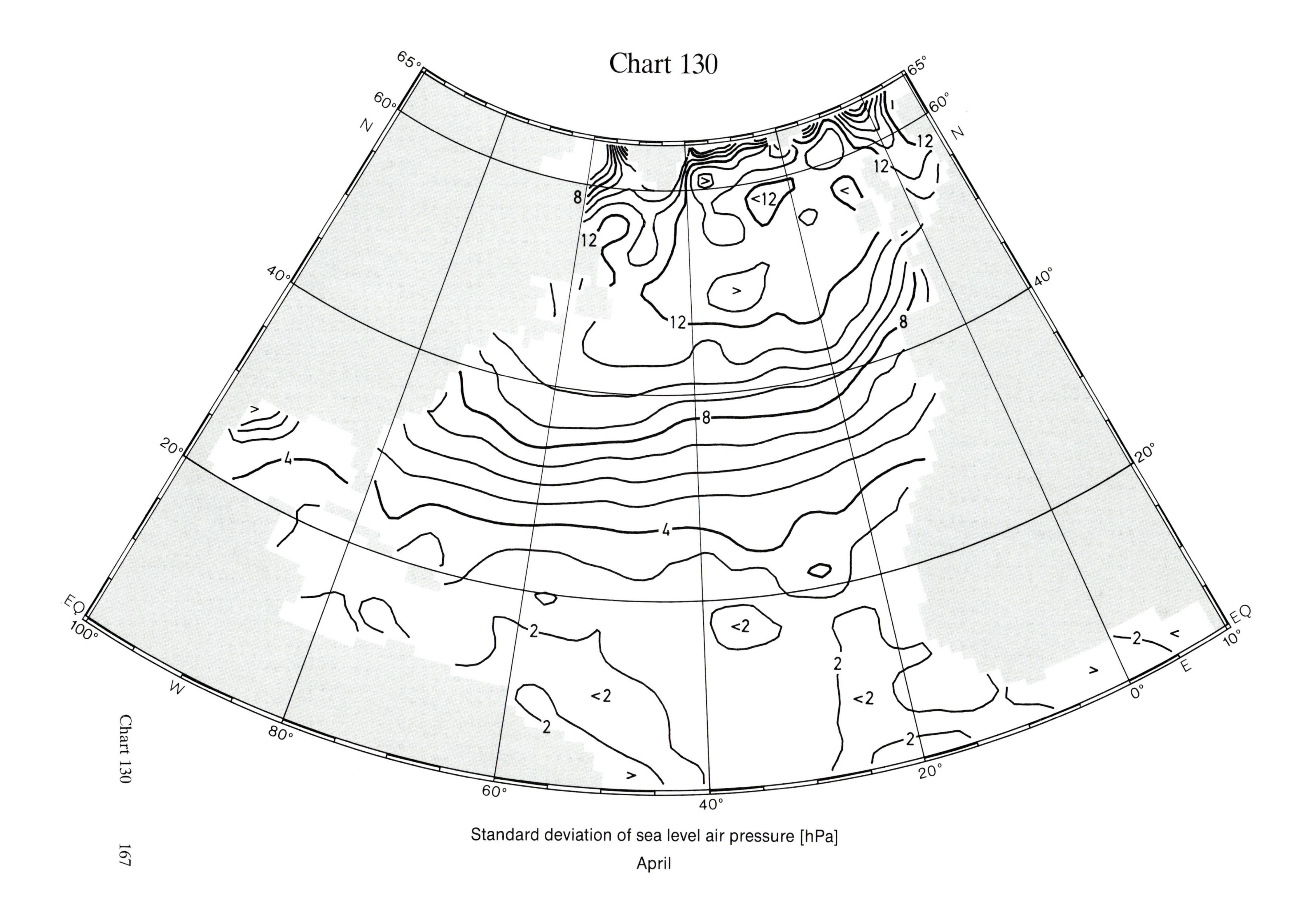

Standard deviation of sea level air pressure [hPa]
April

Chart 131

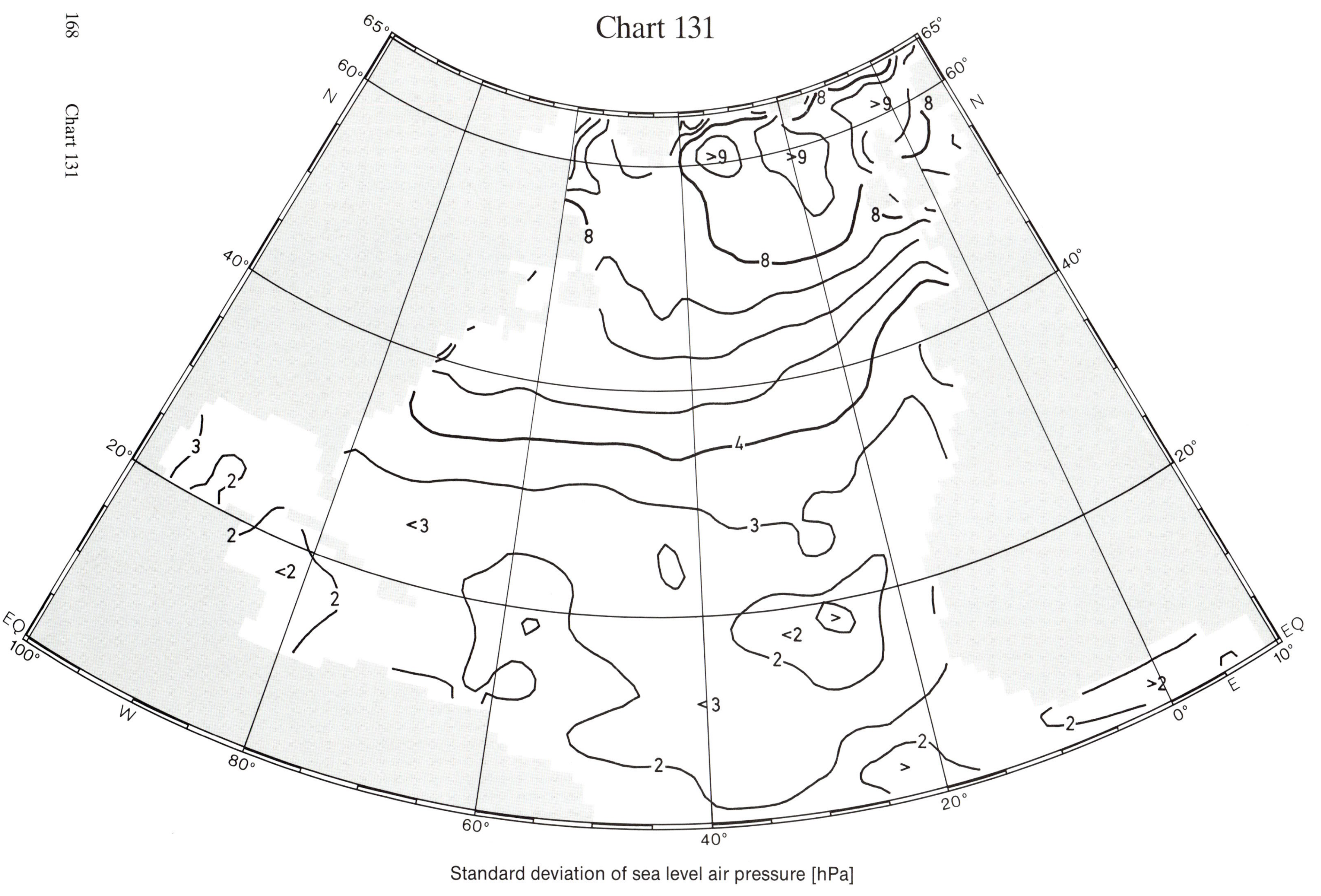

Standard deviation of sea level air pressure [hPa]

July

Chart 132

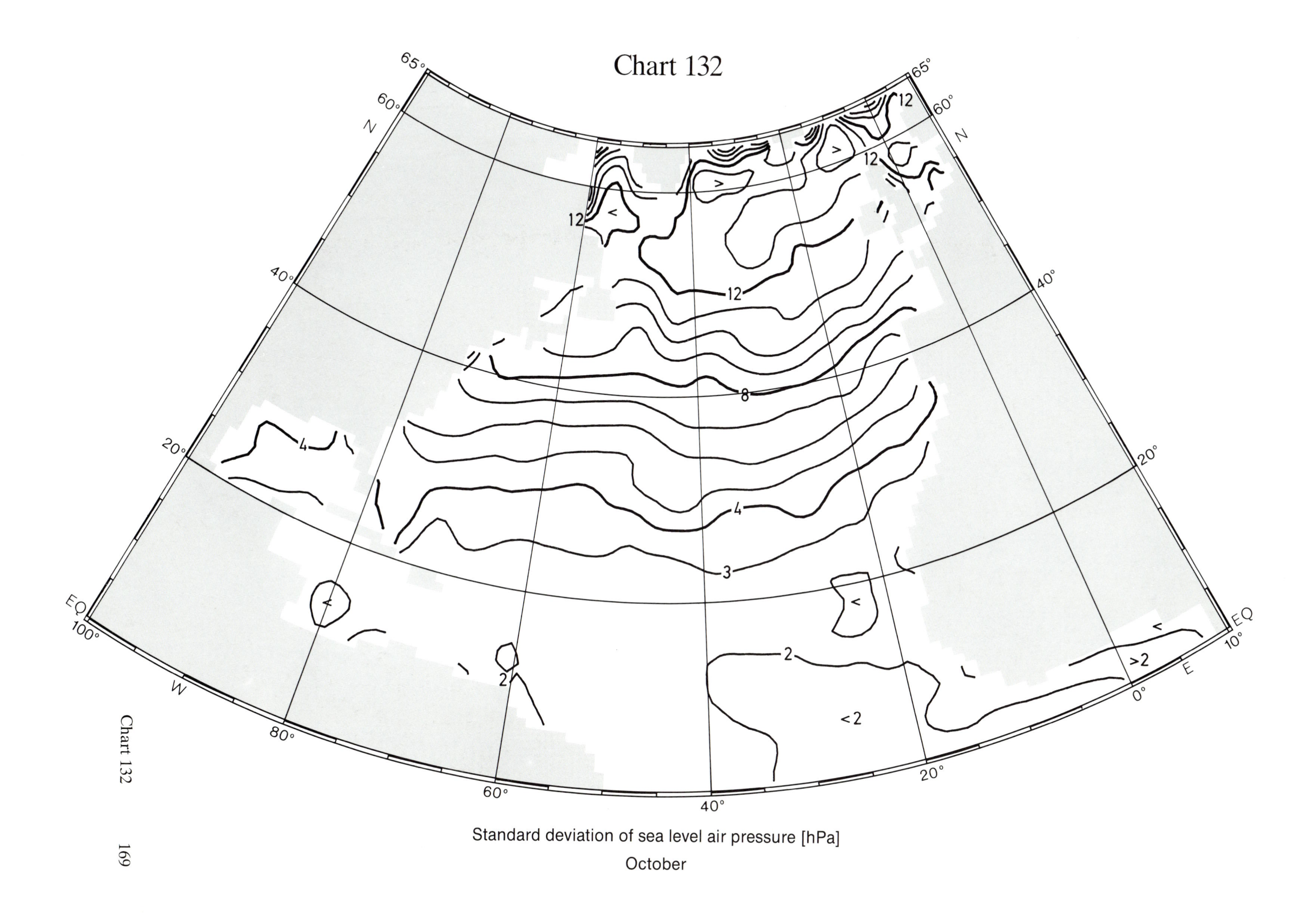

Standard deviation of sea level air pressure [hPa]

October

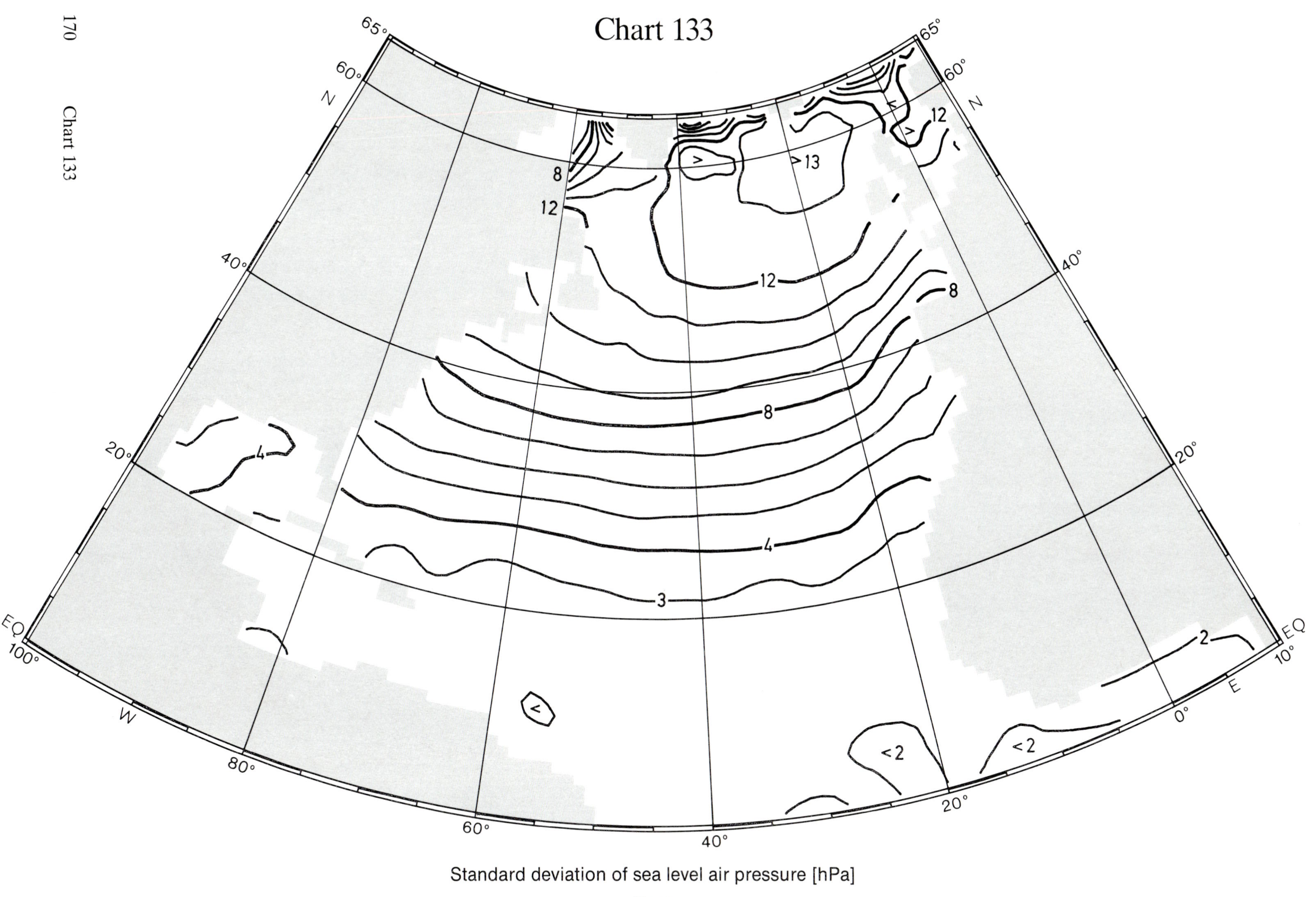

Standard deviation of sea level air pressure [hPa]
Year

Chart 134

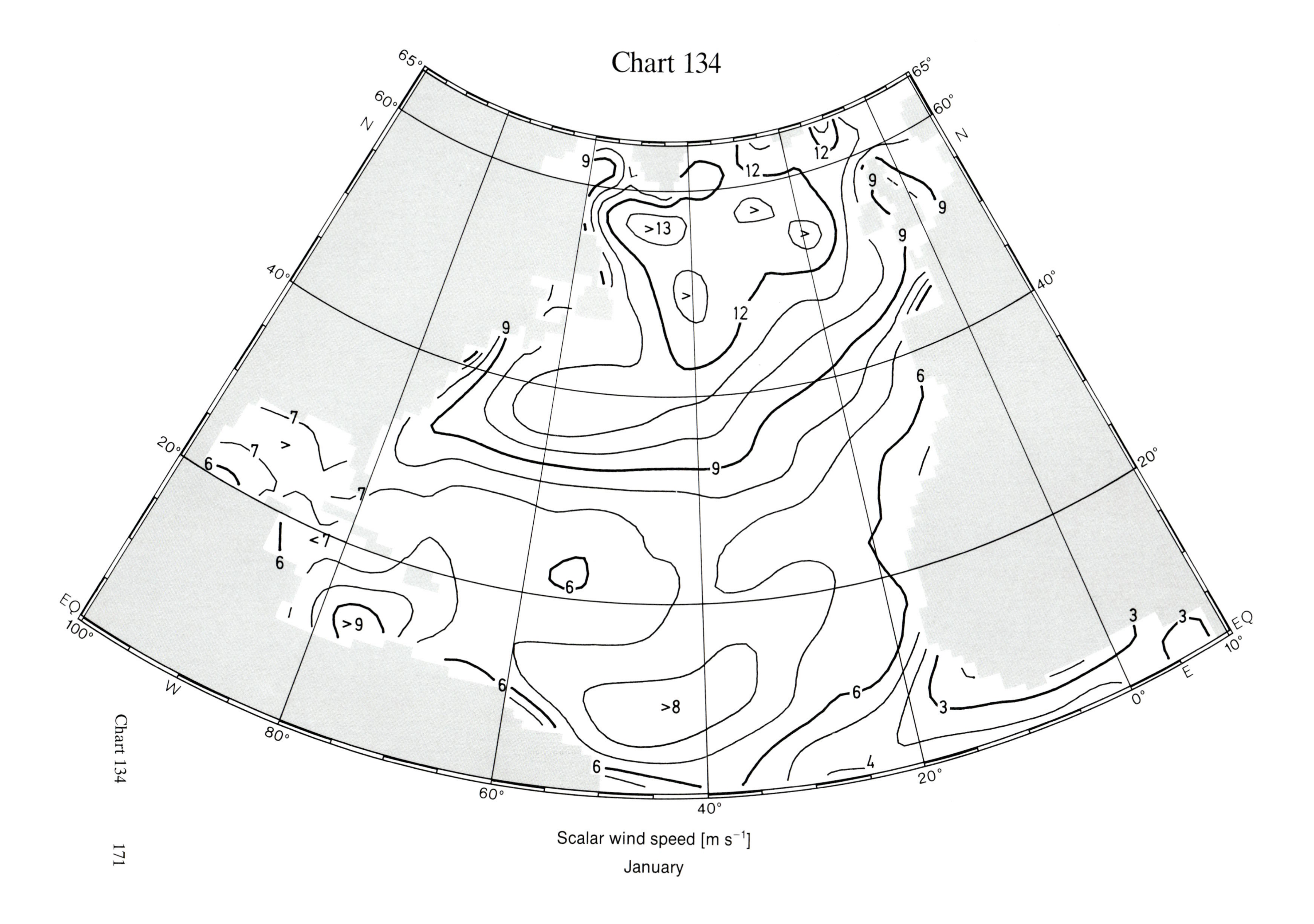

Scalar wind speed [m s⁻¹]
January

Chart 135

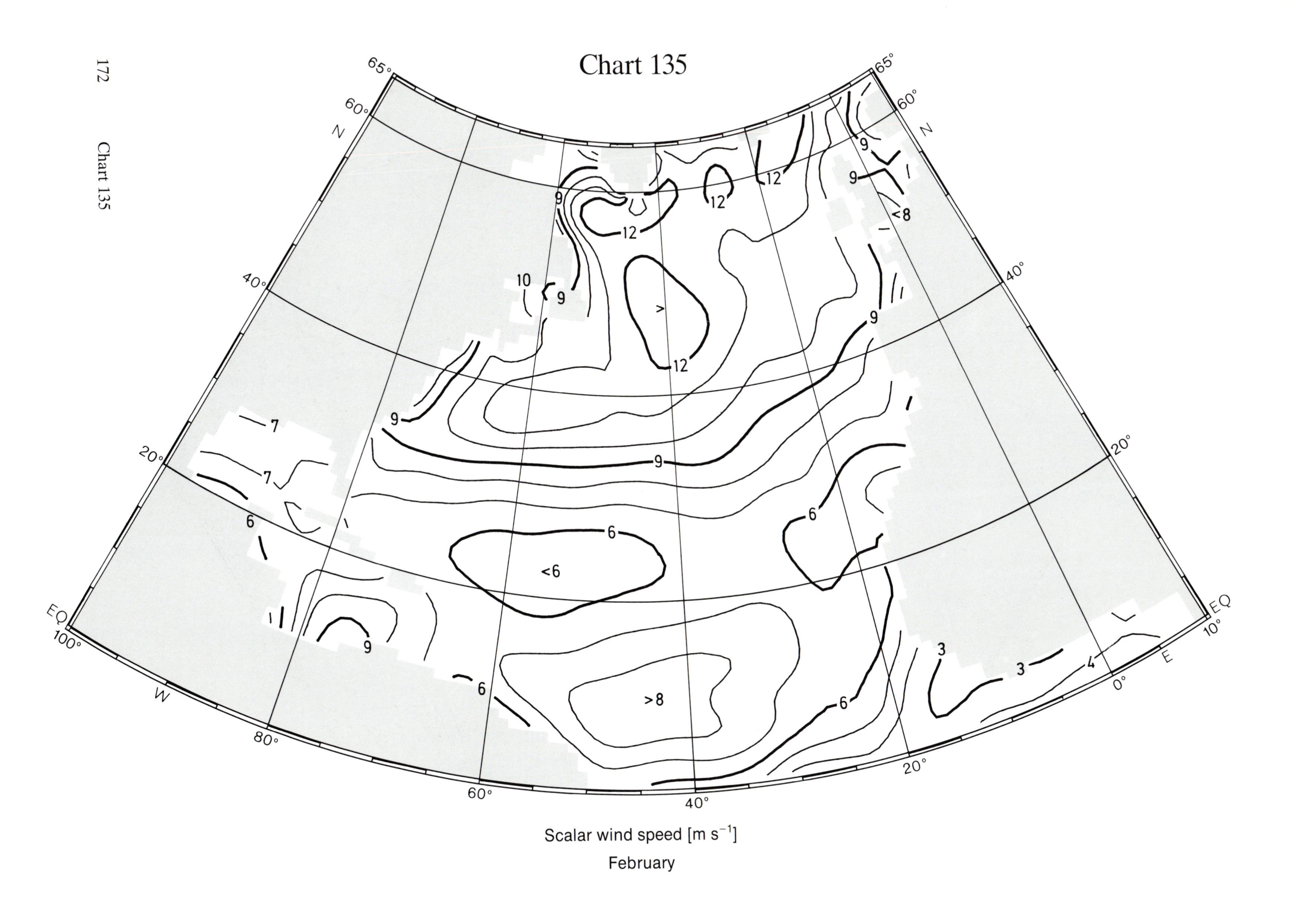

Scalar wind speed [m s^{-1}]
February

Chart 136

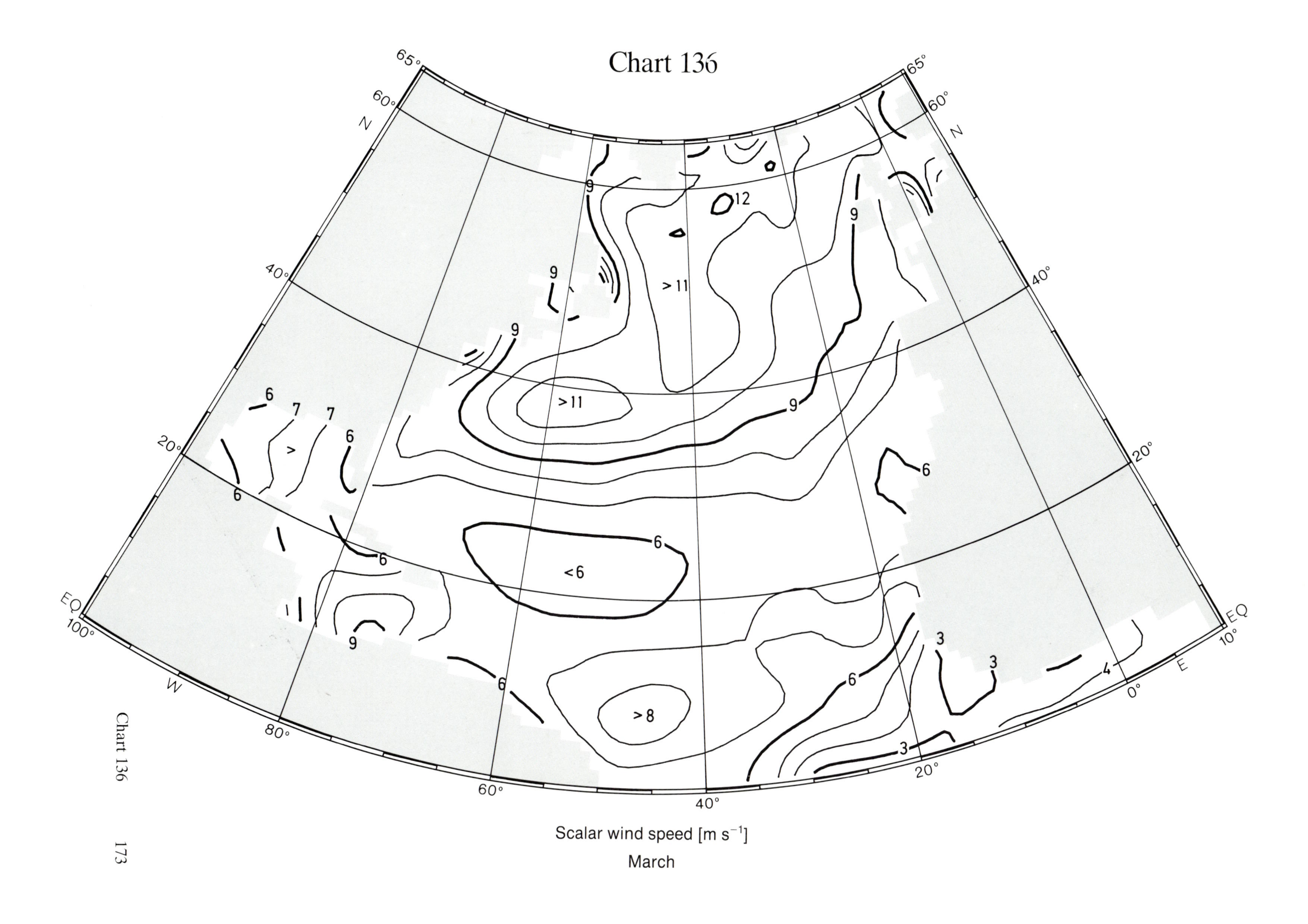
65°
60°
N
40°
20°
EQ
100°
W
80°
60°
40°
20°
0°
E
10°
>11
>11
12
<6
>8
9
6
7
3
4
Scalar wind speed [m s⁻¹]
March

Chart 137

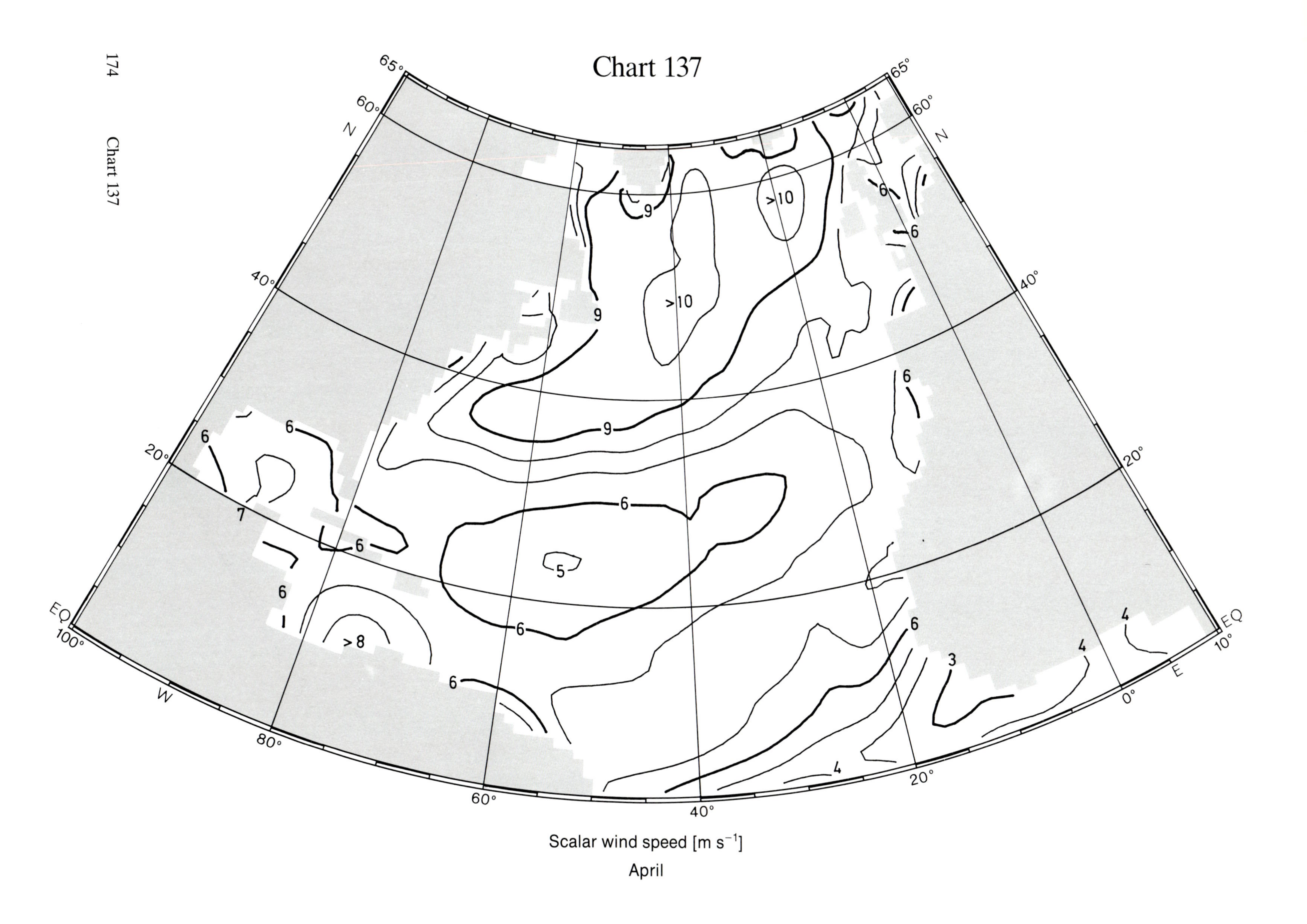

Scalar wind speed [m s^{-1}]
April

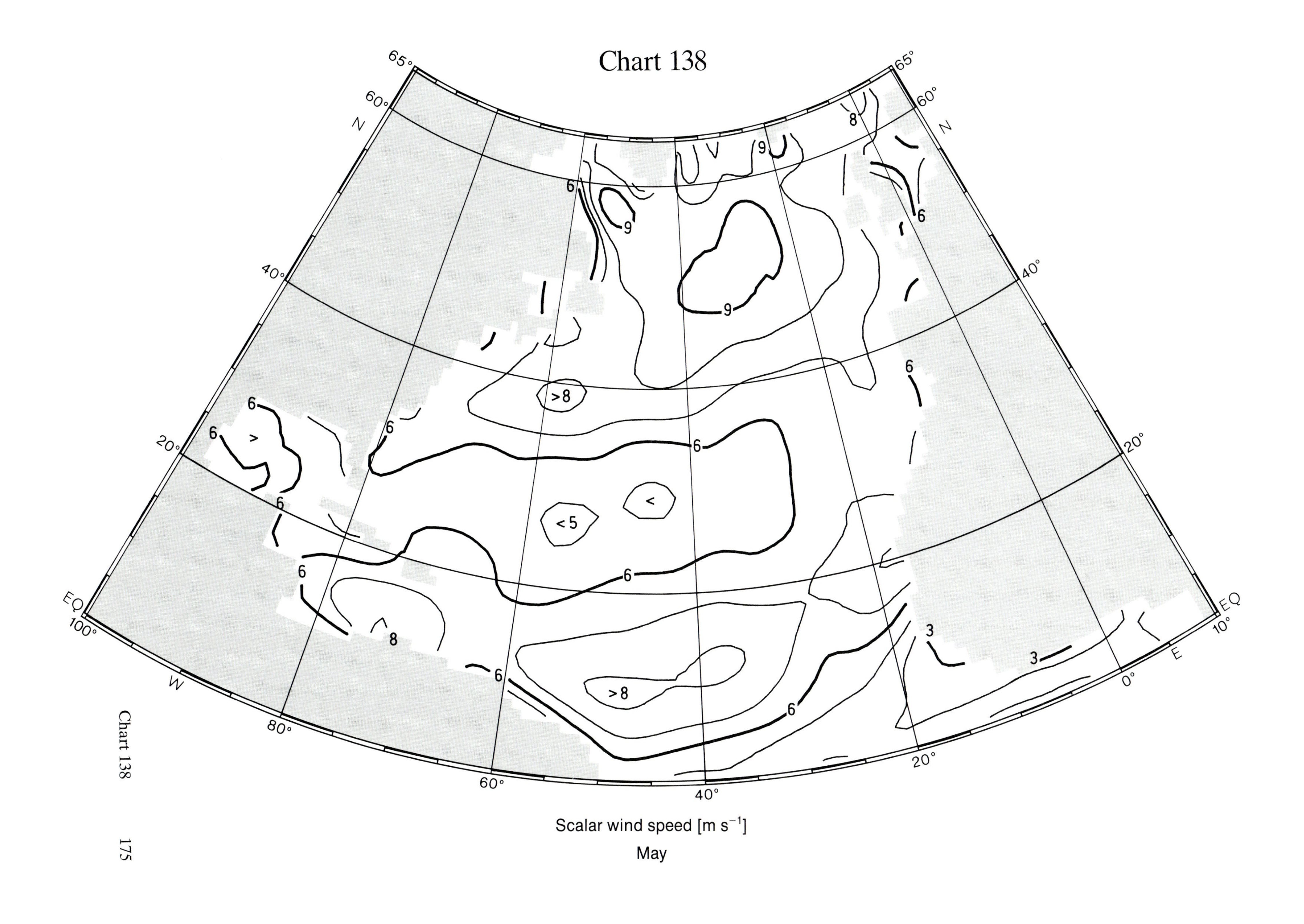
Chart 138
Scalar wind speed [m s⁻¹]
May
65°
60°
N
40°
20°
EQ
100°
80°
60°
40°
20°
0°
10°
W
E
3
6
8
9
>8
<5
<
>

Chart 139

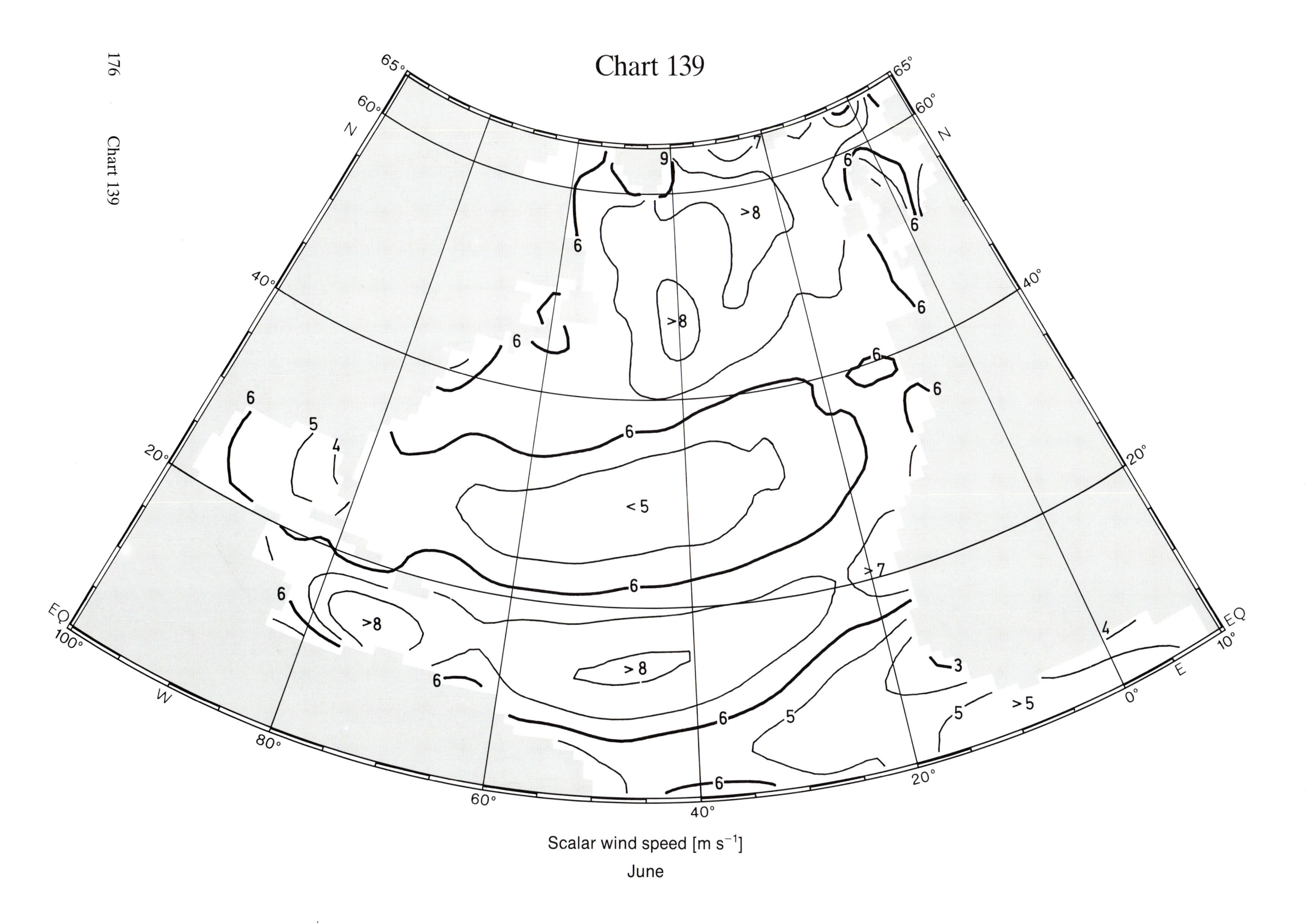

Scalar wind speed [m s^{-1}]

June

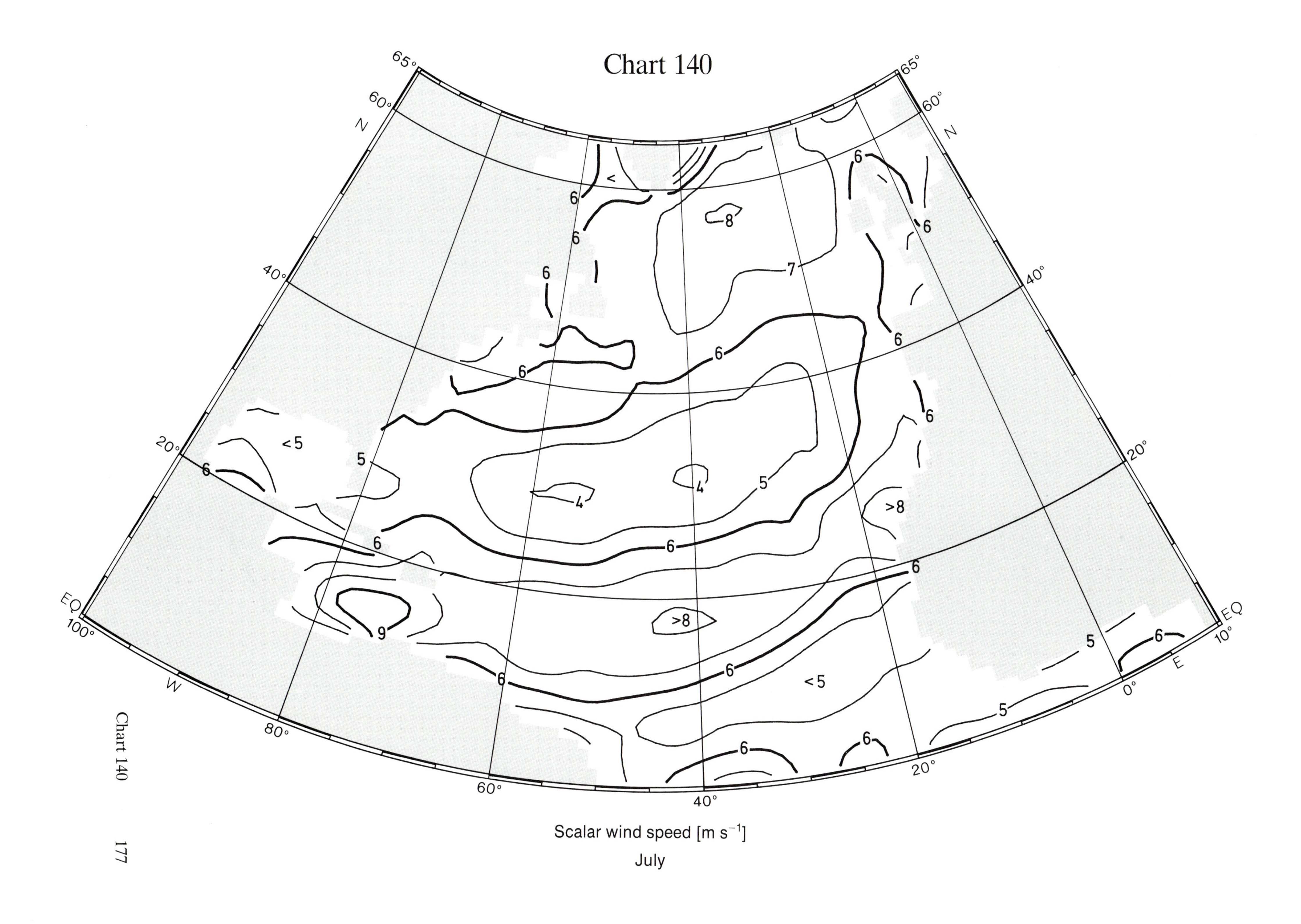
Chart 140
Scalar wind speed [m s⁻¹]
July
65°
60°
40°
20°
EQ
N
W
E
10°
0°
20°
40°
60°
80°
100°
<5
>8
5
6
7
8
4
9

Chart 141

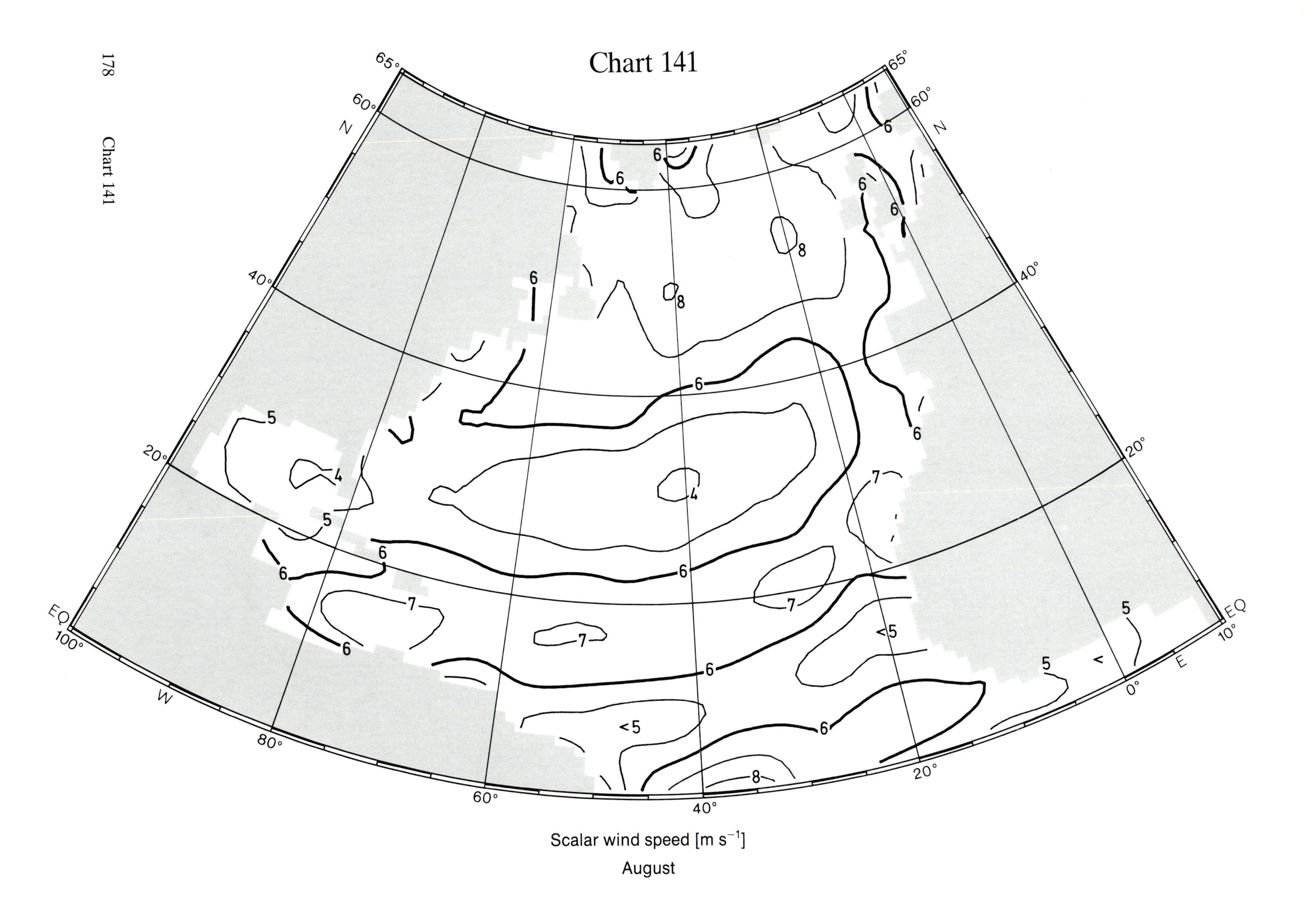

Scalar wind speed [m s^{-1}]
August

Chart 142

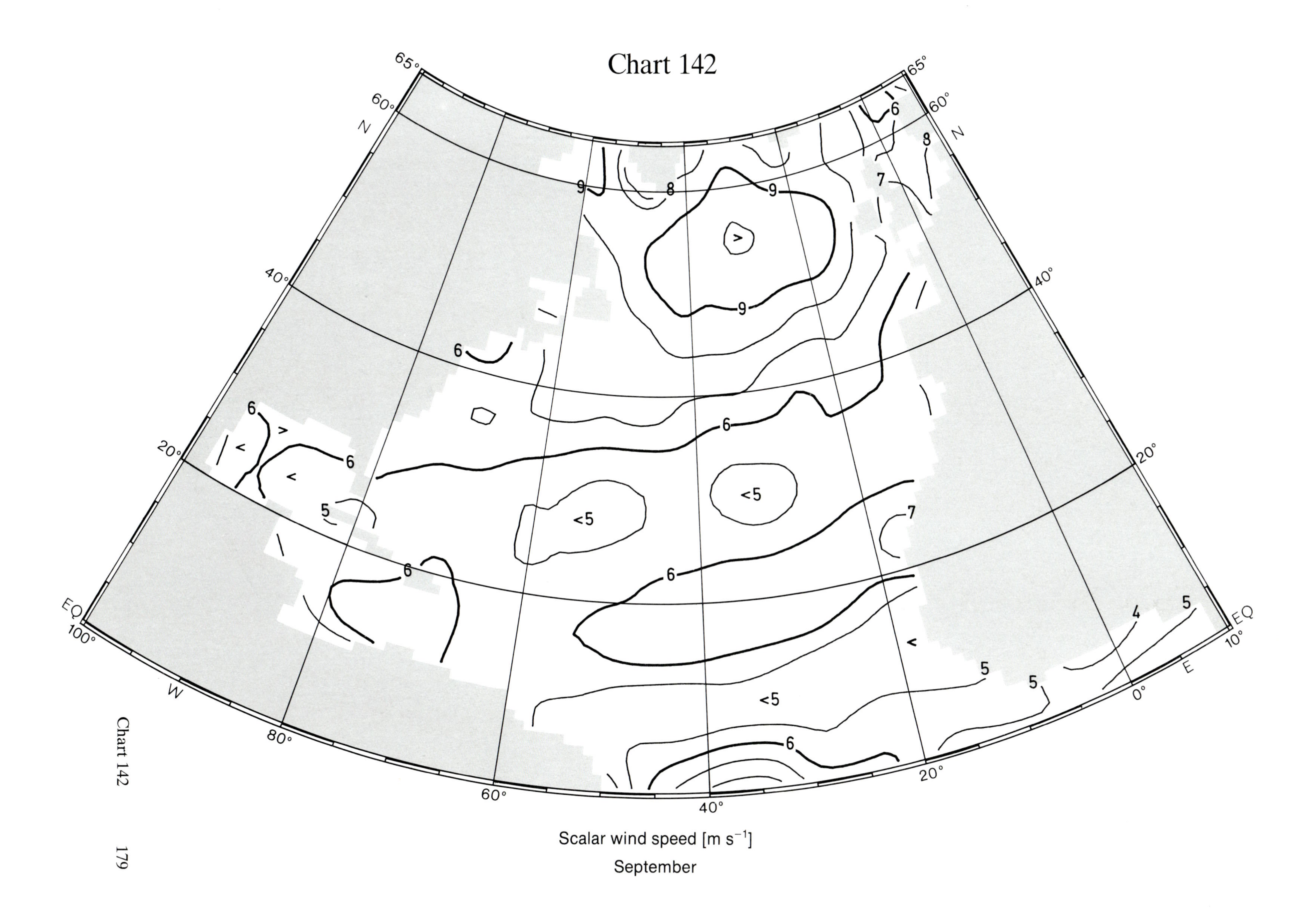

Scalar wind speed [m s^{-1}]

September

Chart 143

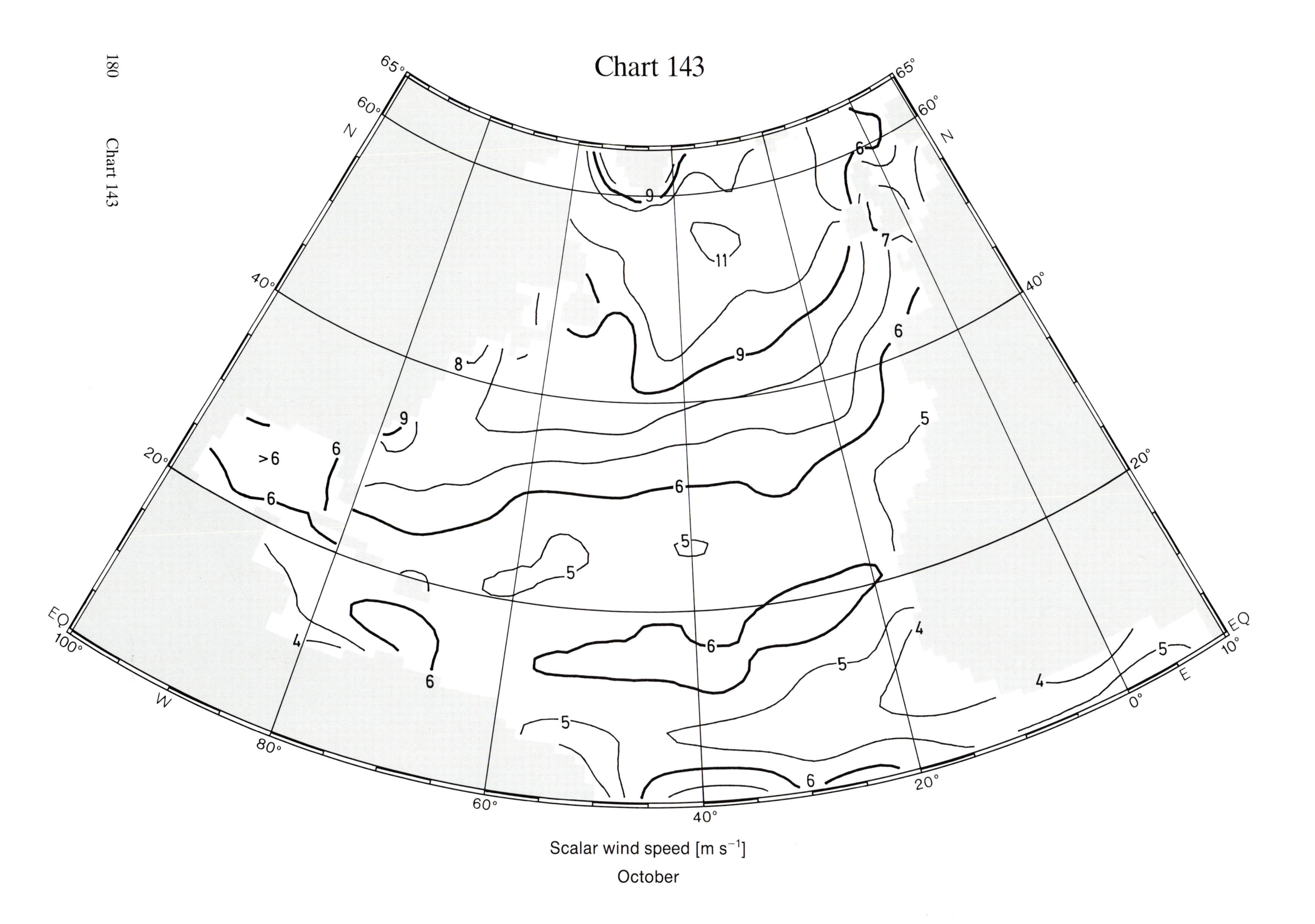

Scalar wind speed [$m\ s^{-1}$]

October

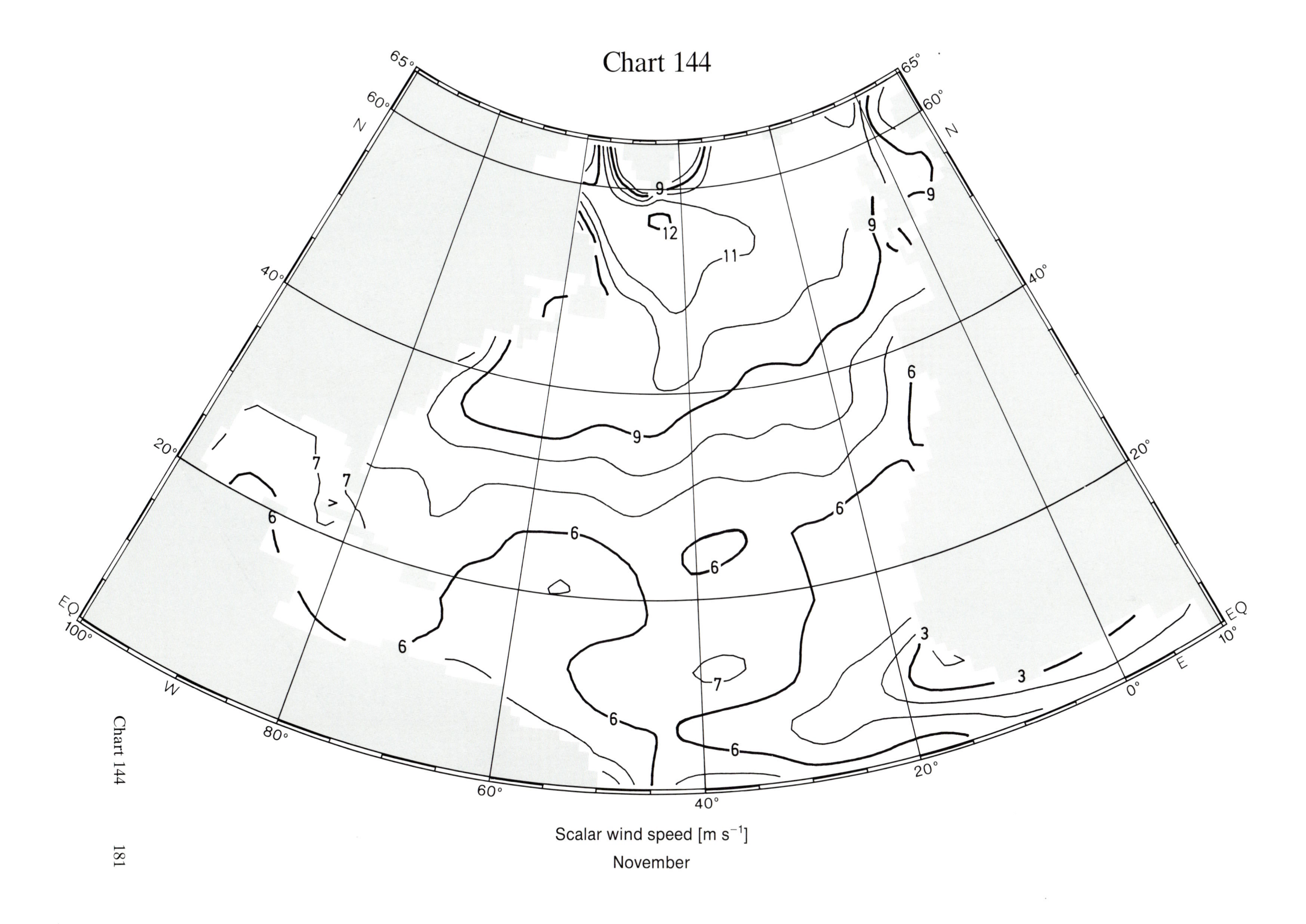
Chart 144
65°
60°
N
40°
20°
EQ
100°
W
80°
60°
40°
20°
0°
E
10°
3
6
7
9
11
12
Scalar wind speed [m s⁻¹]
November

Chart 145

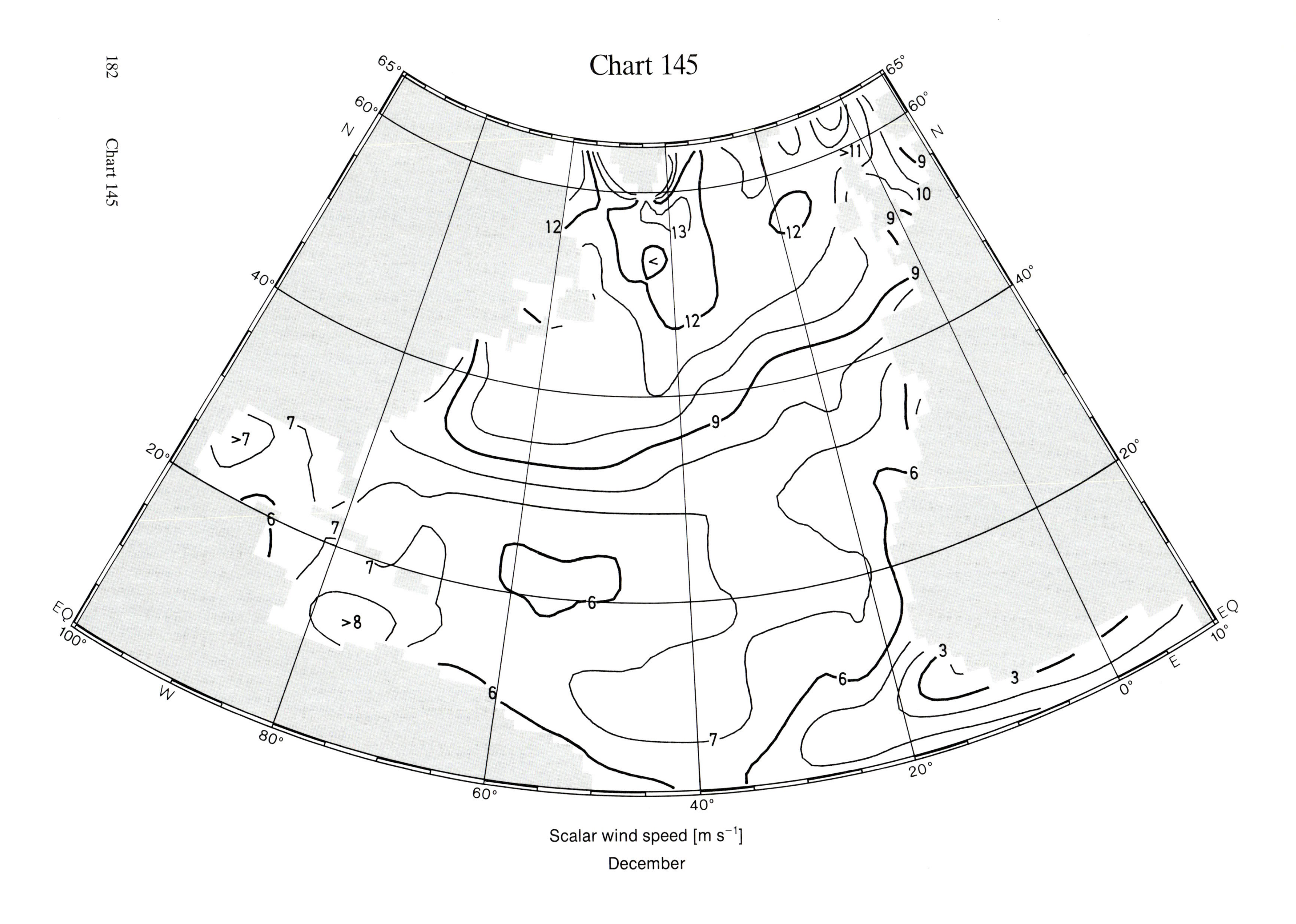

Scalar wind speed [m s^{-1}]
December

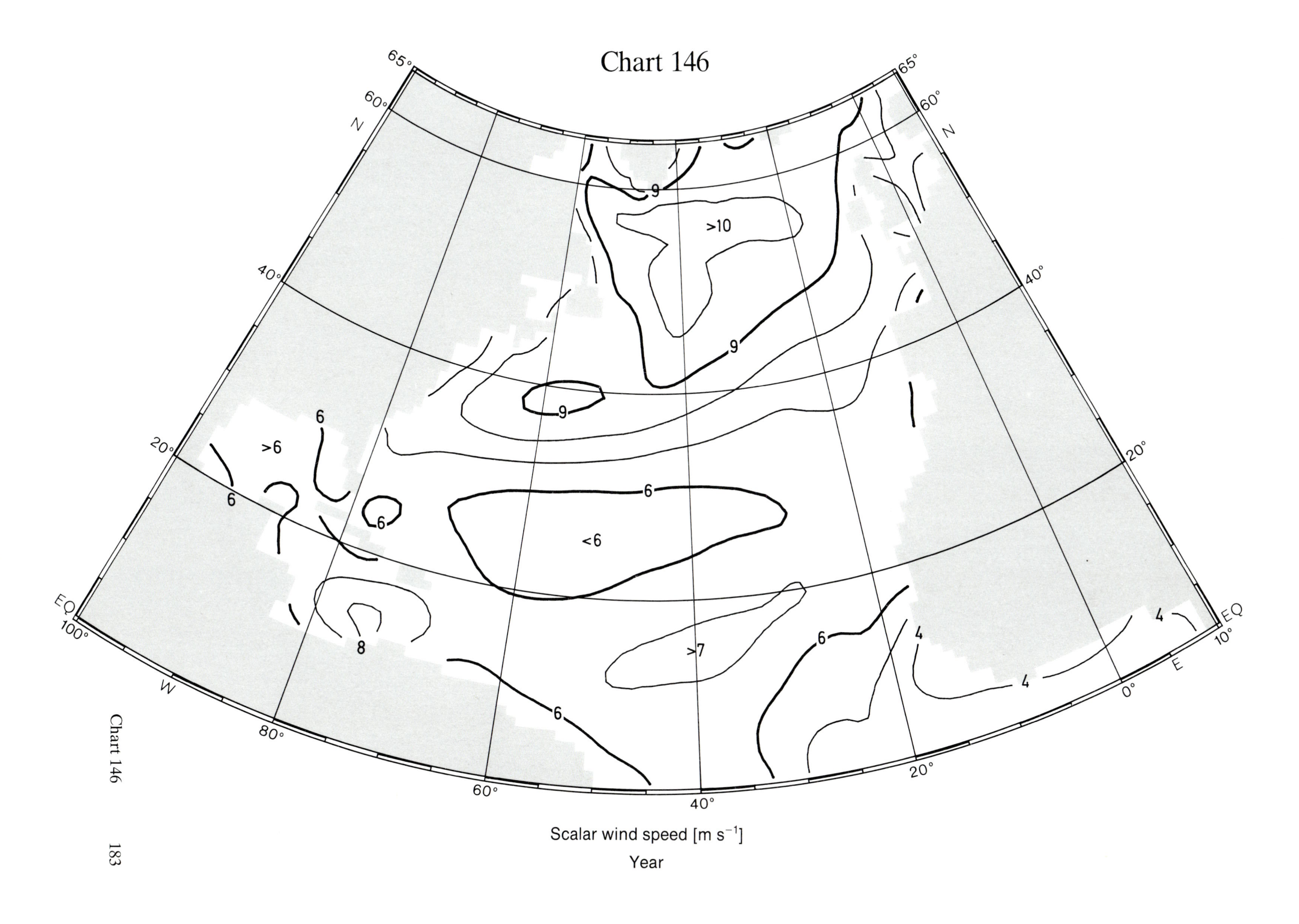

Chart 146
Scalar wind speed [m s⁻¹]
Year
>10
9
<6
>7
8
>6
6
4
65°
60°
40°
20°
10°
0°
80°
100°
EQ
N
E
W

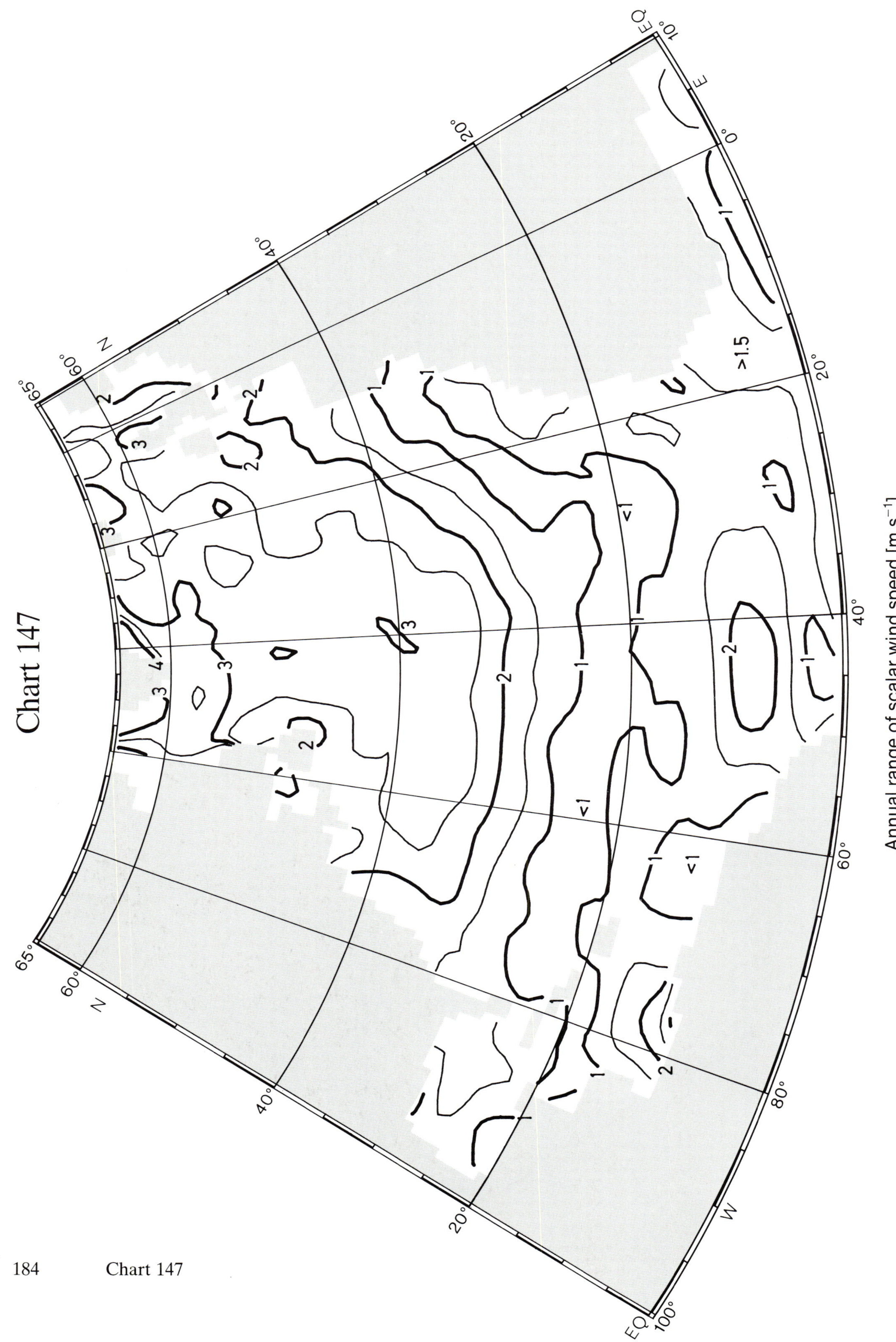
Chart 147
Annual range of scalar wind speed [m s⁻¹]
65°
60°
40°
20°
EQ
100°
80°
60°
40°
20°
0°
10°
>1.5
<1
1
2
3
4

Chart 148

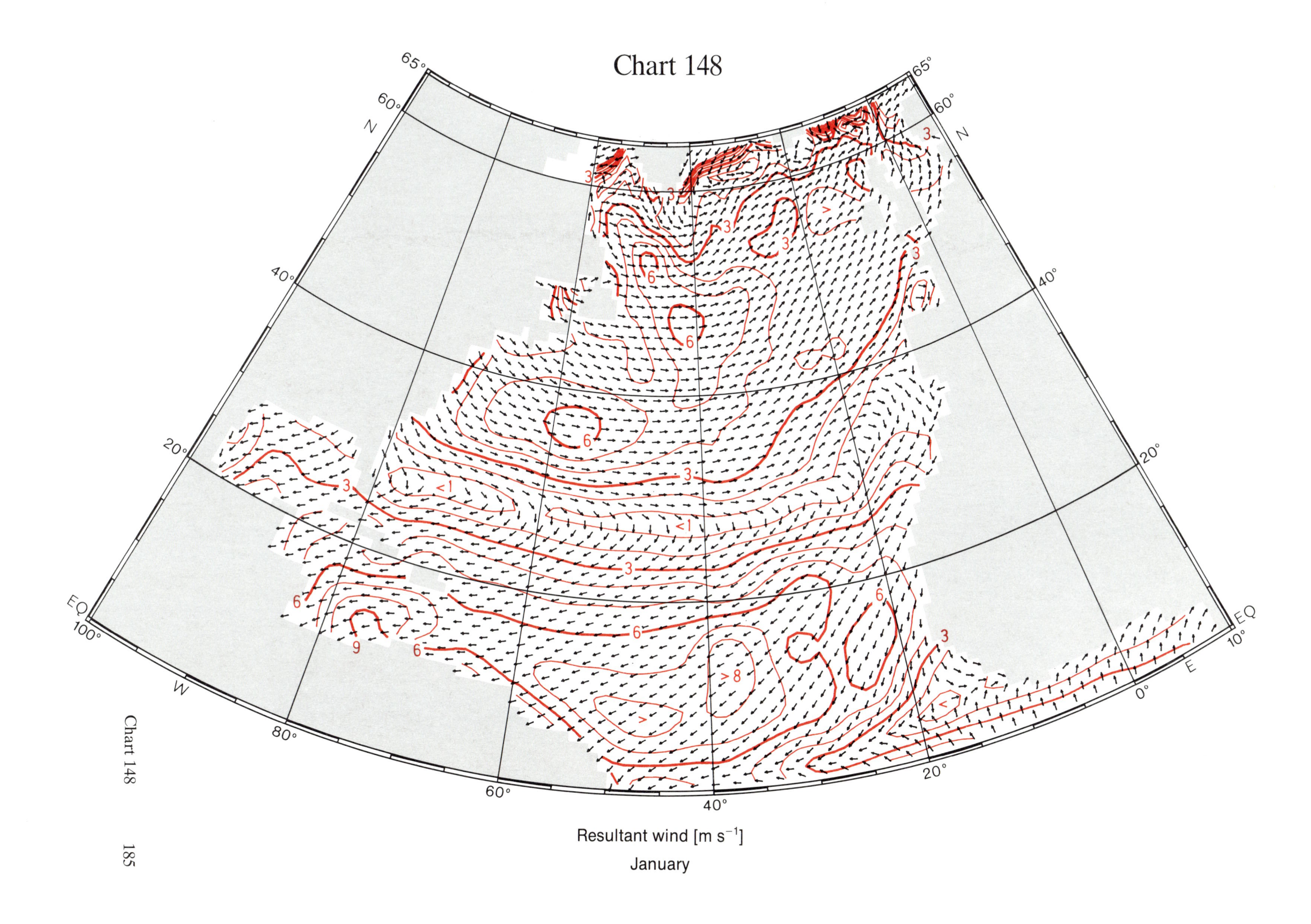

Resultant wind [m s^{-1}]
January

Chart 149

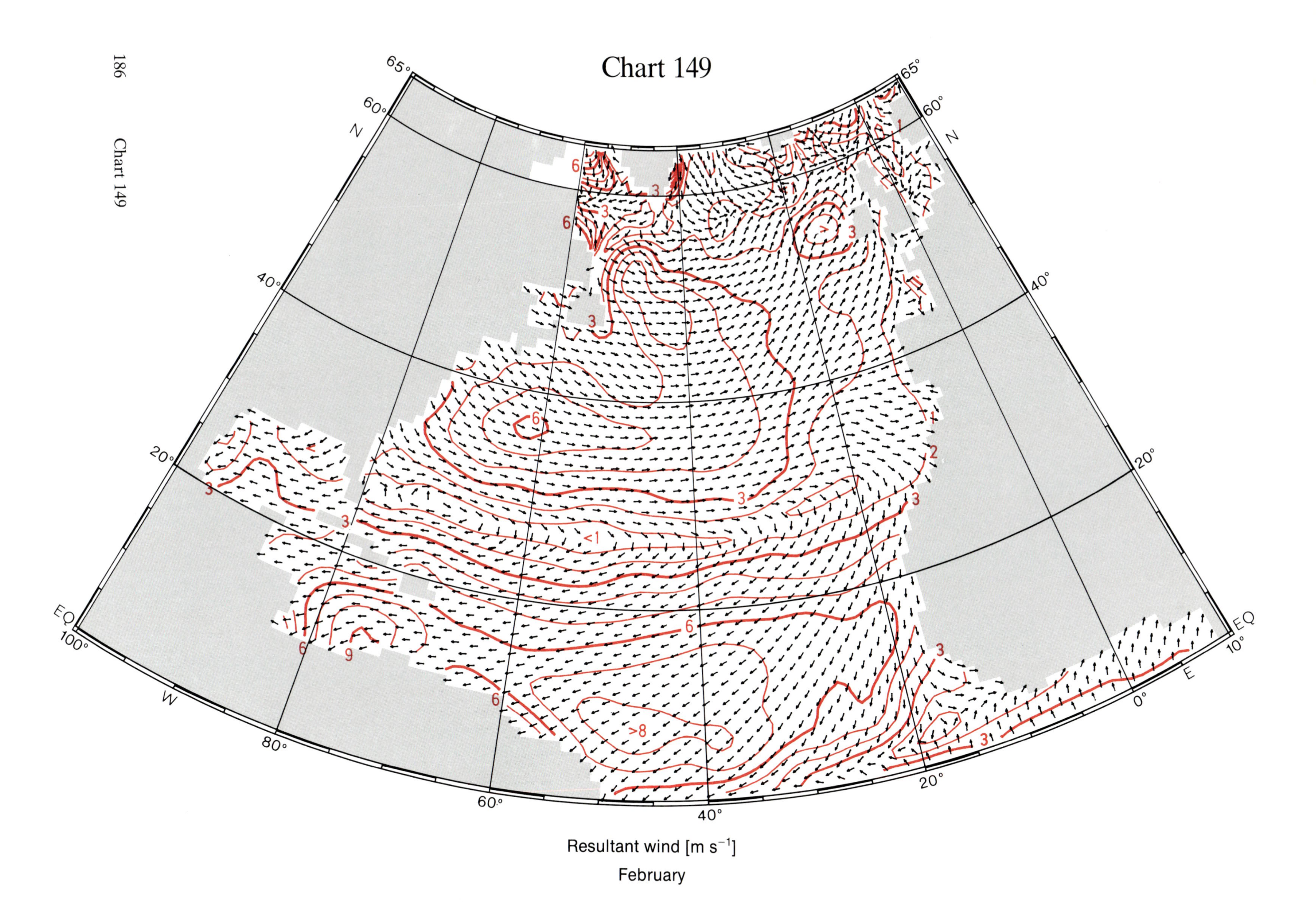

Resultant wind [m s^{-1}]

February

Chart 150

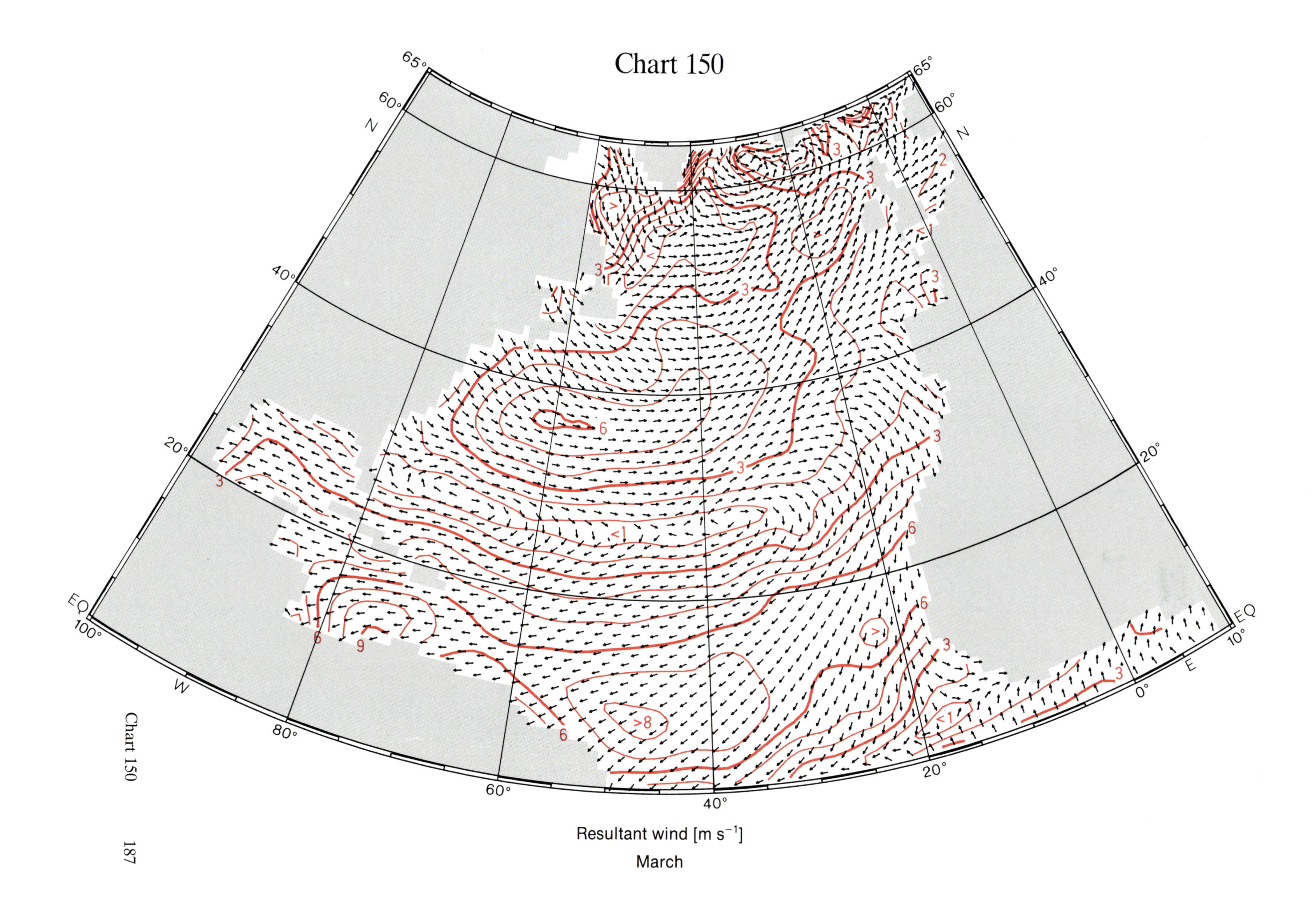

Chart 151

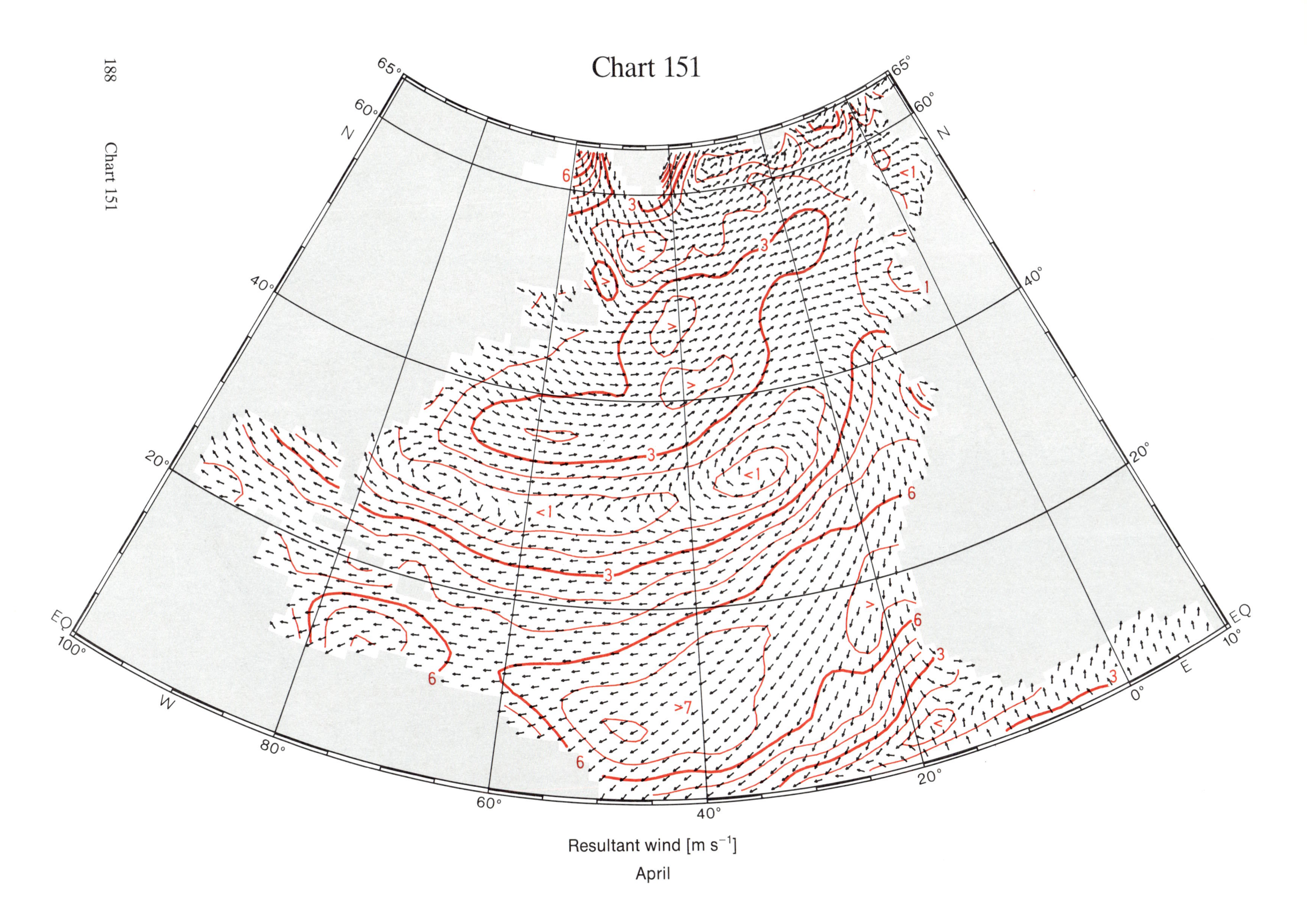

Resultant wind [m s^{-1}]

April

Chart 152

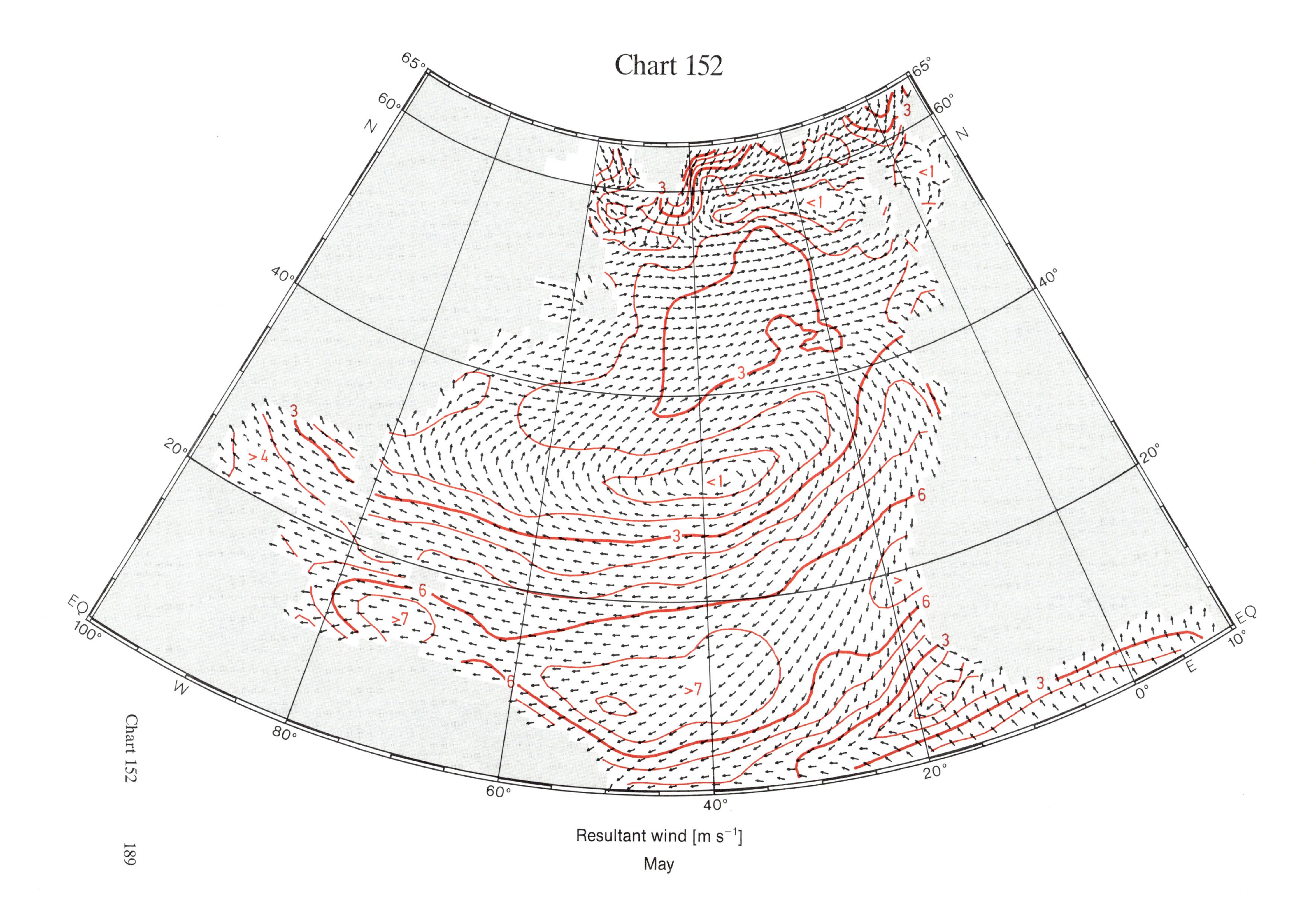

Resultant wind [m s^{-1}]
May

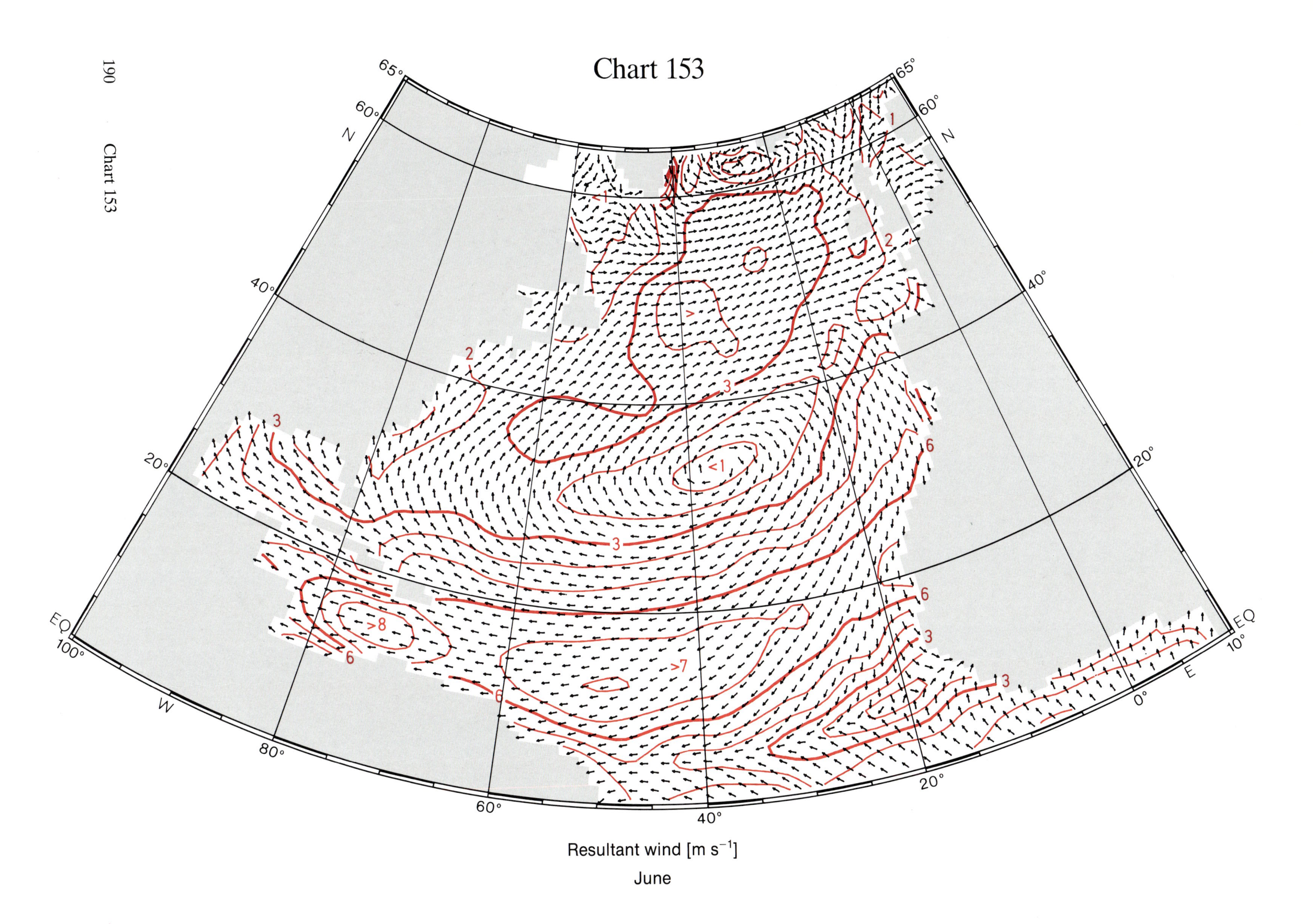
Chart 153
65°
60°
N
40°
20°
EQ
100°
W
80°
60°
40°
20°
0°
E
10°
<1
>7
>8
1
2
3
6
Resultant wind [m s^{-1}]
June

Chart 154

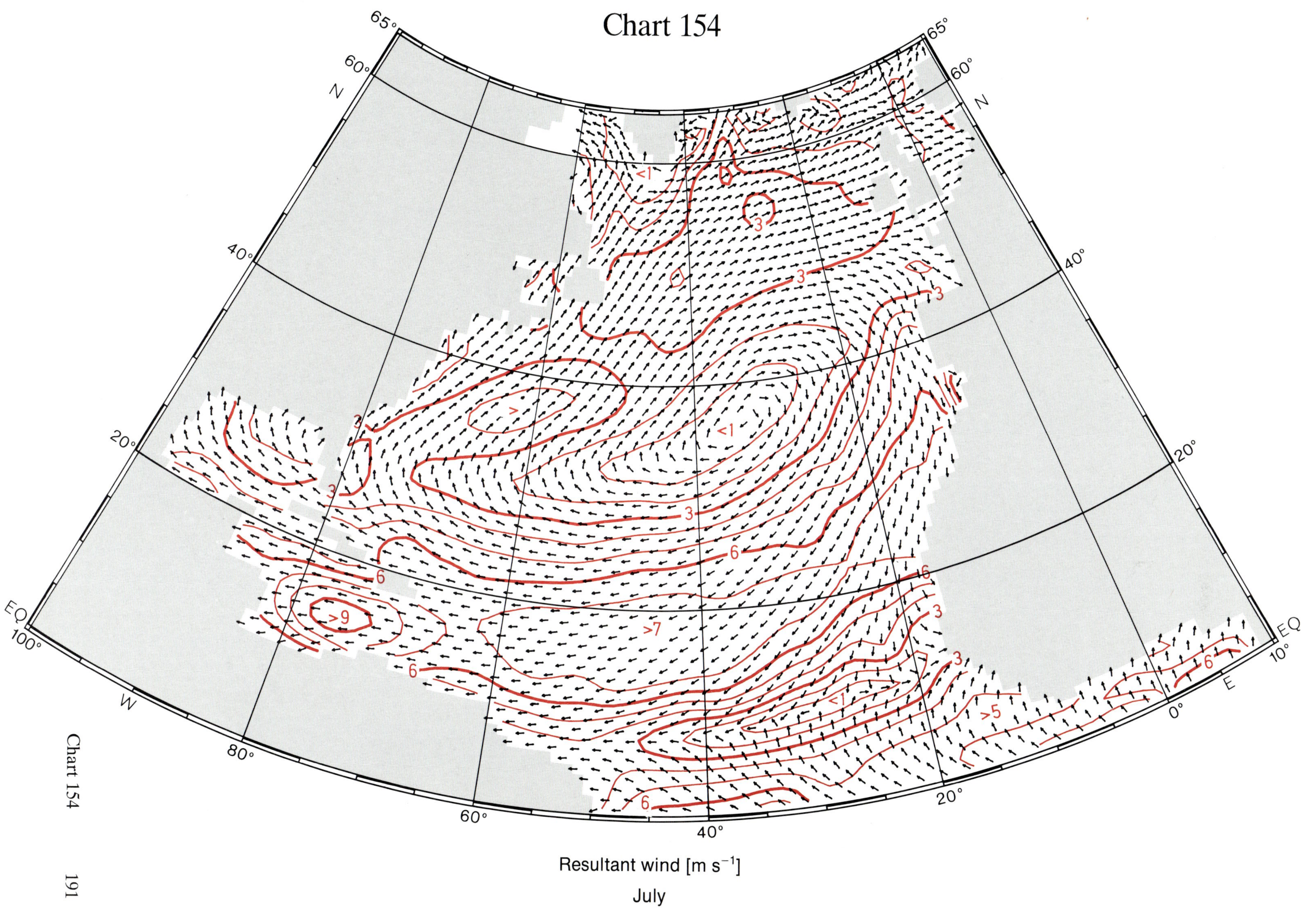
65°
60°
N
40°
20°
EQ
100°
W
80°
60°
40°
20°
0°
E
10°
<1
3
6
>5
>7
>9
Resultant wind [m s⁻¹]
July

Chart 155

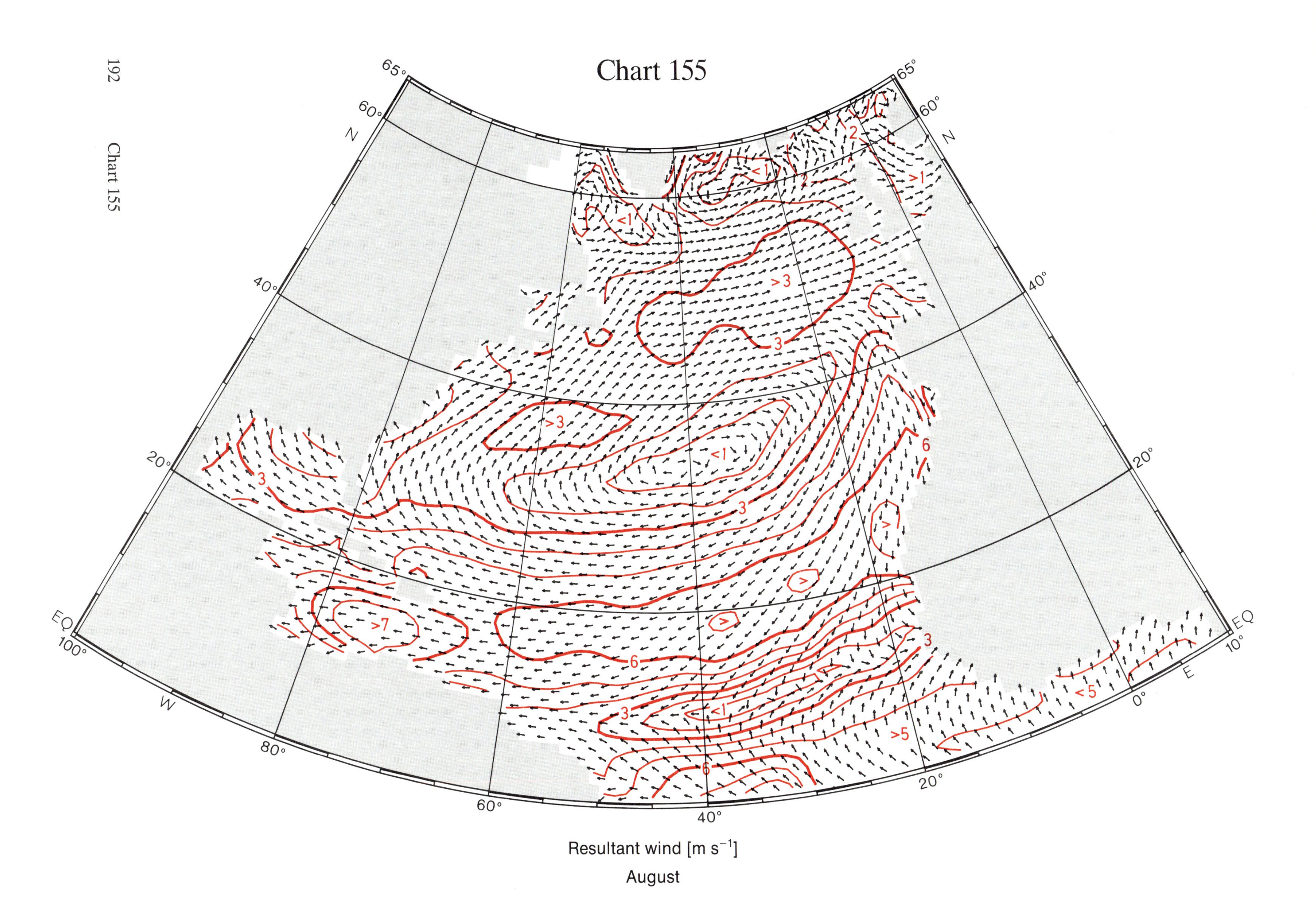

Resultant wind [m s^{-1}]
August

Chart 156

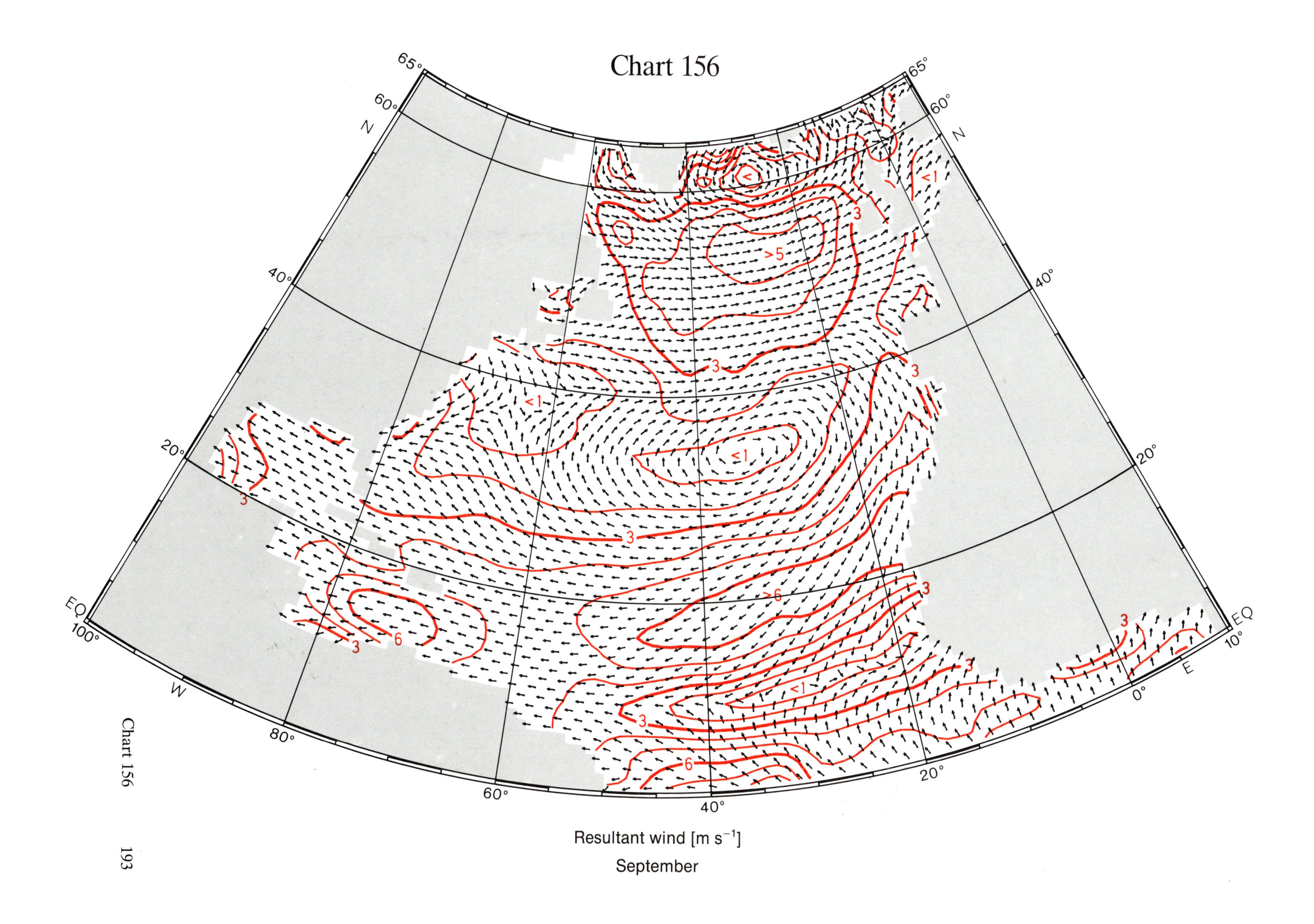

Resultant wind [m s^{-1}]
September

Chart 157

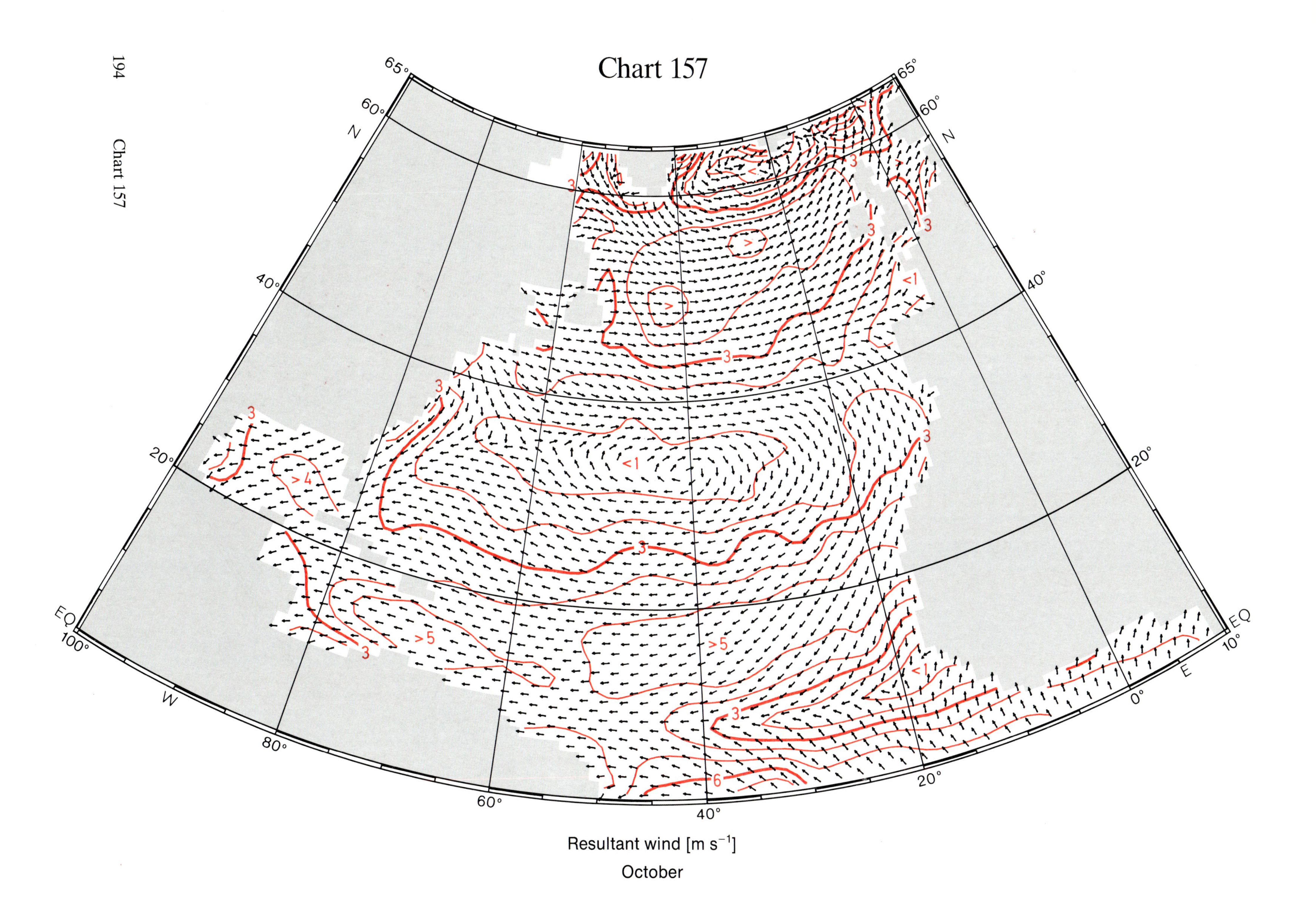

Resultant wind [$m\ s^{-1}$]

October

Chart 158

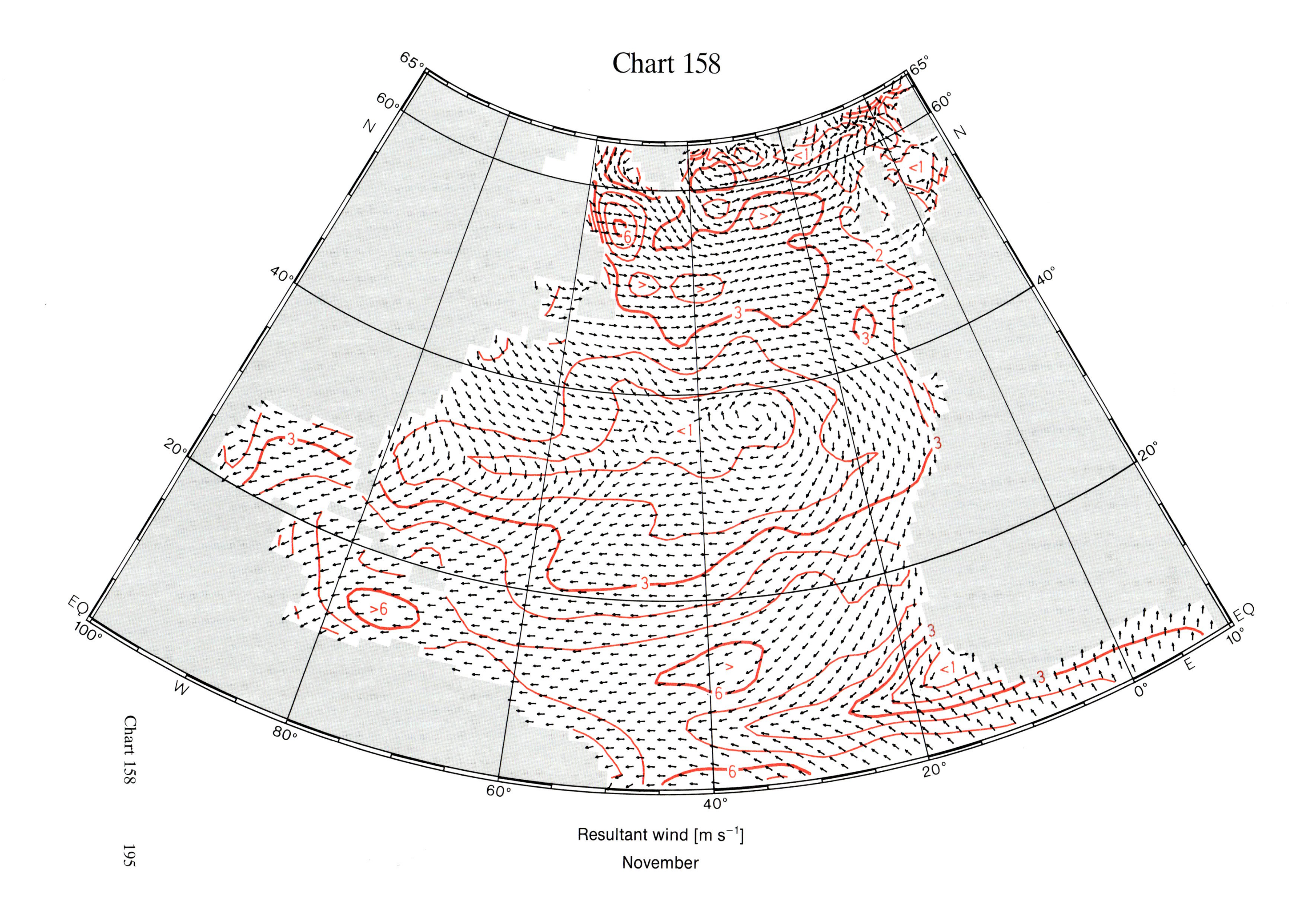

Resultant wind [m s^{-1}]

November

Chart 159

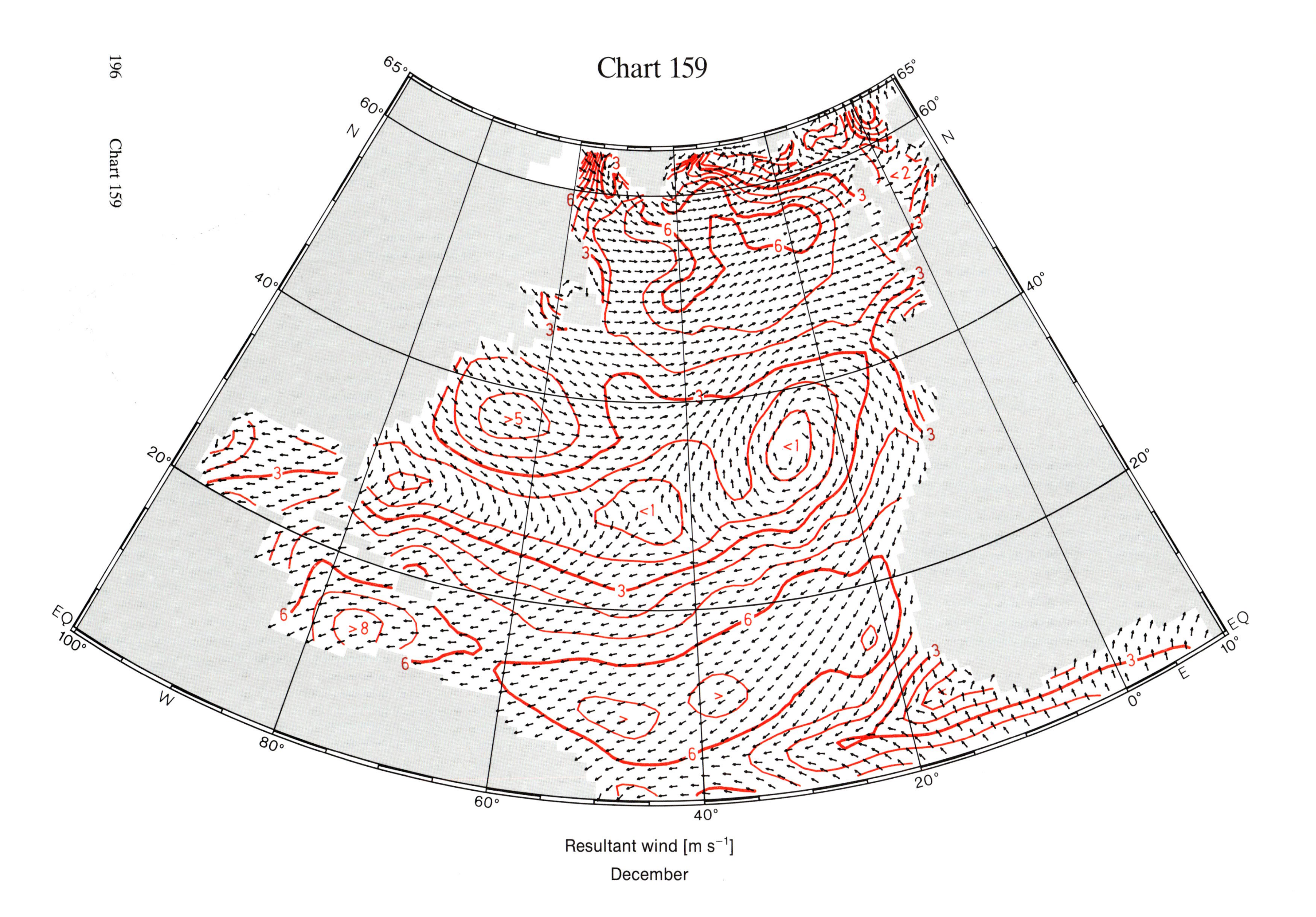

Resultant wind [m s^{-1}]

December

Chart 160

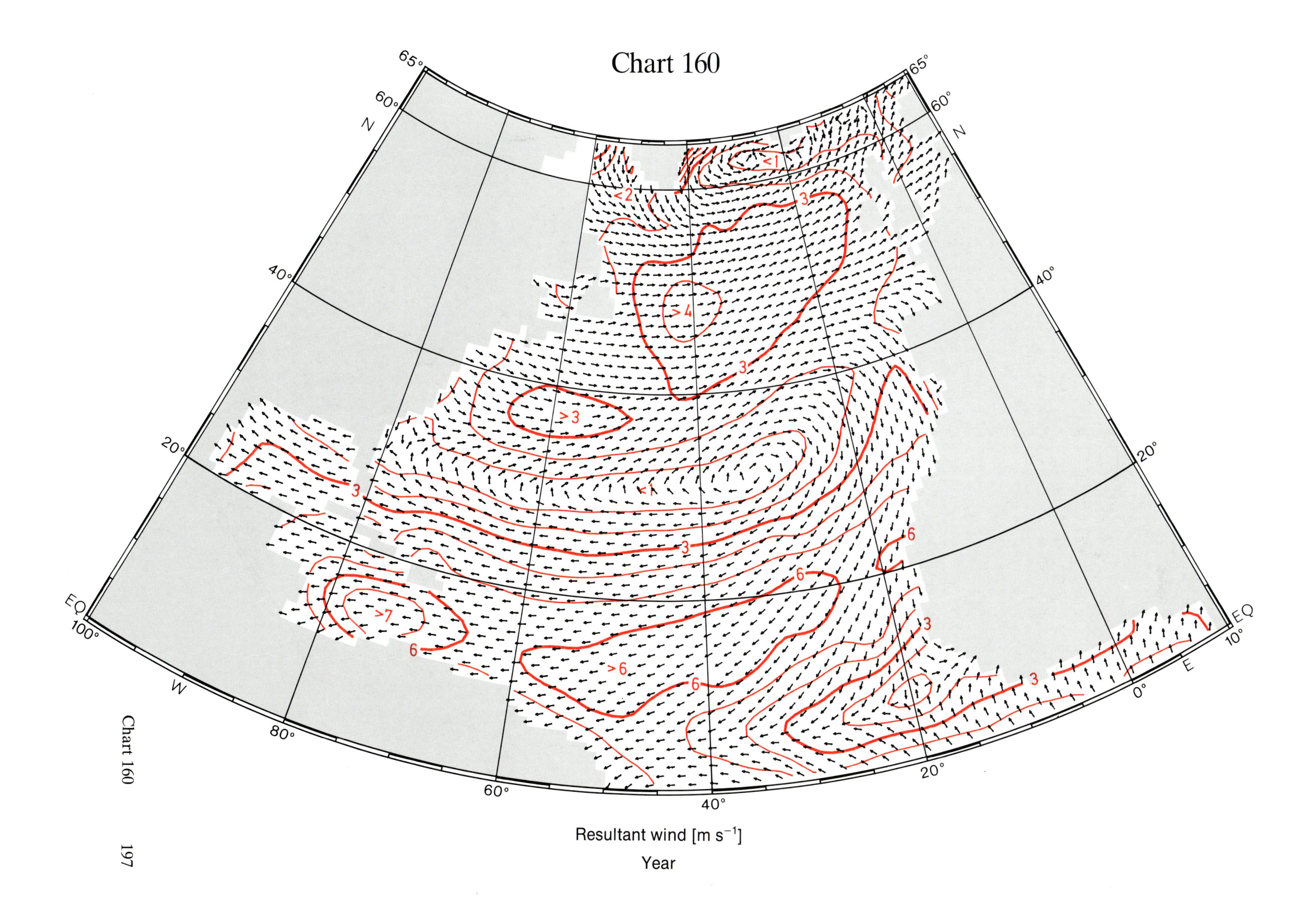
65°
60°
N
40°
20°
EQ
100°
W
80°
60°
40°
20°
0°
E
10°
3
6
>6
>7
>4
>3
<1
<2
Resultant wind [m s⁻¹]
Year

Chart 161

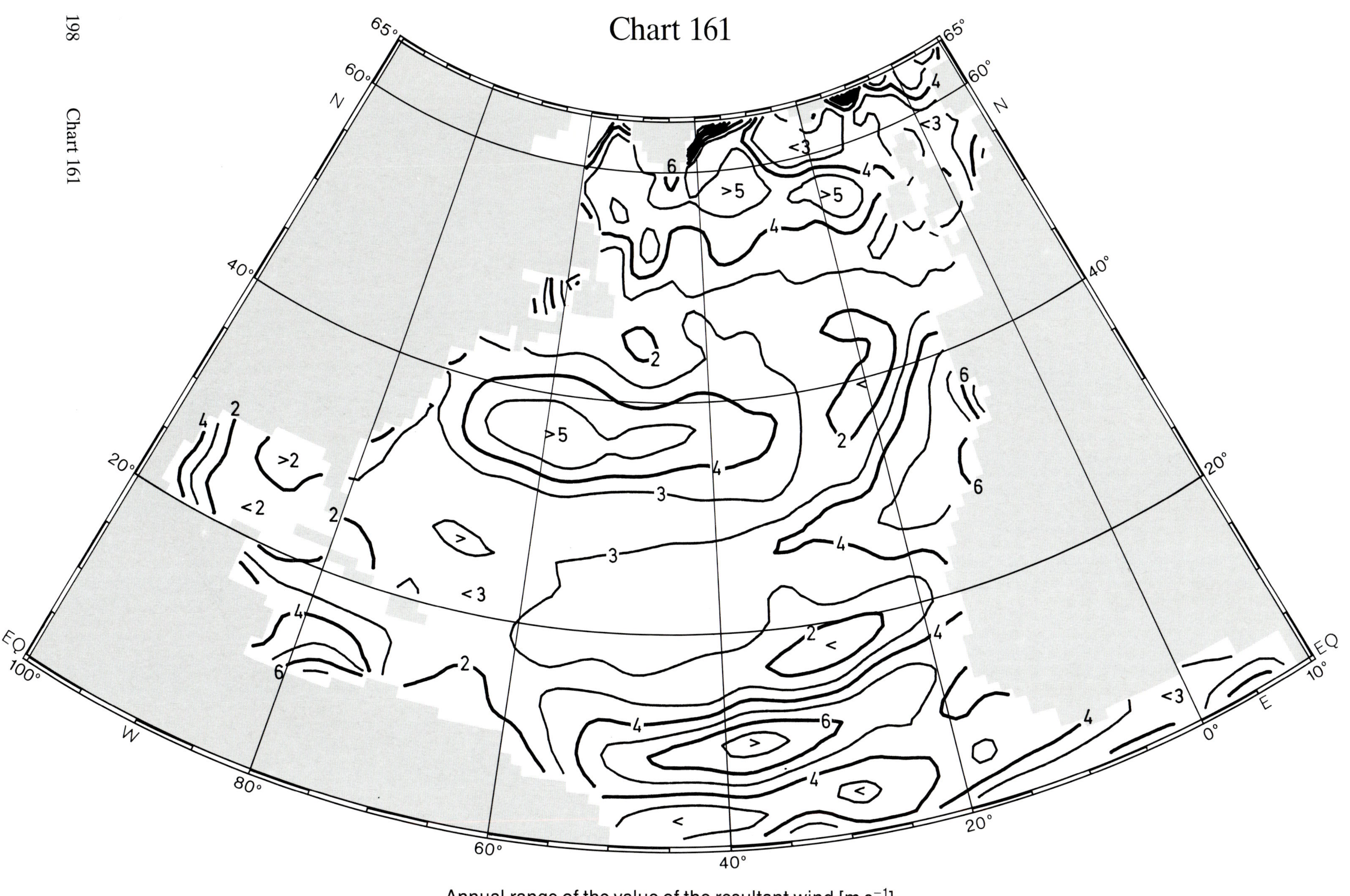

Annual range of the value of the resultant wind [m s^{-1}]

Chart 162

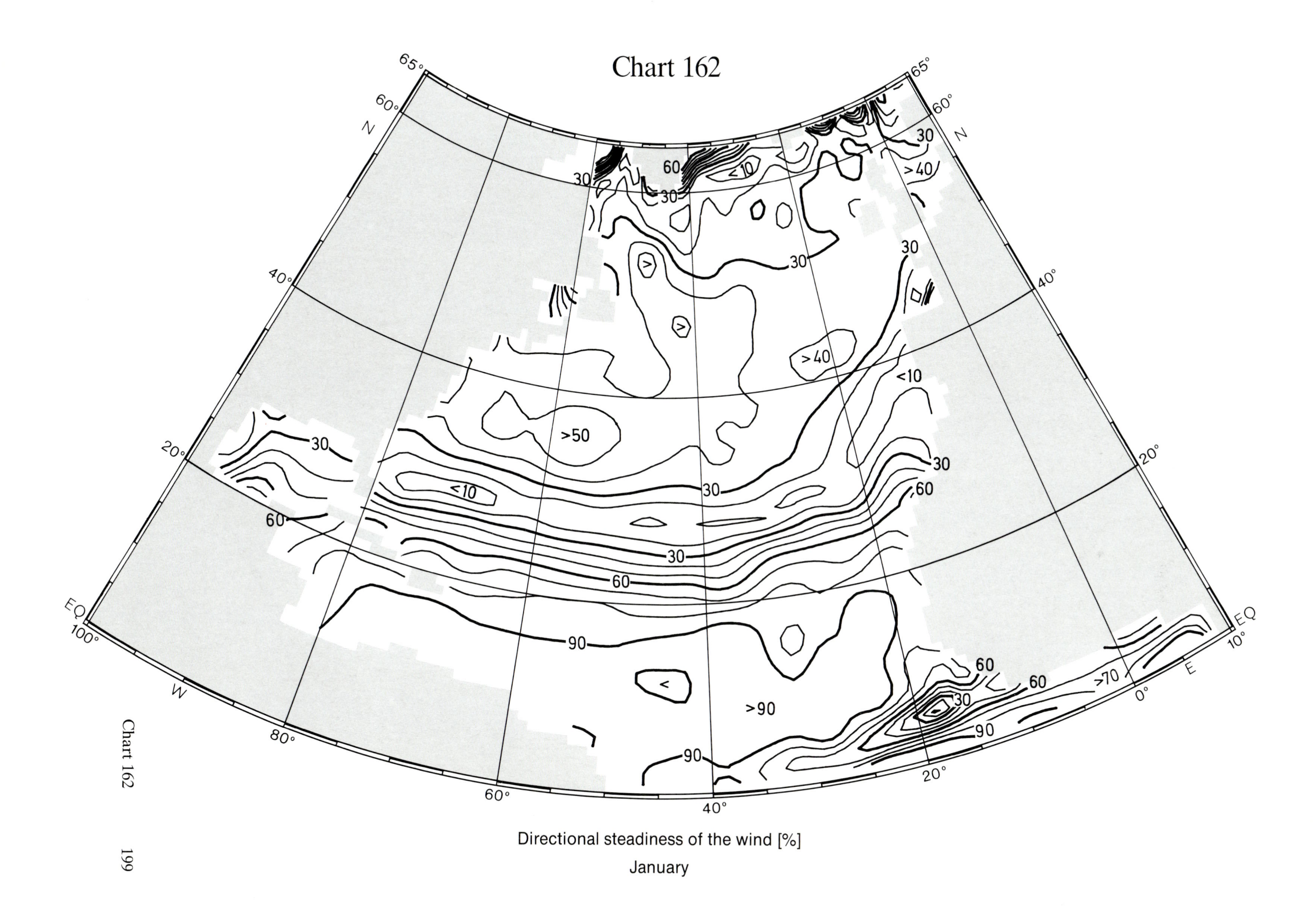

Directional steadiness of the wind [%]

January

Chart 163

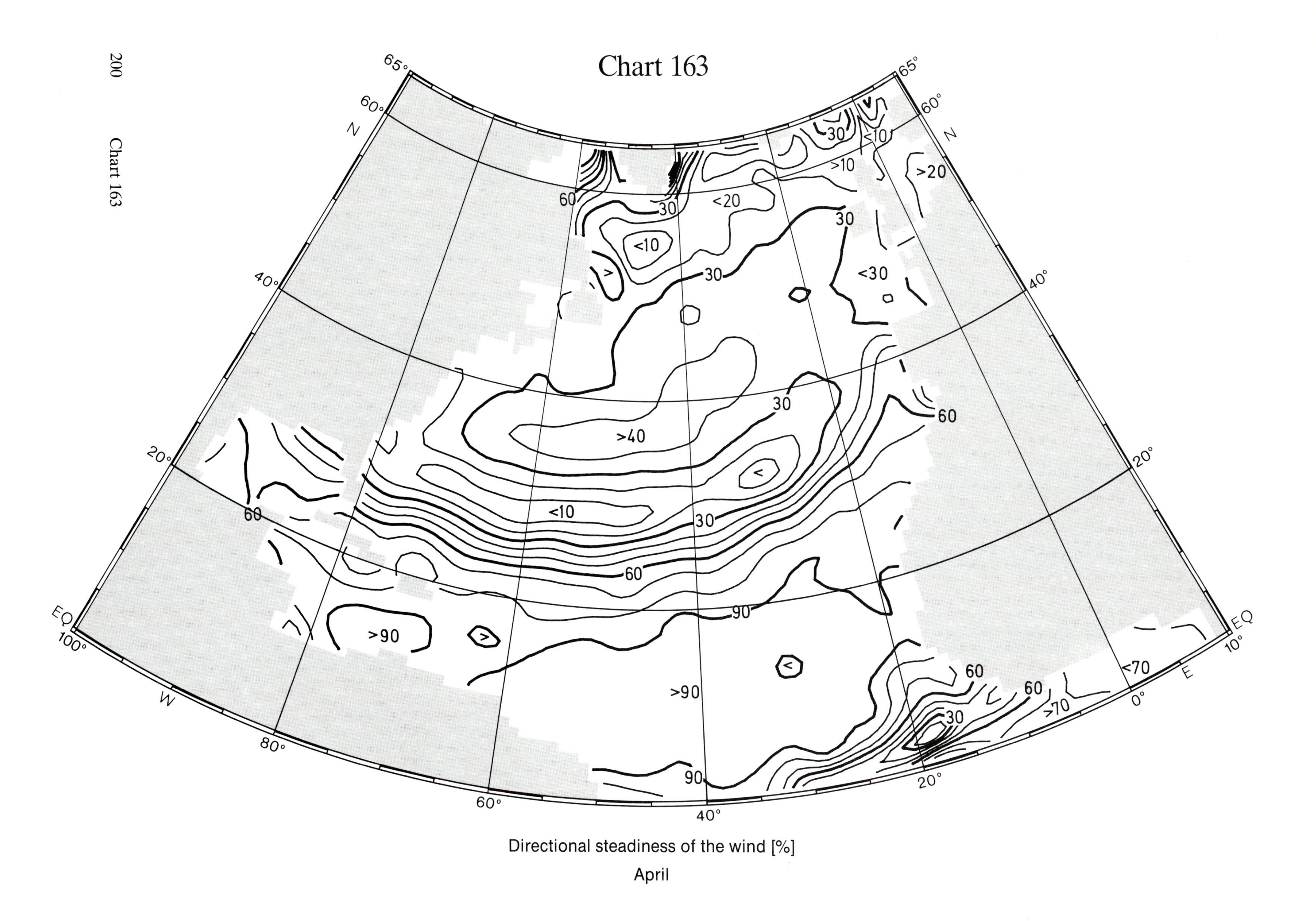

Directional steadiness of the wind [%]
April

Chart 164

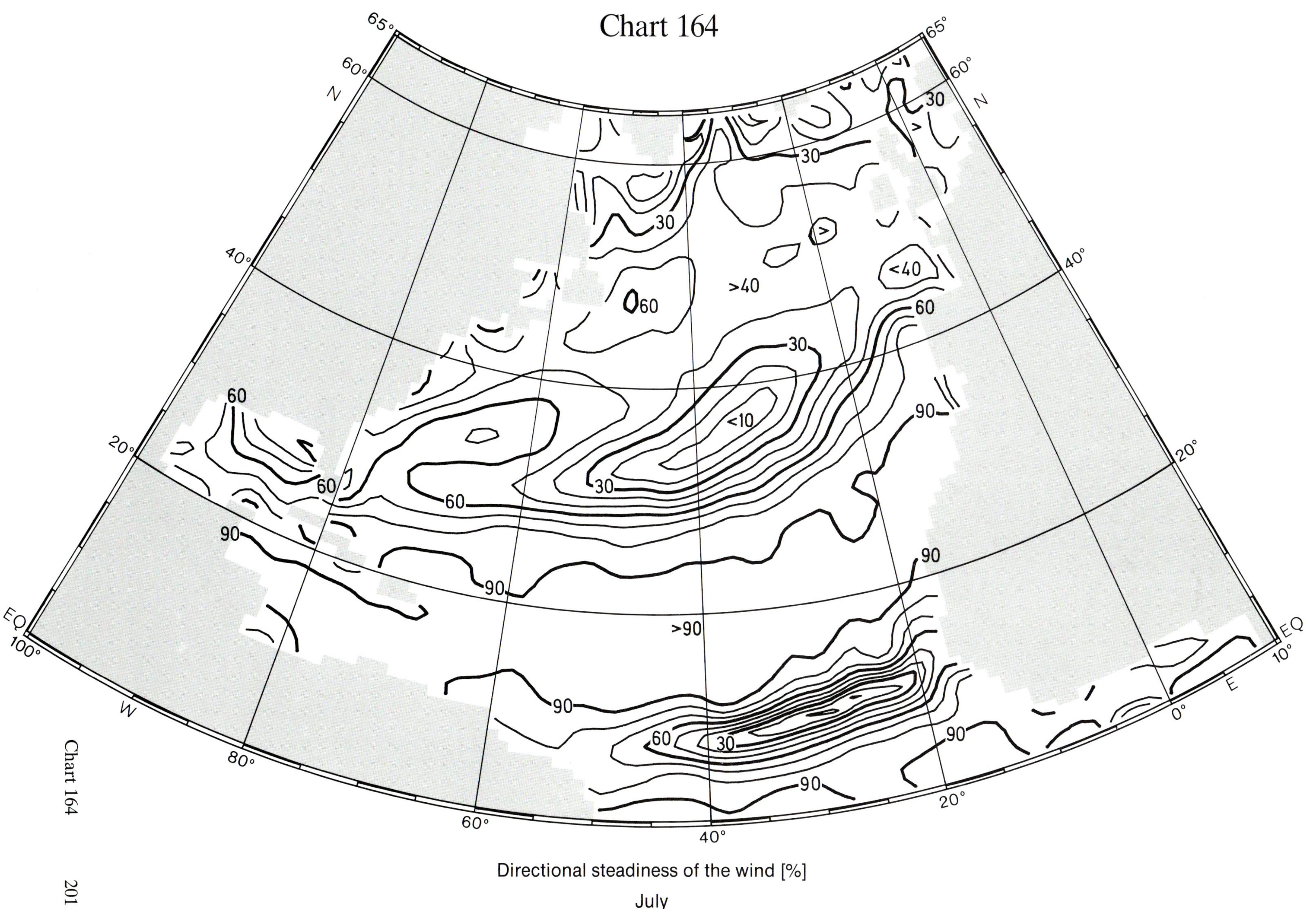

Directional steadiness of the wind [%]

July

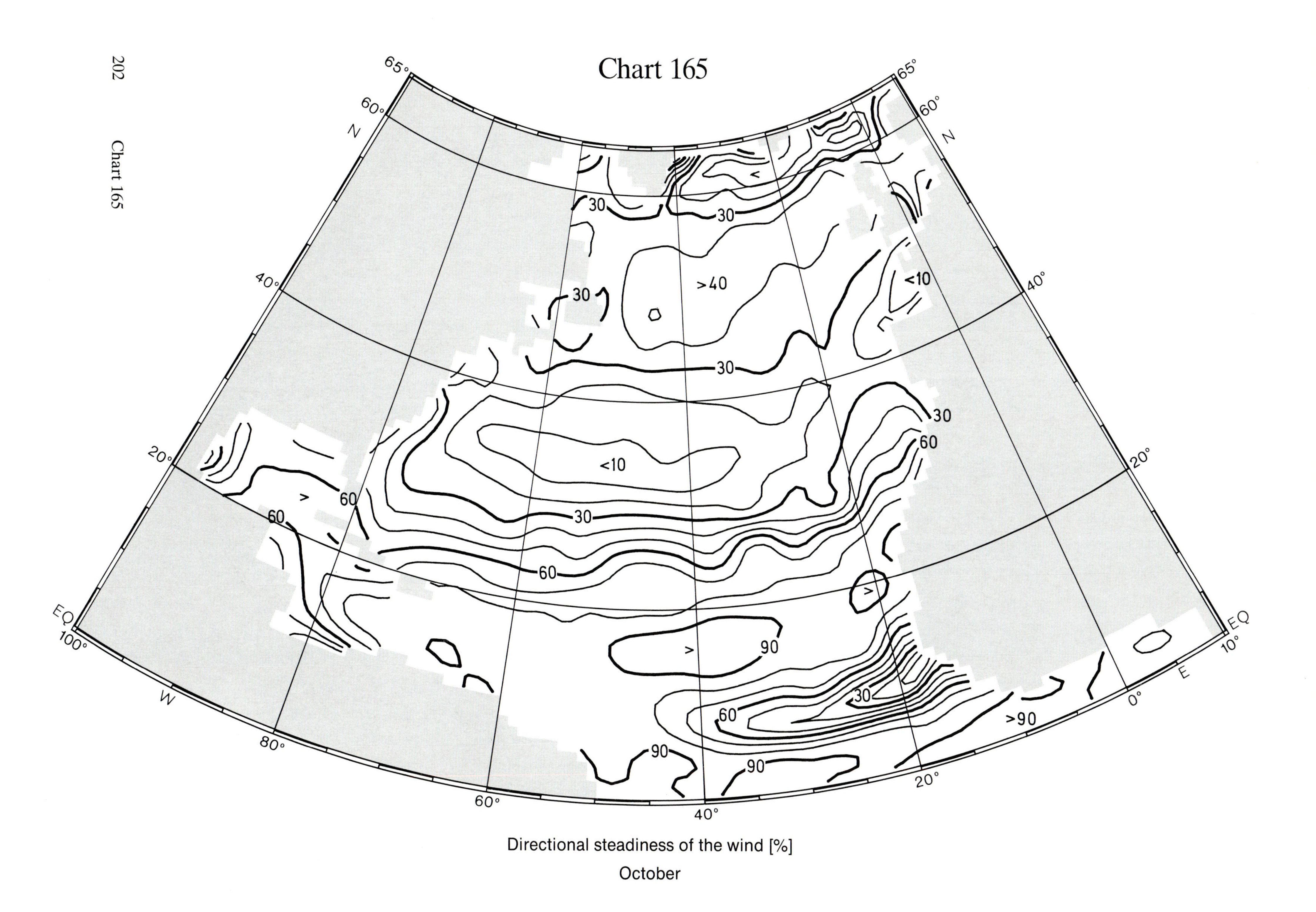
Chart 165
65°
60°
N
40°
20°
EQ
100°
W
80°
60°
40°
20°
0°
E
10°
30
>40
<10
60
90
>90
Directional steadiness of the wind [%]
October

Chart 166

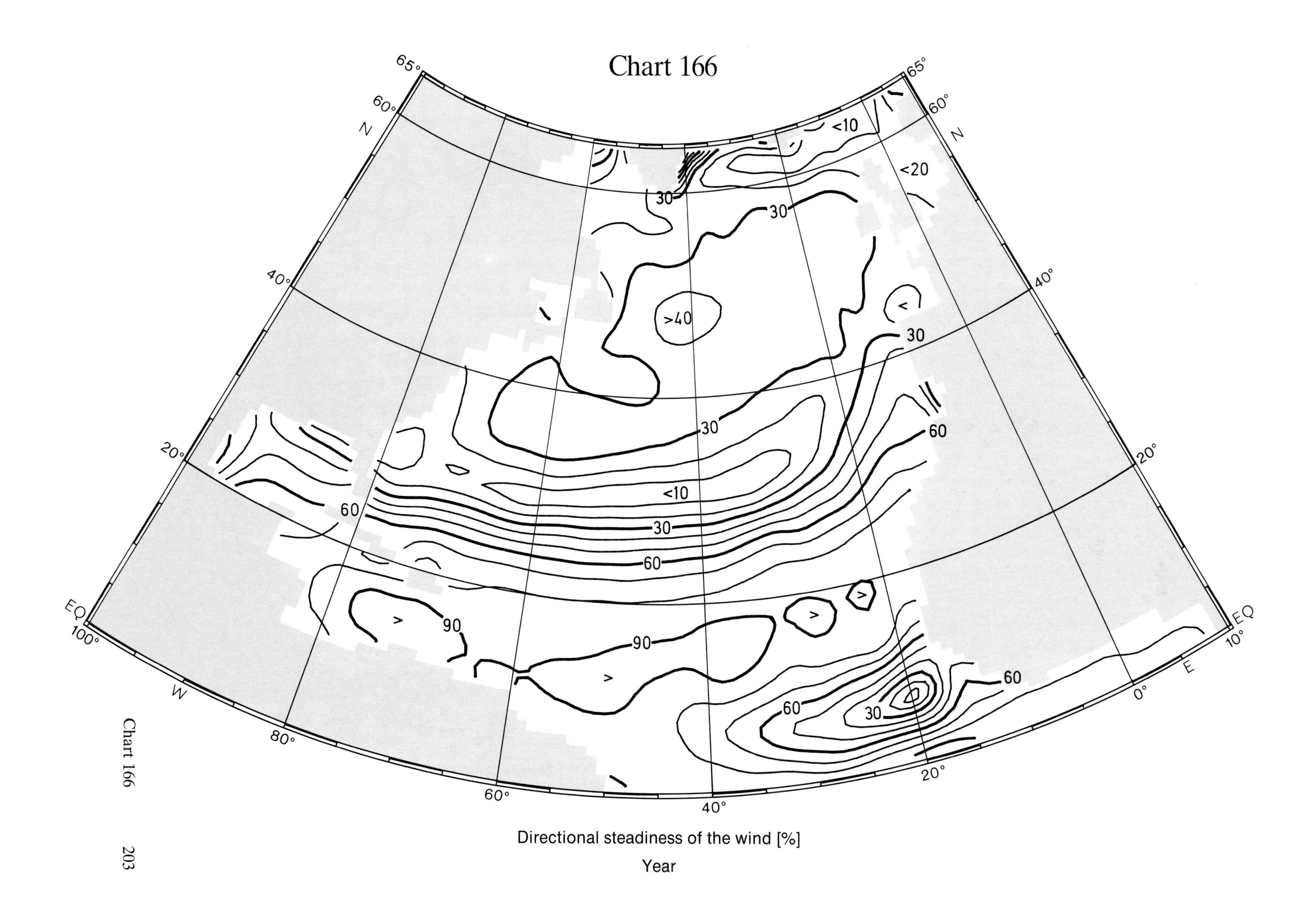

Directional steadiness of the wind [%]

Year

Chart 167

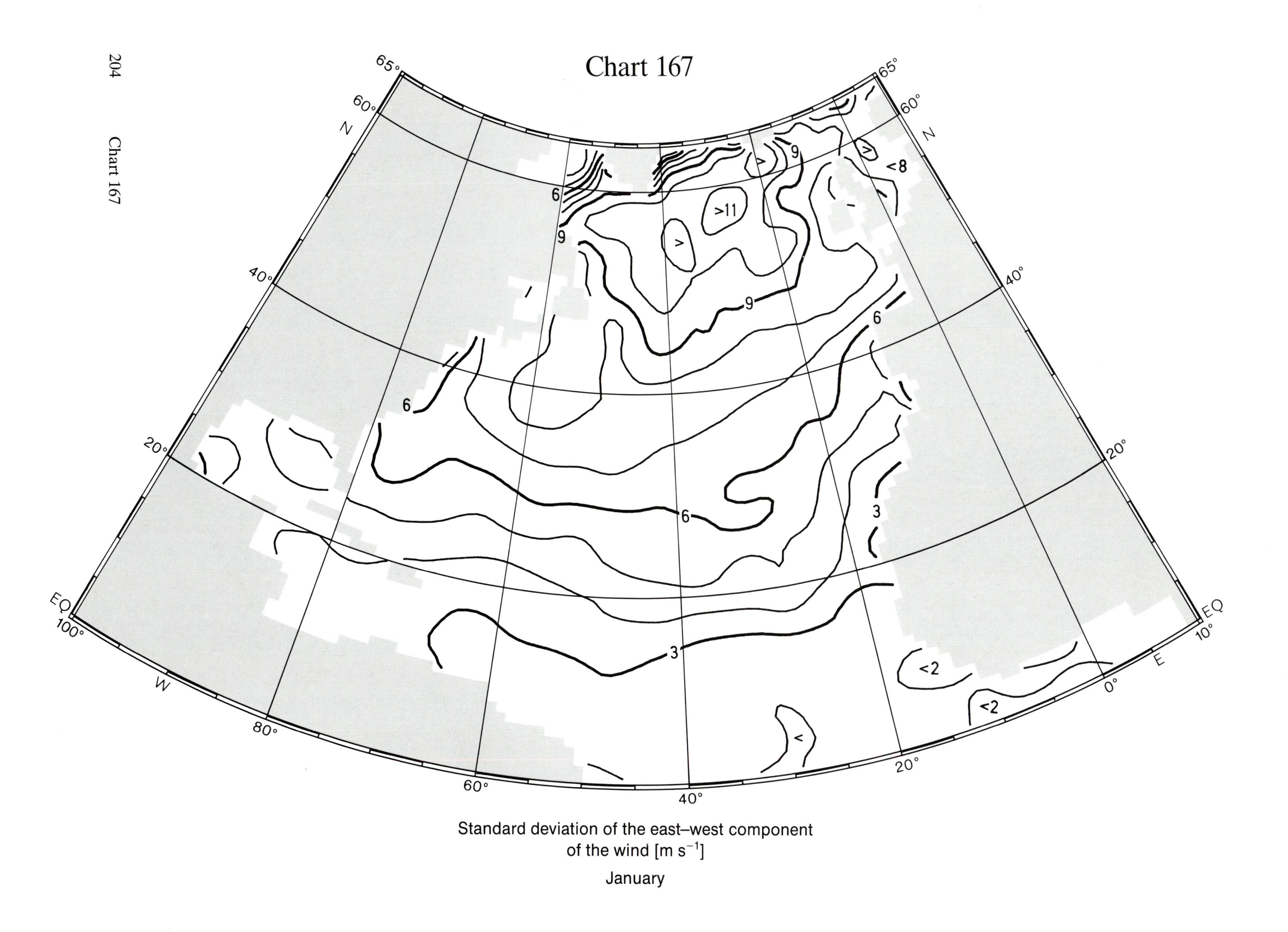

Standard deviation of the east–west component of the wind [m s^{-1}]

January

Chart 168

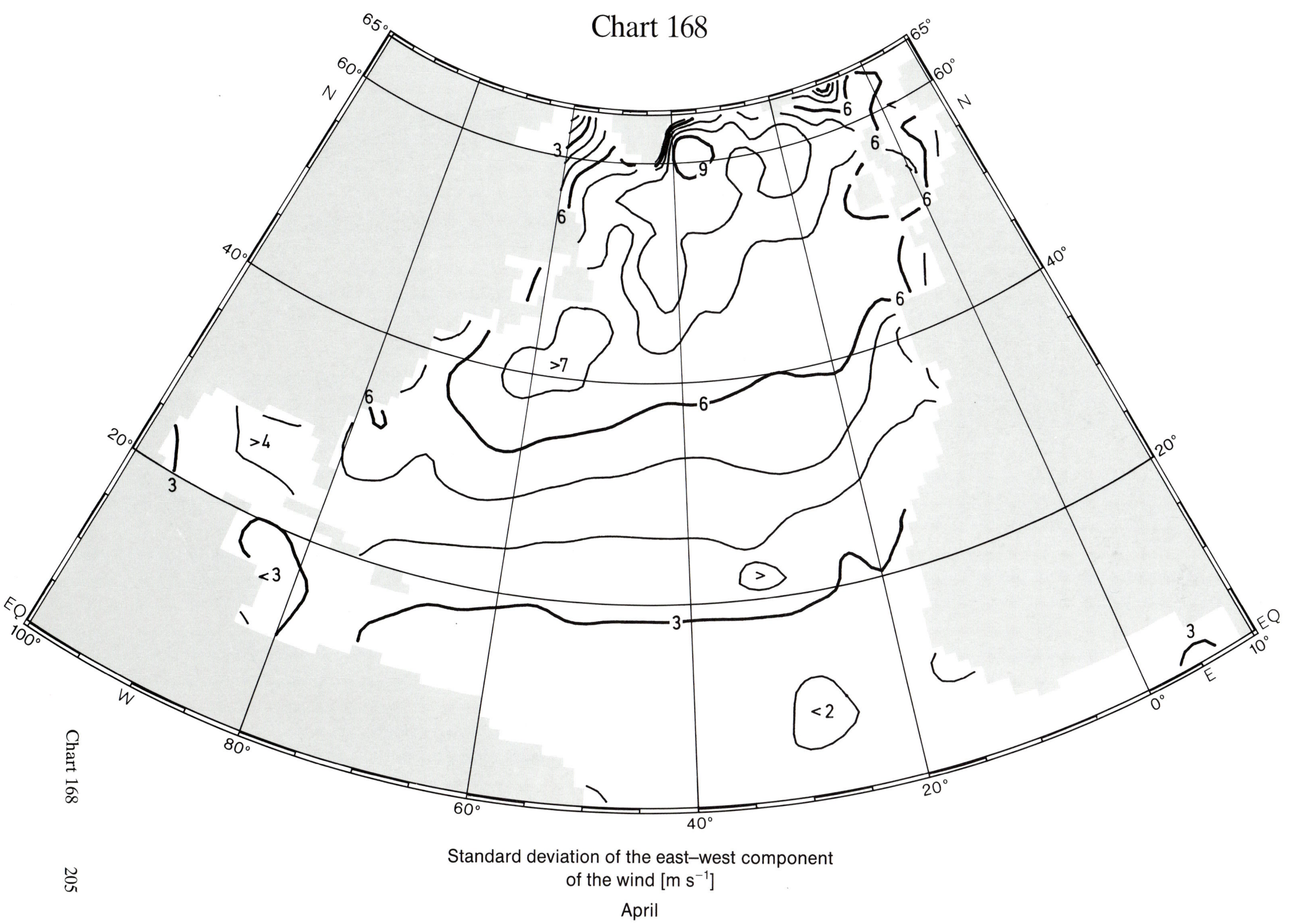

Standard deviation of the east–west component
of the wind [m s^{-1}]
April

Chart 169

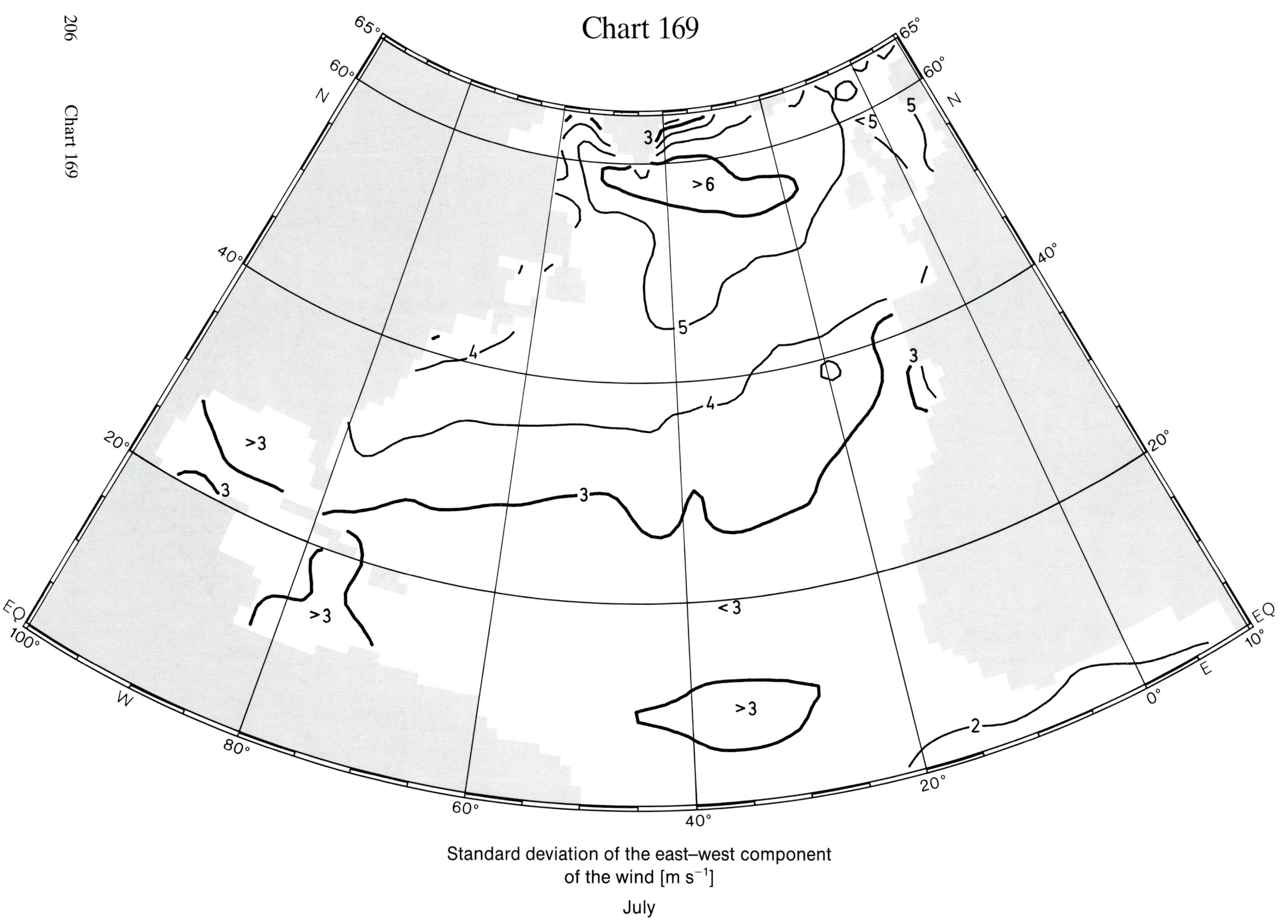

Standard deviation of the east–west component of the wind [m s^{-1}]

July

Chart 170

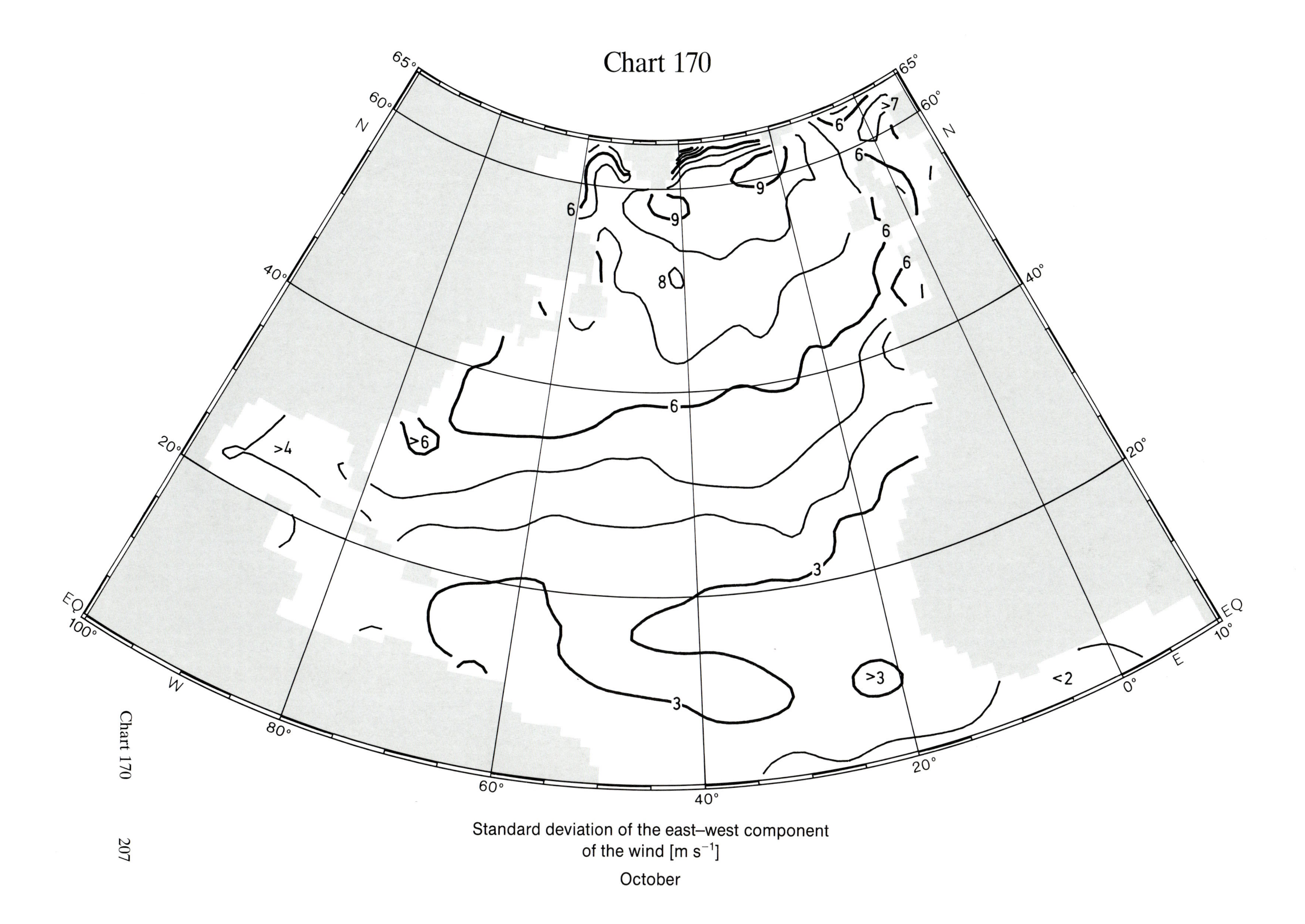

Standard deviation of the east–west component
of the wind [m s^{-1}]
October

Chart 171

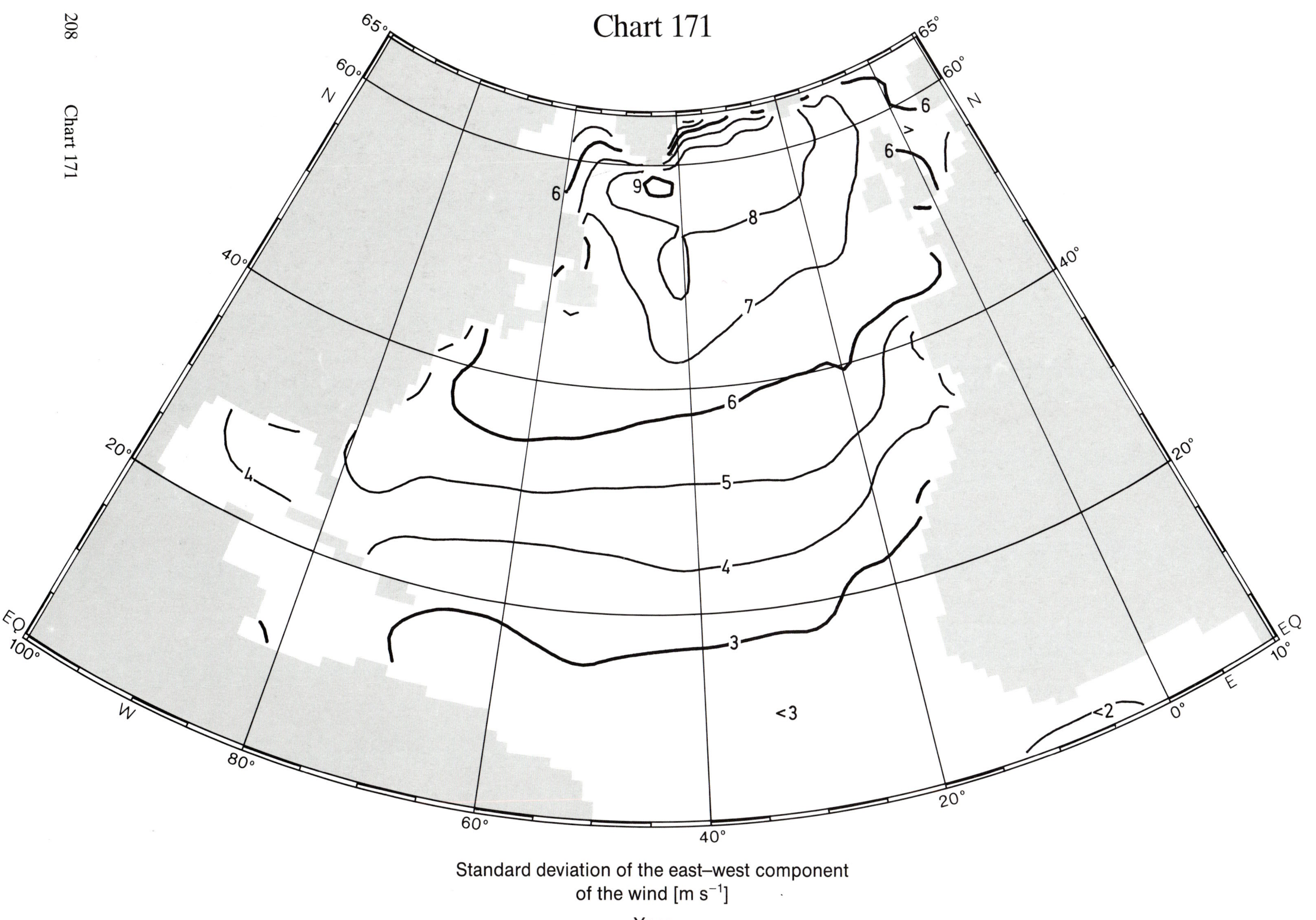

Standard deviation of the east–west component of the wind [m s^{-1}]
Year

Chart 172

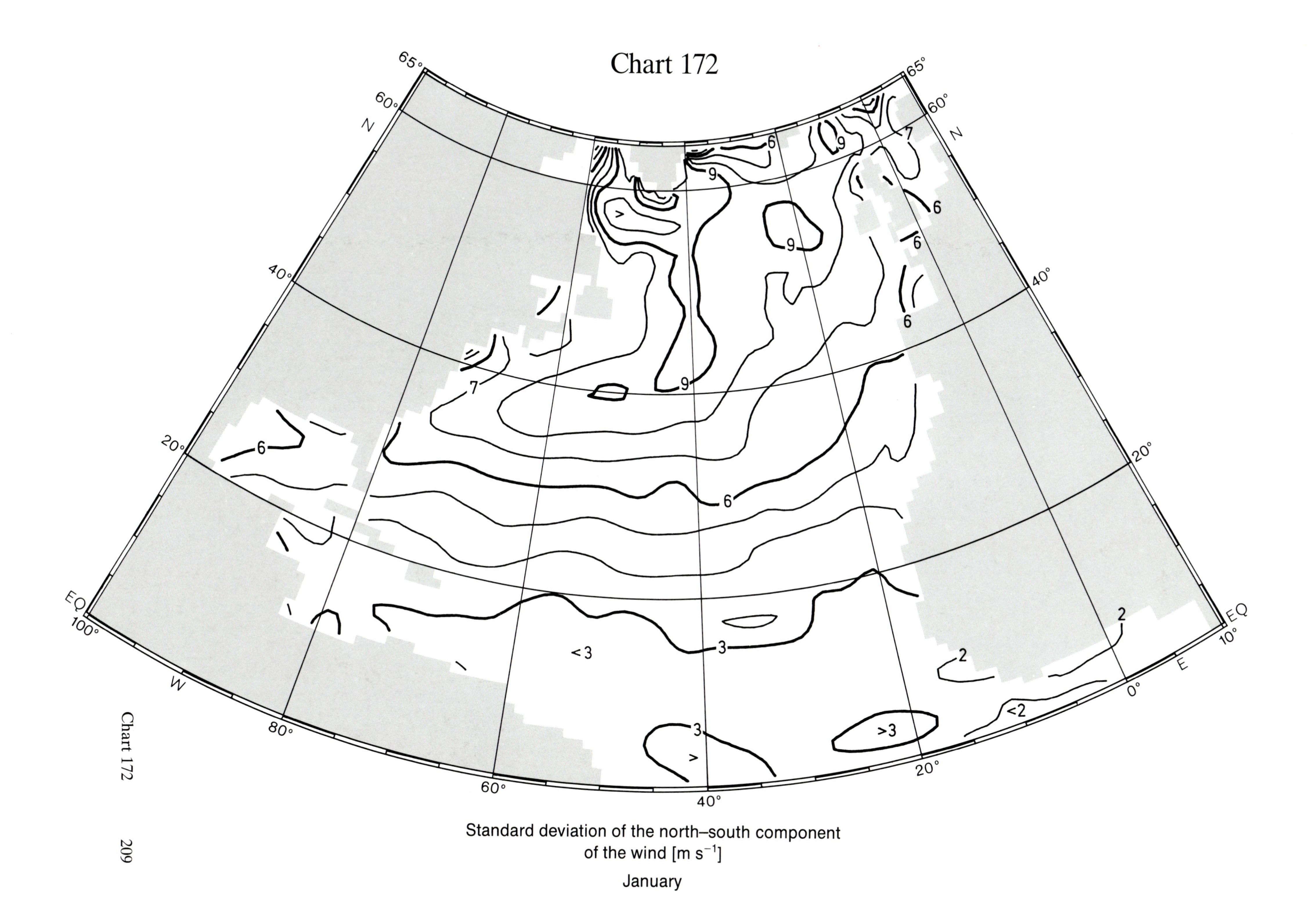

Standard deviation of the north–south component of the wind [m s^{-1}]

January

Chart 173

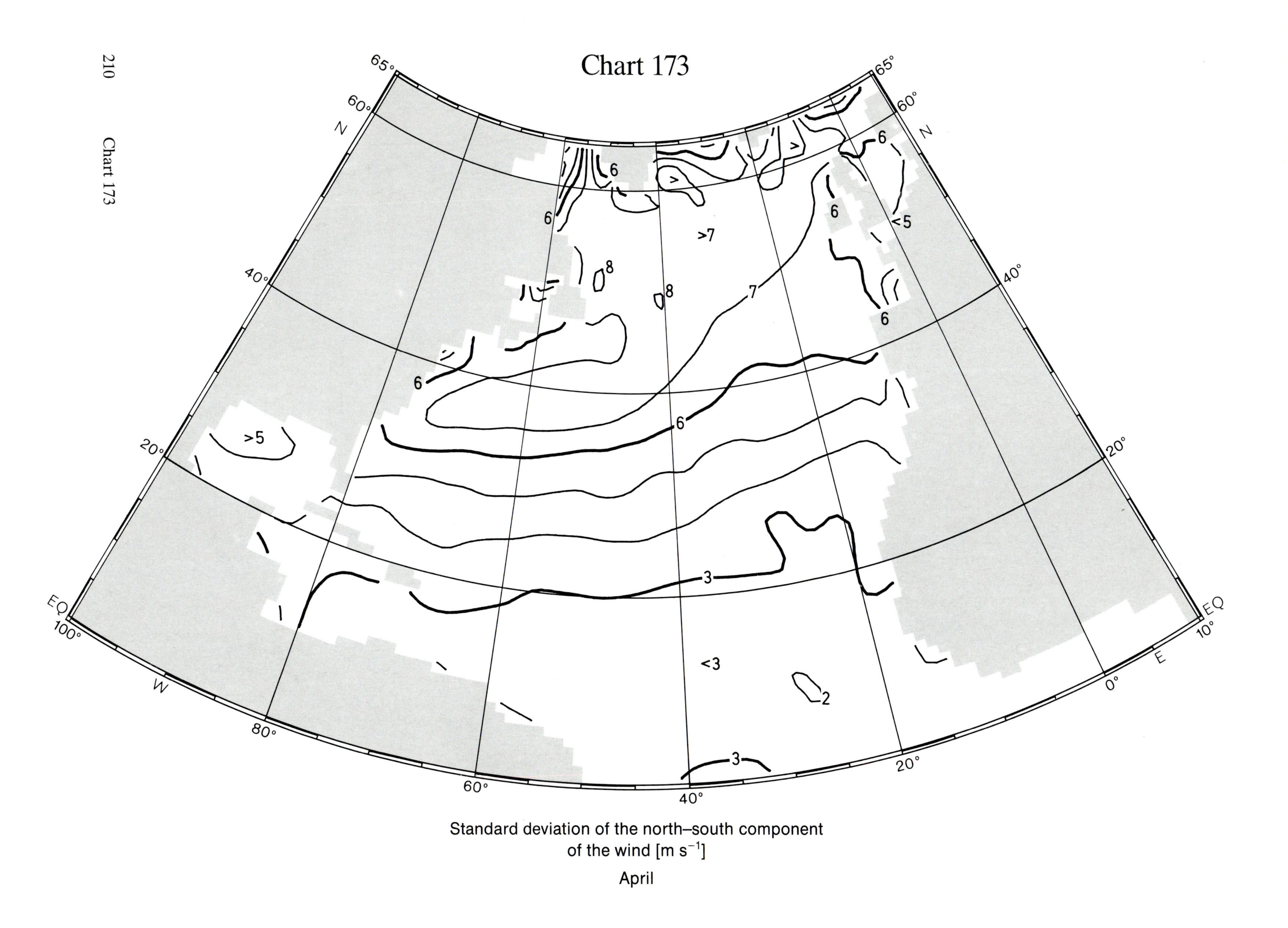

Standard deviation of the north–south component of the wind [m s^{-1}]

April

Chart 174

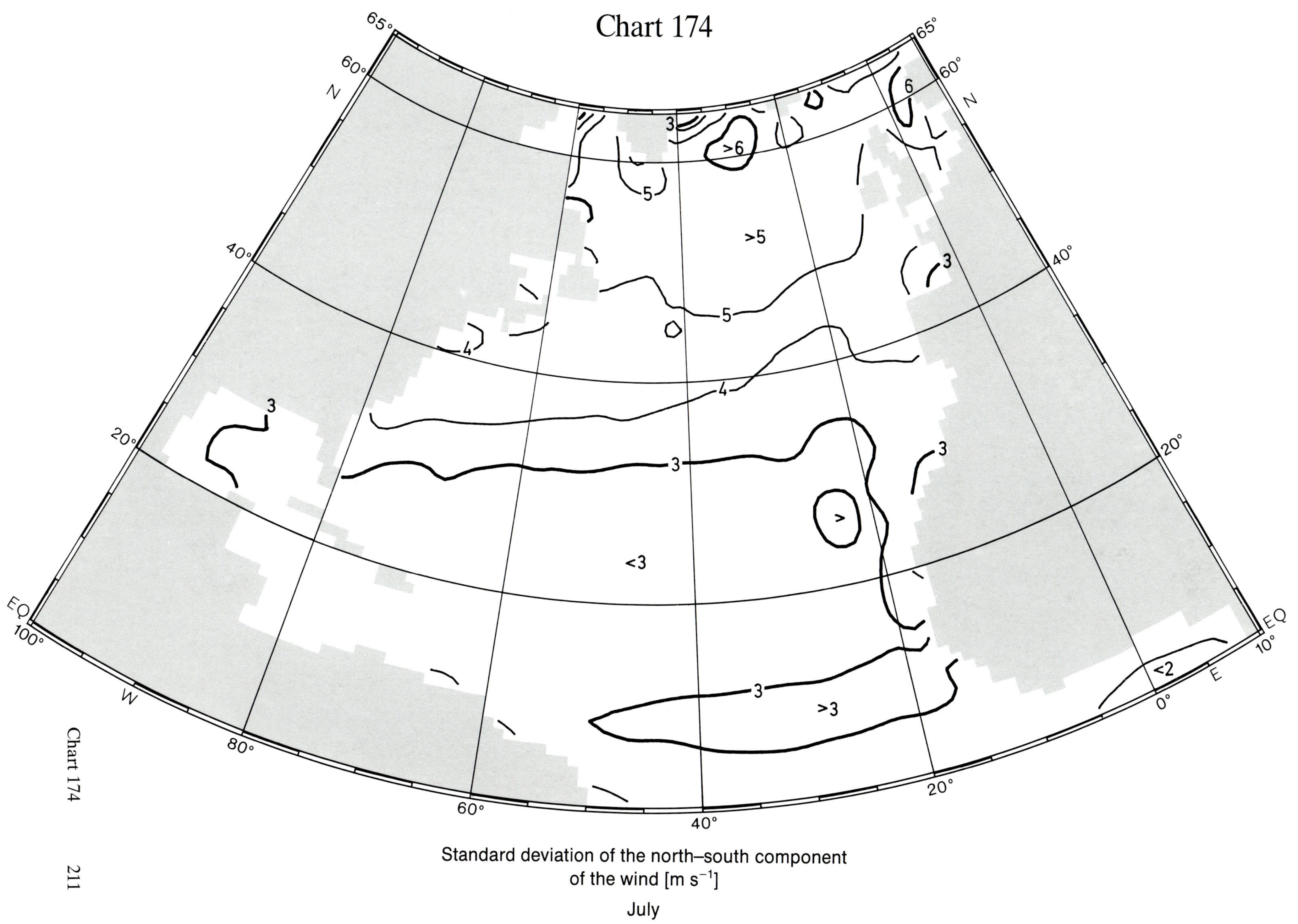

Standard deviation of the north–south component of the wind [m s^{-1}]

July

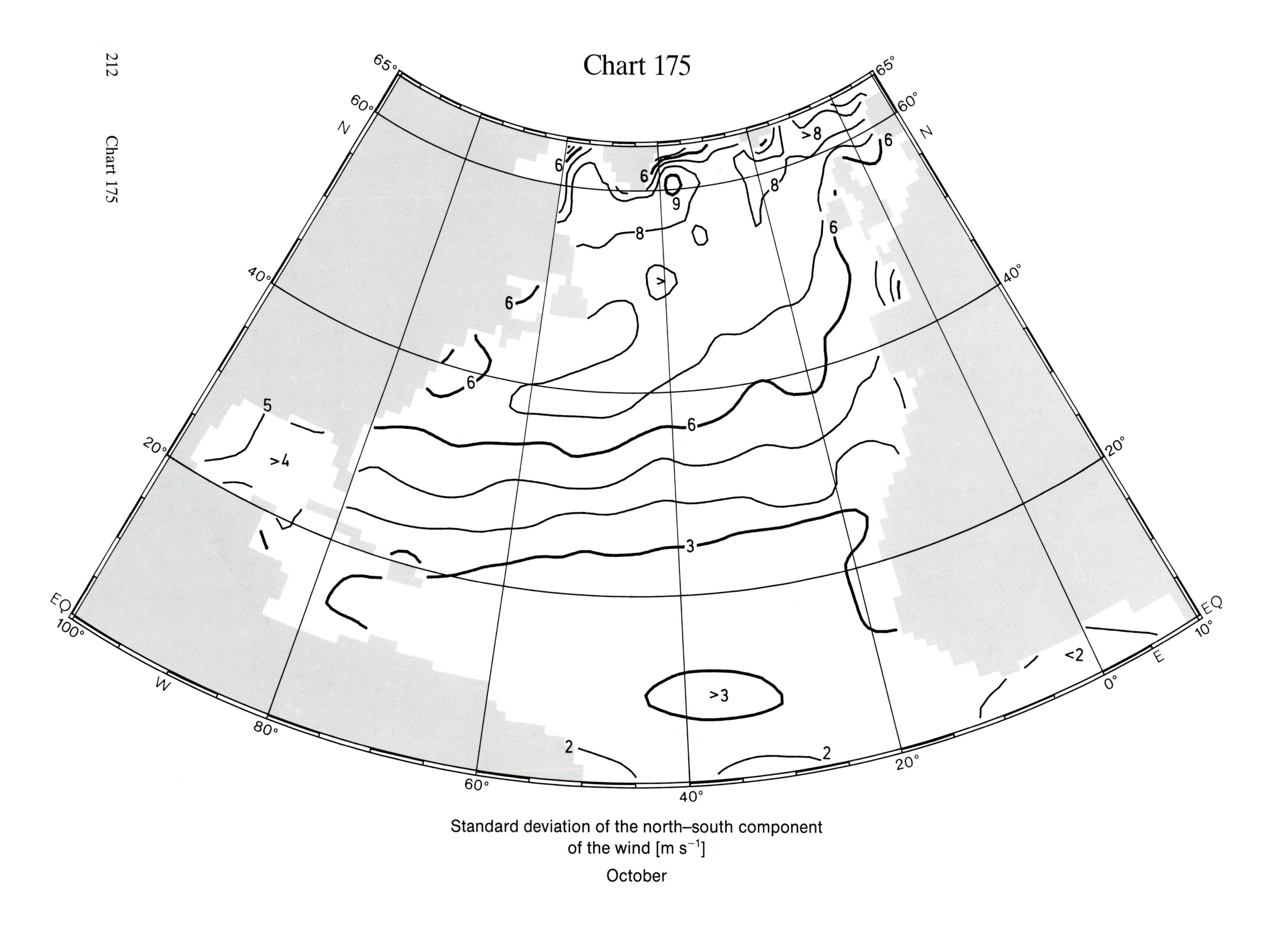

Standard deviation of the north–south component of the wind [m s^{-1}]
October

Chart 176

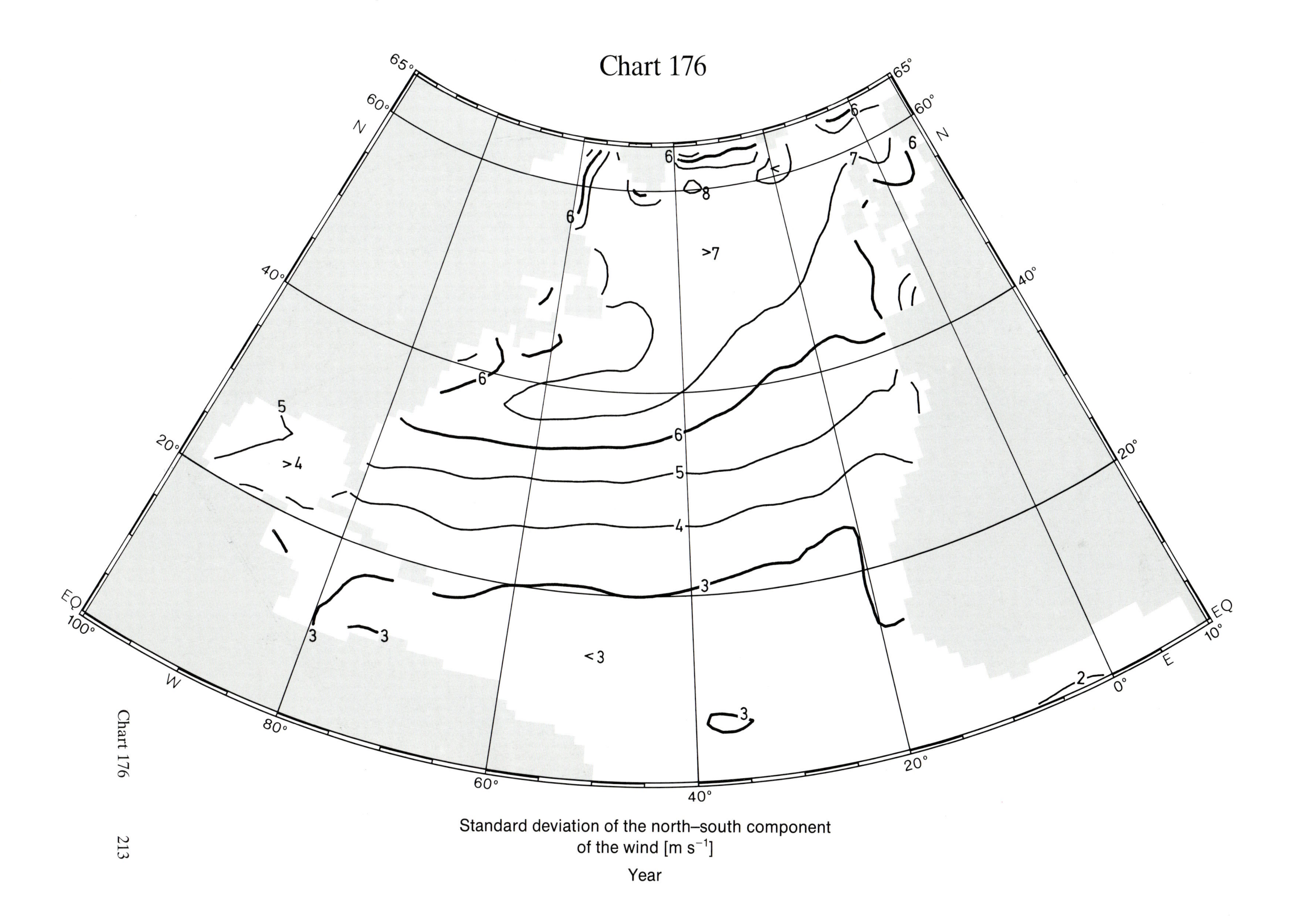

Standard deviation of the north–south component of the wind [m s^{-1}]

Year

Chart 177

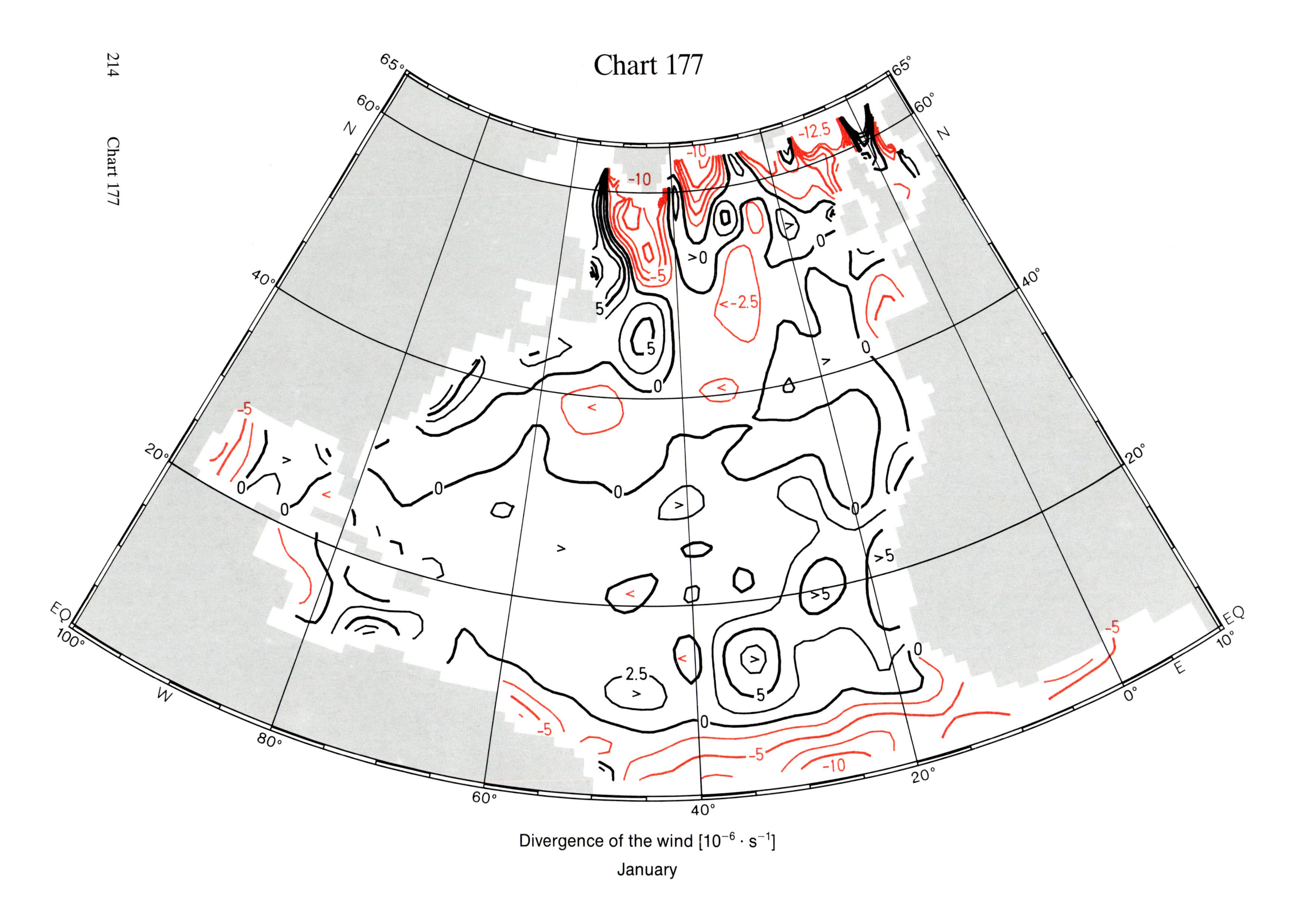

Divergence of the wind [$10^{-6} \cdot s^{-1}$]

January

Chart 178

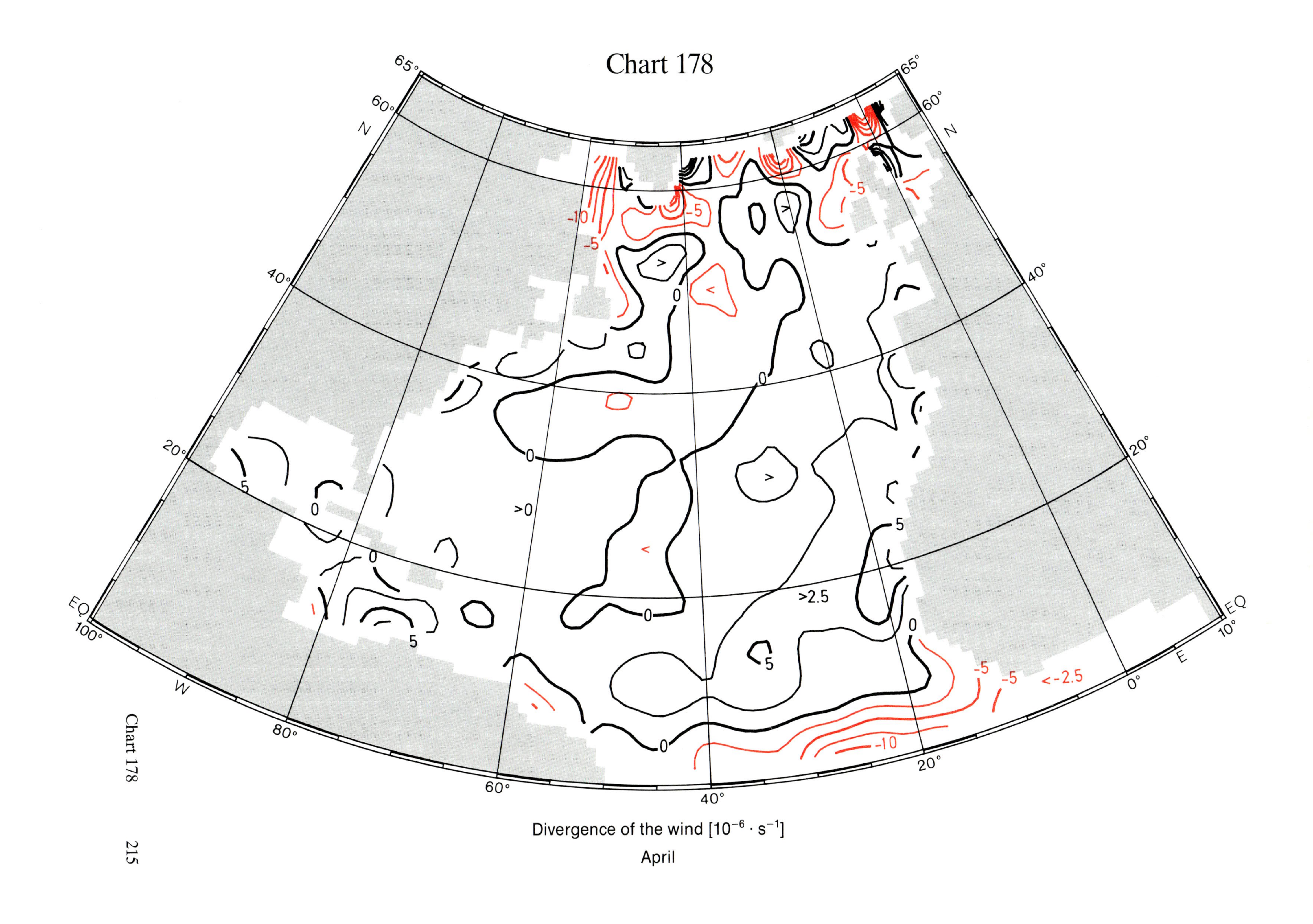

Divergence of the wind [$10^{-6} \cdot s^{-1}$]
April

Chart 179

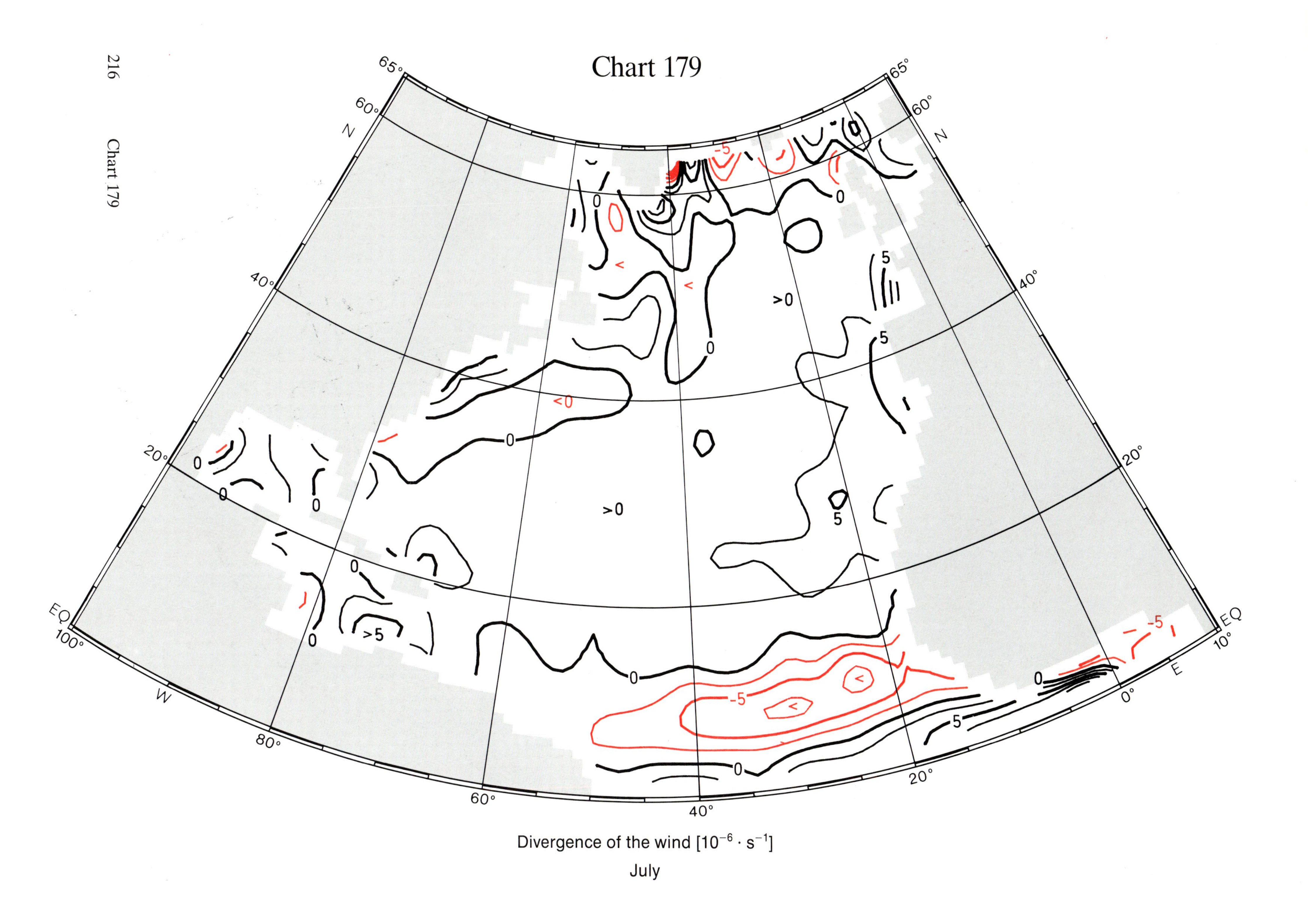

Divergence of the wind [$10^{-6} \cdot s^{-1}$]
July

Chart 180

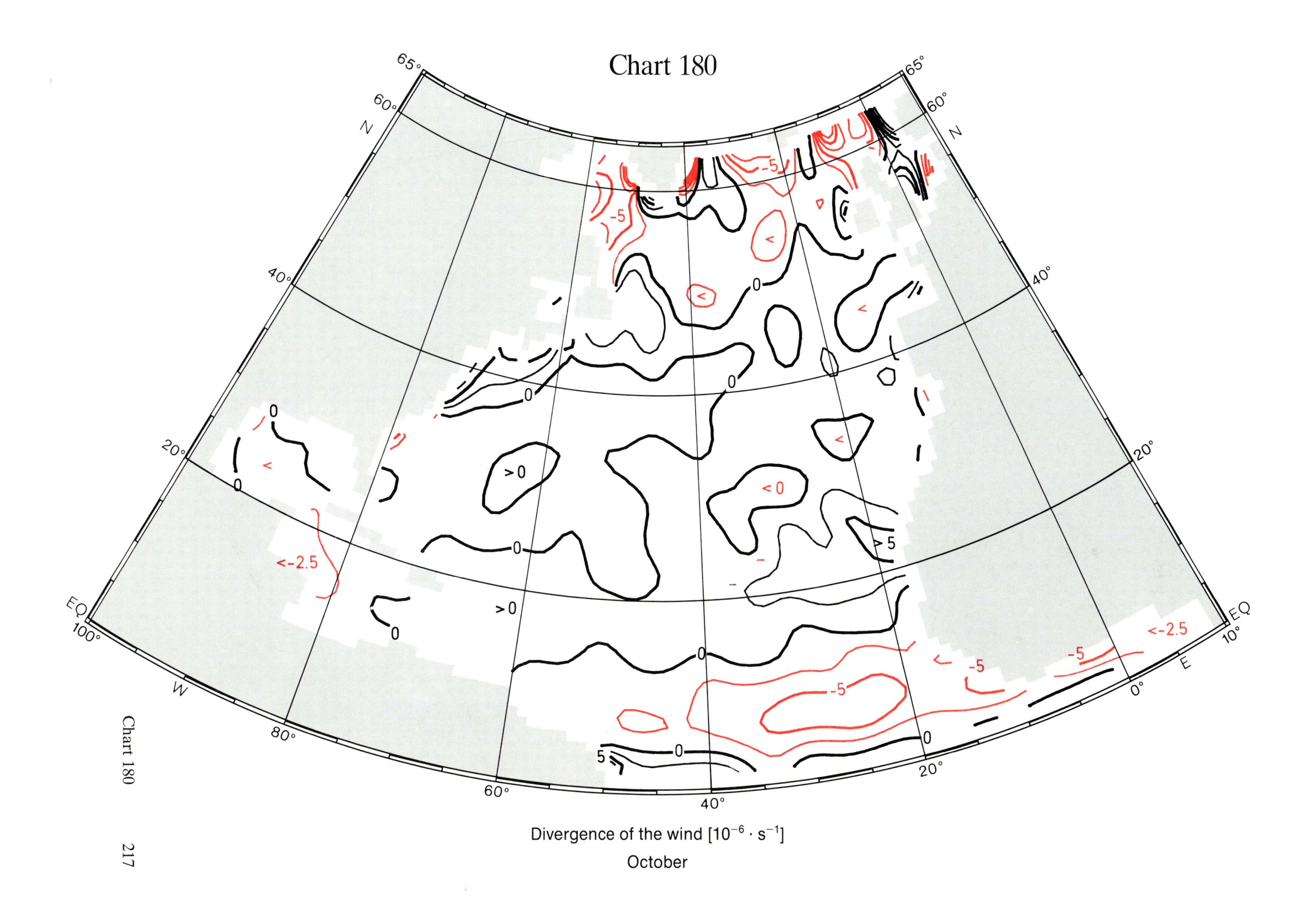

Divergence of the wind [$10^{-6} \cdot s^{-1}$]

October

Chart 181

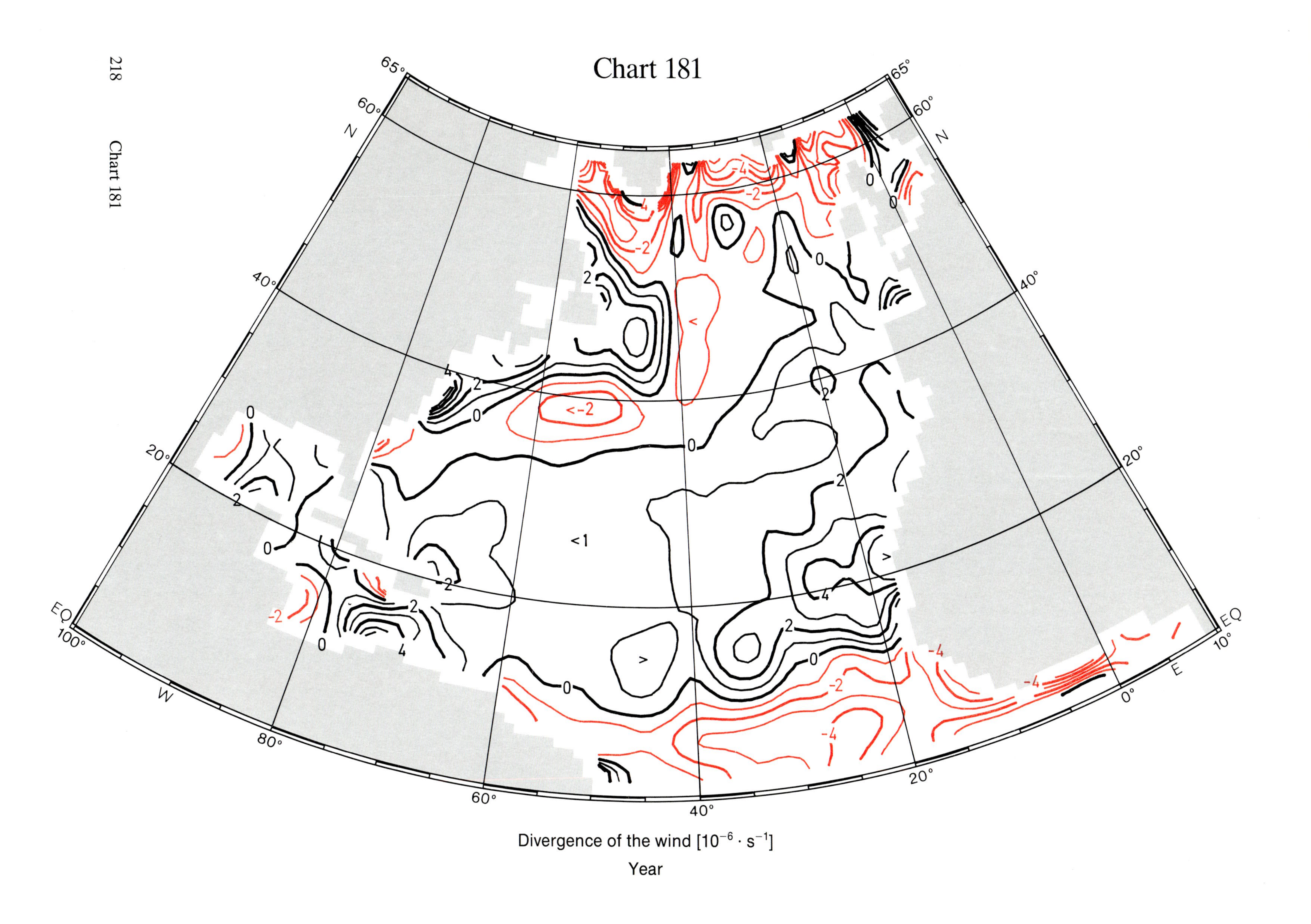

Divergence of the wind [$10^{-6} \cdot s^{-1}$]

Year